Advances in
Space Science and Technology
Volume 6

Contributors

Bryce S. DeWitt
John M. Lowther
Lloyd G. Marts
Saul Moskowitz
Geoffrey K. C. Pardoe
Mitchell R. Sharpe, Jr.
Robert H. Waterman, Jr.
Paul Weinschel

Advances in

Space Science

and Technology

Edited by **FREDERICK I. ORDWAY, III**

*General Astronautics Research
Corporation
Washington, D. C.
and
Huntsville, Alabama*

VOLUME 6

ACADEMIC PRESS NEW YORK and LONDON **1964**

ACADEMIC PRESS INC.
111 Fifth Avenue, New York 10003, New York

United Kingdom Edition published by
ACADEMIC PRESS INC. (LONDON) LTD.
Berkeley Square House, London W.1

Library of Congress Catalog Card Number: 59-15760

Contributors to Volume 6

Numbers in parentheses refer to the page on which the author's contribution begins.

BRYCE S. DEWITT (1), *Institute of Field Physics, Department of Physics, University of North Carolina, Chapel Hill, North Carolina*

JOHN M. LOWTHER (245), *Chrysler Corporation, Cape Kennedy, Florida*

LLOYD G. MARTS (173), *Industrial Economics Division, Denver Research Institute, University of Denver, Denver, Colorado*

SAUL MOSKOWITZ (101), *Navigation and Guidance Staff, Space Division, Kollsman Instrument Corporation, Elmhurst, New York*

GEOFFREY K. C. PARDOE (39), *Space Division, Hawker Siddeley Dynamics Limited, London, England*

MITCHELL R. SHARPE, JR. (245), *University of Alabama Center, Huntsville, Alabama*

ROBERT H. WATERMAN, JR. (171), *Industrial Economics Division, Denver Research Institute, University of Denver, Denver, Colorado*

PAUL WEINSCHEL (101), *Navigation and Guidance Staff, Space Division, Kollsman Istrument Corporation, Elmhurst, New York*

Foreword

Our first chapter assesses the state of human knowledge about gravitation, with particular reference to astronautics. Following an introductory chapter that provides an observational description of the gravitational field, a review is given of experimental tests of the theory of gravitation, including the three "classical" checks (frequency shift, bending of light rays by the Sun, and planetary perihelion precession) and still undefined cosmological tests. The final section deals with measurements in deep space (atomic clocks in satellites, gravitational radiation-measuring antennas, etc.) and speculations on the distant future.

Turning to technological aspects of astronautics, an author intimately connected with the European space program in general and the European Launcher Development Organization (ELDO) in particular describes the myriad problems inherent in the integration of the various stages of space carrier (launch) vehicles. In the United States and in the Soviet Union it is difficult enough to solve the interface and integration problems, but in the multination ELDO effort it sometimes seems miraculous that progress is made at all—the carrier alone has a first stage developed and produced in Great Britain, a second stage from France, and a third stage of German creation. This chapter is by no means built around the vehicle, but the step-by-step treatment inevitably leads into a short discussion of the European carrier. The main body of the essay offers a survey of the dynamic requirements of staging, the development of carriers, the integration of staged vehicles, the installation of payloads, general engineering problems of ground launched carriers, ground support facilities, and the feasibility of integrating spaceship stages in orbit, on the Moon, and on the planets.

In probably the most complete survey in the open literature, two experts write on the intriguing and exceptionally critical subject of navigational instrumentation for accomplishing orbital and deep space missions. In turn, the authors describe the three basic flight missions (orbital, lunar, and planetary), the mission phases (boost to injection, early midcourse, late midcourse, etc.), and the observables in space navigation (for example, the stars, the Sun, the planets) in terms of the flight phase. Navigational concepts are next considered—both explicit, wherein navigational measurements (the observables) are reduced to positional and vector information, and implicit, wherein the true present or future position remains undetermined. The place of man in space navigation receives attention, full account being taken of human limitations and the constraints these pose in space flight. The use of optical instrumentation, man's ability as a computer, and navigational instrumentation and its operational employment are then described. Dispelling any notions that may still exist about the role of man in a spaceship, the authors do not look forward to complete automation in space navigation, asserting that even when "automatic computation is employed, the primary observations must (at a minimum) be initiated by the operator and may possibly be performed in their entirety by him."

One of the most important and widely debated issues of the mid-1960's is the commercial value of the prodigious effort that has been, is being, and will be made in missile, space, and directly related programs. The National Aeronautics and Space Administration (NASA) inaugurated a program to first search out and then publicize the many inventions, innovations, techniques, and methods growing out of efforts that at first glance seemed to offer few tangible benefits to civilization. In order to identify space by-products and beneficial techniques of possible commercial importance, NASA in late 1961 took the important step of approving the initiation of a study, selecting the Industrial Economics Division of the Denver Research Institute to undertake it. The fourth chapter presents the results of this study.

The authors first discuss what they call "technological transfer," a term much broader in scope than the restrictive term "by-product" that is generally employed in this context. In turn, they analyze the major elements of technological transfer from the missile/space programs, namely (1) stimulation of basic and applied research; (2) development or improvement of process and techniques; (3) improvement of existing products; (4) increased availability of materials, testing equipment, and laboratory equipment; (5) development of new products; and (6) cost reduction. They then cite numerous concrete instances of technological transfer, following which they elaborate on its significance and on both the incentives and barriers to commercial applications of space-generated technology.

The final chapter, which is Part 1 of a treatise (to be completed in Volume 7) on world rocket, missile, and space carrier vehicle testing, launching, and tracking operations, describes the major facilities located in the United States. It is the result of several years of painstaking research concerning a field that, as far as the Editor knows, has never previously been discussed so comprehensively in the permanent literature in any language.

The organization of both parts of this chapter is such that comparisons can readily be made between one launch center and another. Following the general introduction, geographical factors are covered, including climate and transportation access. A description of the launching facilities is then given, after which tracking and communications are described. A section on support activities normally terminates the material on a given range.

During late 1963 and early 1964 efforts were made to strengthen the Editorial Advisory Board, providing *Advances in Space Science and Technology* with representation in Africa, Asia, Australia, and South America. Representing South Africa is Dr. F. J. Hewitt, Director of the National Institute for Telecommunications Research of the South African Council for Scientific and Industrial Research. Professor Hideo Itokawa, who joined the Board in October 1963, directs the Itokawa Laboratory of the Institute of Industrial Science, University of Tokyo, and is a member of the National Committee on Space Research in Japan. A month earlier Dr. David F. Martyn became a member, representing Australia. He directs the Upper Atmosphere Section of the Commonwealth Scientific and Industrial Research Organization and is chairman of the Australian National Committee on Space Research. And from Argentina we are honored to have Ing. Teófilo M. Tabanera, President of the Comisión Nacional de Investigaciones Espaciales. On behalf of the publisher and the original members of the Board, we welcome all these distinguished scientists to the Editorial Advisory Board of *Advances in Space Science and Technology* and look forward to many years of close and fruitful cooperation.

Special thanks are due to Janice Duffy Lounsbery for her assistance in editing portions of the volume.

FREDERICK I. ORDWAY, III

Huntsville, Alabama
July 1964

Professor Dr.-Ing. Eugen Sänger

On the 10th of February 1964 our German representative on the Editorial Advisory Board died most unexpectedly at the Technical University of Berlin. Professor Dr.-Ing. Eugen Sänger was not only one of the world's outstanding authorities on astronautics and among the few true pioneers of rocketry and space flight theory, but an old and close friend of the Editor and of most of his fellow members on the board. His loss will be acutely felt by all of us and by the entire scientific community.

Professor Sänger was born in Pressnitz, Germany on 22 September 1905 and, from 1912 until 1923, attended elementary and high schools in Austria and Hungary. From 1923 to 1929 (when he graduated) he studied at the Technische Hochschule at Graz and Vienna. In Vienna he later received his diploma in structural engineering and his doctorate of engineering science.

From 1936 to 1943 he collaborated in the establishment of the rocket flight experimental station of the Deutsche Luftfahrtforschungsanstalt Hermann Göring, where he tested liquid propellant rocket engines up to chamber pressures of 100 atmospheres and where he inaugurated ramjet development in Germany. From there he moved to the Deutsche Forschungsanstalt für Segelflug, E. V. in Bavaria, working on ramjets at speeds up to Mach 0.6 and altitudes up to 7 km. During his two years at that location he and Dr. Irene Bredt—who after the war became his wife—conceived the awesome antipodal, or rocket boosted glide, bomber concept. Their famed top secret report, *Über einen Raketenantrieb für Fernbomber,* was published in August 1944 by the Zentrale für wissenschaftliches Berichtswesen der Luftfahrtforschung and distributed to 70 organizations and individuals.

After the war Sänger joined the French Direction Technique et Industrielle du Ministère de l'Armement outside Paris, providing services also to the Société Matra and Nord Aviation, all the time continuing his work on ramjets and rockets. He returned to Germany in 1954 and established the Forschungsinstitut für Physik der Strahlantriebe, Europe's first space research institute. Concurrently, he was a consultant for ramjets at Daimler-Benz and a faculty member of Stuttgart's Technische Hochschule. In 1960 he lectured at the University of Cairo and a

year later became technical advisor at Gollnow-Werke AG (later Avia-
test) and research scientist at Junkers Flugzeug und Motorenwerke AG.
Finally, in 1963, he was named full professor of the Department and of
the Institute of Flugtechnik IV (Elements of Space Flight Technology)
at the Technische Universität Berlin. Since 1950 he had been active in
the International Astronautical Federation, being instrumental in its organization, serving at one time as its president, and assisting in the creation and development of its journal, the *Astronautica Acta*. During his
later years he became internationally known for his theoretical work on
photon rockets. Professor Sänger was widely honored in Germany and in
many countries throughout the world.

Dr. Irene Sänger-Bredt, Professor Sänger's widow, long-time co-
worker, and renowned physicist, has graciously consented to fill her late
husband's place on the Editorial Advisory Board. In April 1964 she
wrote that "I consider it as a great honor and a duty to my dear husband
to accept this, although I never will be able to replace him really. In
any way, I will try to do my best in his memory."

Contents

Gravity

BRYCE S. DeWITT

Integration of Payload and Stages of Space Carrier Vehicles

GEOFFREY K. C. PARDOE

Navigational Instrumentation for Space Flight

SAUL MOSKOWITZ AND PAUL WEINSCHEL

Space-Related Technology: Its Commercial Use

ROBERT H. WATERMAN, JR., AND LLOYD G. MARTS

Progress in Rocket, Missile, and Space Carrier Vehicle Testing, Launching, and Tracking Technology
Part 1: Facilities in the United States

MITCHELL R. SHARPE, JR., AND M. LOWTHER

Contents of Previous Volumes

Supplementary Monographs

Supplement 1: O. H. LANGE AND R. J. STEIN
Space Carrier Vehicles: Design, Development, and Testing of Launching
  Rockets, 1963

Supplement 2: N. A. WEIL
Lunar and Planetary Surface Conditions, *in preparation*

Gravity

BRYCE S. DeWITT

Institute of Field Physics
Department of Physics, University of North Carolina
Chapel Hill, North Carolina

I. Observational Description of the Gravitational Field

A. Geodetic Deviation

We know that the Earth possesses a gravitational field because the Earth holds us to its surface. With the aid of the most rudimentary devices, we can measure at the surface of the Earth the ubiquitous force which keeps us and all our familiar surroundings in place and which renders the lifting of heavy payloads into the sky and into space such an expensive undertaking. The word *field* is essentially a synonym for *force*, or rather,

1

for the "set of all forces at different locations." As a concept, it describes how the force varies from place to place.

Is that all it is, just a conceptual, or mathematical, bookkeeping system? The answer to that depends on our point of view—on what we want the notion of "field" to do for us. If we want it to suggest new ideas, the answer is no, but if we are using it merely to correlate observational data, the answer is yes.

What are these observational data? In the case of gravitational forces this is a good question, because the answer is not as evident as one might think, and in the process of determining it we find that the concept of "field" becomes transformed, through several degrees of sophistication, into something quite different from the naive bookkeeping system with which we started. We can then say that the field idea is no longer purely a mathematical construct. Although it is expressible in mathematical language, it leads a conceptual as well as apparently a physical life of its own. This is a metaphysical point, perhaps, but important.

Observations are performed, supposedly, on something that really exists. The first difficulty we encounter with gravity is the *equivalence principle:* gravitational forces differ in no way from inertial forces. This principle is certainly true in first approximation, and in view of the high accuracy to which the Eötvös experiment (designed to determine just this question) has been pushed, it is an extremely good approximation for localized matter.

How do we know, then, that the Earth is not just a giant spaceship which has been accelerating upwards under us over the ages and keeping our feet on the ground? Is the answer to be that we would long ago have begun to detect a relative motion with respect to the stars? If so, then how do we know that the Earth, the Solar System, and all the stars and galaxies are not accelerating together? One obvious answer is that the world is round, and if the acceleration were positive on one side, it would have to be negative on the other!

Evidently, therefore, gravity must truly be a force in its own right, concentrated in and generated by matter. Indeed, as the Cavendish experiment shows, gravity is concentrated in quite small bits of matter (infinitesimal on the astronomical scale). The gravitational force between any two objects is always found to be attractive. Moreover, not only is the force exerted on a given body proportional to the mass of the body (the equivalence principle) but so also is the force which that body exerts on other bodies. It is even possible to demonstrate the inverse square distance law of this force in the laboratory.

A suspicion remains, however, that something is still missing from this picture. Suppose we pick up the laboratory and let it drop. The force which the Earth exerts is then suddenly gone as far as the experimenter inside is concerned. No other force behaves in this way. An electrically charged experimenter falling freely in an electric field would still feel the full force of the electrical pull. His neutral components would tend to tear themselves away from his charged components, and these in turn would tend to separate from one another as the charge which each carries varies from place to place. (We neglect their mutual interactions.) Why is the gravitational field felt only when there is direct physical contact?

This leads us back to the equivalence principle again, and to two questions. What does the equivalence principle mean? Can it be violated? In order to answer them, let us remove the direct physical contact with the Earth, blast off into space, and go into orbit. Suppose we do even more than that. Suppose we close all the ports, shut off radio contact, and go into complete isolation. Can we then still tell that the Earth is there?

The answer is that we can if our spaceship is big enough and our laboratory equipment sensitive enough. Imagine two little steel balls ("test bodies" they are sometimes called) floating freely inside the cabin. It may happen that they are in line with the center of the Earth. If so, then the orbit of one has a slightly greater radius than the other. Their rotation periods will consequently differ, and no matter how careful we are to place them initially at rest with respect to one another and with respect to the cabin, the space between them will eventually begin to widen at a steadily increasing rate. The same thing will happen, to a greater or lesser degree, in other orientations as well, both when the spaceship is in a circular orbit and when in a highly elliptical orbit, although sometimes the separation will tend to decrease rather than to increase. It should be emphasized that the "test bodies" have been assumed to be placed far enough apart so that their mutual attraction may be neglected. Similarly it is assumed that the cabin has been designed in such a way that its own gravitational effect is negligible (e.g., uniform spherical shell).

From this imaginative experiment we learn that the equivalence principle is only a local principle. Over a sufficiently large domain (i.e., over a region big enough to distinguish a gravitational phenomenon from an inertial phenomenon), it is violated. The tendency of the test bodies to diverge from or converge toward one another provides a measure of this violation. The exact law which describes the violation is called the law of *geodetic deviation*. For the purposes of this chapter the mathematical statement of the law will not be needed, although it is quite simple. It will be sufficient to know simply the reason for its existence and how it manifests itself.

We may now ask two questions. First, does the geodetic deviation effect manifest itself in any already familiar way? And second, does it by itself uniquely determine the gravitational field and hence the properties of the body which produces the field, insofar as these properties (e.g., the mass) relate to gravitation? The answer to both questions is yes. The ocean tides, for example, are a manifestation of geodetic deviation. By means of the tides the Earth would know that the Sun and the Moon exist even if they were optically invisible. Moreover, if it were possible to adjust the radius of the Earth's orbit over a measurable continuous range, the gravitational field of the Sun could be accurately mapped out and the Sun's mass determined simply by observing the variation in the height of the tides.

Even if the oceans did not exist, the Sun and the Moon could be detected by the strains which they set up in the Earth. In the imaginary experiment described above, the test bodies actually can be omitted. The strains in the spaceship itself are by themselves sufficient, in principle, to detect the Earth. That such strain-producing forces can sometimes be quite strong is illustrated by the innermost moons of Saturn, which have crumbled into rings.

The idea that the real gravitational forces, i.e., those which remain after contact forces are removed, are strain-producing forces is an important one. Gravity and strain are, in fact, intimately related. Let us observe, however, what this idea has done to our field concept. We began with the conventional idea that the strength of the gravitational field is measured in units of force per unit mass, i.e., acceleration. But, recognizing that if this were the whole story, gravitational forces would differ in no way at all from inertial (i.e., acceleration) forces, we have come to the more sophisticated view that the gravitational field really should be measured in units of force gradient per unit mass, that is, in reciprocal square seconds.

From this point of view the basic observational data, which provide a direct measure of the real gravitational field, are the geodetic deviation effects. We may ask, however, whether these data are really the most handy to use. Can we not, in principle at any rate, map out the Earth's gravitational field in three dimensions by building scaffolds into the sky and using ordinary gravity meters? No, we cannot. The use of static gravity meters prevents us from measuring certain components of the gravitational field which arise from the rotation of the Earth in consequence of the laws of general relativity, much as magnetic forces are special relativistic consequences of the motion of electric charges [1]. It is true that the magnetic components of the Earth's gravitational field are only inferred; they have never been measured. But the use of static gravity meters will never reveal them to us.

The possibility remains of simply filling the sky with a lot of high speed projectiles and mapping the gravitational field by observing their orbital motion. Relativistic as well as static components can, in principle, be detected in this way without the use of pairs of test bodies undergoing geodetic deviation. A crucial factor which makes this possible, however, is that we have the "fixed" stars to serve as a reference frame for determining the projectile orbits.

B. The Frame of the "Fixed" Stars

Suppose the only visible celestial objects were the Sun, the Moon, and the planets. How would celestial mechanics have fared under these circumstances? Certainly the problem of unraveling the puzzle of planetary motion would have resisted solution longer than it did, and it is likely that some branch of physics other than astronomy, e.g., elasticity or hydrodynamics, would have fathered the development of modern mathematics and modern science. However, in time the true laws of planetary motion would have been discovered, for the Earth itself provides us with a useful reference system. The fact that the Earth's axis is not perpendicular to the plane of the ecliptic would make it possible for us to distinguish the absolute rotation of the Earth from its daily rotation with respect to the Sun. However, even with the use of this reference system, the agreement between theory and experiment would have been far from perfect, because the Earth's axis suffers a general precession through the sky, with a period of about 25,800 years, which impairs its usefulness as a gyrocompass. Eventually, this too would have been discovered, with the aid of modern computing machines, perhaps; and a least squares adjustment based on Newton's theory.

Actually a least squares analysis is useful even when the stars are visible, and it has in fact been performed [2]. Our ability to see the stars permits us, in one leap, to start our calculations at a much higher level of accuracy than can our imaginary and less fortunate brethren to whom the stars are invisible, but we must still correct for the motion of the stars as they partake of the general galactic rotation. That is, even the fixed stars are not "fixed."

The goal that one strives for in solving the problem of planetary motion is simplicity. The Copernican theory is simpler than that of Ptolemy; the Newtonian-Keplerian theory is simpler than that of Copernicus; and, if economy of means be the chief criterion, the Einstein theory is simpler than that of Newton in explaining the anomaly in Mercury's perihelion precession. Even if the stars were invisible, the Earth always presented the same face to the Sun, and the planetary orbits were arranged randomly at arbitrary angles to the ecliptic; or worse still, if there were several small suns, and the planets formed merely a statistical swarm, we would still strive to discover that elusive absolute frame in which the laws of gravity and motion become simple.

Here, however, there might be some differences of opinion as to what constitutes simplicity, particularly if the planets constituted a random swarm. In this case the raw observational data would certainly not be declination and right ascension. Assuming we have radar, they would instead be relative distance and relative velocity, and this leads us back, with a vengeance, to the concept of strain in its purest and simplest form. The members of the planetary swarm define, at any one instant, a coordinate system. At the next instant these coordinates have already begun to undergo a deformation, a change of shape, in short, a strain. The object of experimental research and theoretical inquiry would become that of establishing the laws which relate, not the rate of change of right ascension, but the rate of change of strain, and its rate of change in turn (i.e., the geodetic deviation) to the distribution of matter which the swarm of planets and suns provides.

The astronaut, randomly suspended in his tumbling space capsule, can easily appreciate the primacy of relative position and relative velocity as data for getting one's bearings. He is perhaps our first true relativist. Yet even the astronaut may not be prepared for the shock of learning that the very concept of distance is elusive, that the Euclidean manner in which he uses the concept in the small confines of his cabin, or even in interplanetary space, may lead him into contradictions on the grand scale of the universe. It is only because man's effort to conquer the stars is still puny that this difficulty is not likely to bother him in a practical way for many years to come.

To his counterpart in the planetary swarm, it would probably come as no shock at all to learn that space may not be Euclidean. Having become used to curvilinear coordinates at an early age, the latter would accept philosophically the minor modification this would entail in his description of the world. His shock would come at the advent of the theory of special relativity when he learns that time too is involved in the breakdown of Euclidean ideas and that one cannot speak of space without speaking simultaneously of time. Having passed this hurdle, however, he would presumably be just as willing to keep a suspicious eye on his clocks as he already does on his meter sticks.

The lesson to be learned from all this is that although over a limited region of space and time one may by hard work find a simplest coordinate system—Euclidean, inertial, or what have you—in which to correlate observational data on gravitational (or any other) phenomena, it may not really be the best. Even the fixed stars do not provide a safe frame; in the vastness of the universe they are ephemeral. Ultimately, no frame is absolute. To obtain absolutes we must be content with restricting our attention to the local scene and fashioning our theory of the world in such a way that it accepts the raw data directly as they come to us. Therefore, although certain frames may be the best for practical purposes, that theoretical frame is best which accepts all coordinate frameworks equally.

II. Theoretical Description of the Gravitational Field

A. Curvature and Riemannian Geometry

Having renounced the comforting security of the fixed stars, we may well ask if there is anything left to live for. Are we not then condemned to an existence without absolutes? If one curvilinear coordinate system is as good as another, how can one have any standards at all? Is this not scientific depravity?

To judge these questions it is useful to do two things. First, we should take note that the subversive ideas of general relativity were not just plucked out of thin air. Nature herself is equally guilty with Einstein in propounding them. The equality of accelerated frames in the absence of gravity to static frames in the presence of gravity is already suggestive enough. Add to this the inescapable fact that even the most carefully constructed empirical frames are ultimately unstable, and the demolition of the absolute is complete.

Second, we should consider the humble potato. If I ask you to draw for me on the surface of a potato the best network of lines to serve as a grid for recording the results of a microscopic examination of the skin, I think you will have some trouble. If I had asked you to do this on the surface of a globe, or an inflated inner tube, or even a light bulb, I think you might have given me a reasonable answer—but on a potato, no. There is no best coordinate system on a potato. And yet the potato is not without distinctive features—a bump here, a depression there, and a couple of wrinkles somewhere else. What absolute property does each potato possess that makes it distinguishable from its neighbor? The answer, in a word, is curvature.

The curvature of space-time is a property no less absolute than that of the potato. It varies from point to point, a little greater in one place and a little less in another; it is only harder to visualize. Furthermore—and this is what is really noteworthy—the concept of curvature fits hand in glove with that of strain, so much so that a theory of gravity can be built on it, which in practice is distinguishable from Newton's theory only with great effort. Mathematically, the connection between the two concepts arises from the similarity of the so-called metric tensor of Riemannian geometry to the well-known strain tensor of the theory of elasticity, and to the rate-of-strain tensor of hydrodynamics.

Riemannian geometry is the theory of distance and angle relationships on a curved surface or in a curved space, and the metric tensor is the basic object of the theory. It is not necessary to use the mathematical language of the theory to be able to appreciate the connection between geometry and strain. Let us imagine that we have a small square piece of well-greased rubber membrane pressed tightly to a smooth but irregularly shaped object. As we slide the membrane from place to place over the surface of the object, its shape undergoes various distortions, remaining nearly square in regions of slight curvature but becoming quite distorted when the curvature is pronounced. That is to say, it undergoes a strain which is directly correlated with the curvature. This strain is of exactly the same type as that which the Earth suffers in the gravitational field of the Sun.

A still better picture of the relationship of Riemannian geometry to the physical world can be obtained by imagining that we have a group of little steel balls somehow constrained to remain in contact with the irregular object but otherwise able to roll freely on its surface. If at a relatively flat place on the surface the group is set in motion in such a way that all the balls are moving parallel to one another with equal velocity, they will retain their relative configuration only so long as they remain in the flat region. As soon as they enter a curved region, they will begin either to converge toward or to diverge from one another according to whether the curvature in that region is positive or negative. This is precisely the phenomenon of geodetic deviation.

In the real physical world we are constrained to remain in the four dimensional universe of space and time. Moreover, we cannot avoid motion in that universe, for we are forever hurtling forward in time. Time is, in fact, the important new element in the picture. It turns out that although space is curved where a gravitational field is present, there is a *curvature in time* which is much more noticeable. This is because of the high value of the velocity of light, which is the magnitude relating the scale of space to that of time. In metric units the curvature of space is measured in reciprocal square centimeters.[1] In any planetary gravitational field this curvature is so slight as to be undetectable by means of static measurements, and yet our headlong rush in time, which converts it into reciprocal square seconds through multiplication by the square of the velocity of light, is so fast that in dynamical situations the curvature becomes noticeable, much as a slight bump on the salt flats of Utah, which a pedestrian would take no notice of, becomes a menace to a high speed racing car. That is why, although the space in our immediate vicinity appears Euclidean (i.e., flat) to a high degree of accuracy, we can nevertheless see the curvature of space-*time* every time we throw a ball into the air.

[1] The curvature of a curve in space is measured in units of reciprocal length. However, the curvature tensor of Riemannian geometry, which is the quantity of direct dynamical significance in the theory of gravitation, has the dimensions of reciprocal area.

With this view of the world, our concept of "field" has reached a higher level of sophistication than ever. The words *curvature* and *gravitational field* are now synonymous. In the process space-time has become endowed with a structure. It is no longer merely an arena in which things happen. Its variable curvature gives it a texture and character that are far more absolute than anything which a mere flat manifold possesses. As Einstein himself pointed out more than once, it goes a long way toward resurrecting the old concept of the ether, but in a form that is now entirely acceptable.

B. The Principle of General Covariance

In order to complete the world picture which the transition from Euclidean to Riemannian geometry opens up to us, it is still necessary to determine the dynamics of the gravitational field itself and how it interacts with systems other than test bodies. That is, we need to find equations which express the law by which the gravitational field is itself produced and propagated in the presence of matter, electromagnetic fields, and all other fields of force. In this we are guided by the *principle of general covariance:* all our fundamental equations should look the same in every curvilinear coordinate system. One might wonder whether it is possible to find any equations which satisfy such a restrictive principle as this. Actually there are an infinite number, each of which could, in principle, serve as a basis for a theory of gravity. The covariance principle by itself is not sufficient to determine the theory; a principle of simplicity is needed as well. It turns out that if the simplest[2] set of equations is chosen, the resulting theory is indistinguishable from Newton's theory when the motion is slow and the fields are no stronger than planetary.

It is to be emphasized that the principle of general covariance is an addition to and not a consequence of the equivalence principle. Although strongly suggested by any view which regards curvature as a primary element of the universe, it cannot be regarded as experimentally proved in the same sense that the equivalence principle has been verified. Its attractiveness lies in the powerful consequences which it implies when it is combined with the so-called *strong principle of equivalence.* The strong principle of equivalence asserts not merely that gravitational and inertial forces are indistinguishable but that all physical phenomena have the same local appearance, whether at rest in a gravitational field or in a state of equivalent acceleration in gravity free space—or, alternatively, that they have the same local appearance in a frame undergoing free fall as they would have in an inertial frame in the flat space-time of special relativity.

---

[2] In this case the word *simplest* can be given a precise mathematical meaning.

This assertion, in combination with the covariance principle, uniquely determines the laws of interaction between the gravitational field and all other dynamical systems. It is completely consistent with the law of geodetic deviation. In the limit in which the extra dynamical systems reduce to an ensemble of test bodies of negligible mass, the motion of each body is found to be geodetic—i.e., along a shortest path in space-time—and the ensemble as a whole exhibits the geodetic deviation effect.

C. Alternatives to Curvature

1. Scalar Theory

No other theory of gravity is so intellectually compelling as Einstein's; nevertheless, since the Riemannian character of space-time has not yet been established as an incontrovertible experimental fact, it is necessary to investigate other possible theories. In selecting these, we may be guided by the principles of special relativity and quantum mechanics which, in their appropriate domains, have been established beyond all doubt. The simplest alternative theory is that which merely takes the Newtonian theory in its ordinary form and converts it from a theory of action-at-a-distance to a theory which satisfies the demands of special relativity. The fundamental field variable of the Newtonian theory is the gravitational scalar[3] potential. This scalar potential is made to satisfy a wave equation instead of Laplace's static equation, so that gravitational effects will be propagated with the speed of light rather than instantaneously. For slow motion the two theories are again indistinguishable.

A characteristic and interesting by-product of relativistic requirements is that the scalar theory predicts that the mass of a body—as measured by applying nongravitational forces—will vary from place to place in a gravitational field. Since no experiment of sufficient sensitivity to test this prediction has yet been performed, it is impossible to deny on this basis the possibility that gravity may actually have a scalar component.

It is possible to show from experiment that gravity cannot be purely scalar. One of the cornerstones of any physical theory is the law of conservation of energy and momentum. Because of the mass-energy relationship this law is of particular importance in gravitation theory. The mathematical object which, in a relativistic theory, describes the distribution of mass-

---

[3] A scalar is a quantity having only a single component which, at a given physical point, is unchanged by a coordinate transformation. Thus, unlike vectors and tensors which have components projected along the coordinate axes, its numerical magnitude does not depend for its definition on the presence of a coordinate system.

energy and momentum in any dynamical system is the *energy-momentum tensor*. Only this tensor, and no other, can serve as the source of the gravitational field. In order for it to serve as the source for a scalar gravitational field, it must be taken in its contracted form. The interaction between gravity and matter then takes place via this contracted form. It is well known, however, that the contracted energy-momentum tensor for light waves vanishes, and therefore light cannot interact with a scalar gravitational field. On the other hand, observation shows that light rays passing close to the Sun are slightly bent. There is no plausible agent other than gravity which can be invoked to explain this bending. Therefore, gravity cannot be wholly scalar.

2. TENSOR THEORIES

A satisfactory vector theory of gravity is very unlikely for two reasons: (1) it would require a vector source, and such a source cannot be constructed from the energy-momentum tensor; and (2) bodies would repel one another rather than attract. This is illustrated by the electromagnetic theory, which is a vector theory. A well known law of electricity is that like charges repel one another. In a vector theory of gravity, only negative masses would be attracted to the Earth.

Tensor theories, on the other hand, can be constructed by the dozen. In these theories the full energy-momentum tensor is used as the source. Source and field each have ten components, so that gravity becomes a more complicated phenomenon than Newtonian theory anticipates.

A few simple and even logically necessary requirements can drastically reduce the number of possible tensor theories. In brief, these requirements are: (1) the gravitational field equations should be Lorentz invariant (a requirement of special relativity) and of the second differential order; this is the simplest nontrivial possibility and remains closest to Newton's theory; and (2) the equations must describe what is known as a *field of long range*. The effects of the field must be felt over enormous distances, and the inverse square force law must hold with a high degree of precision. In the quantum theory, fields of this type are known as *massless fields*. It is an interesting fact that this requirement, taken together with the preceding ones, almost uniquely determines the theory. It completely determines it in first approximation if we at the same time require that the energy contained in the gravitational field shall be always positive. The positive energy assumption is perhaps the weakest link in the inductive chain. It becomes truly essential only when the laws of quantum mechanics are invoked, and quantum phenomena have never been observed to play a role in gravitation. However, if we accept the pertinence of the quantum theory, then the energy must be positive, for otherwise it can be shown that the very fabric of space-time, i.e., the vacuum itself, would become unstable and decay into a meaningless chaos.

The energy question is a very central one to the gravitation theory, for energy is the source of the gravitational field. To continue our line of reasoning, let us now consider this source; more precisely, let us consider the energy-momentum tensor. In the absence of a gravitational field, this tensor satisfies a differential identity which expresses the laws of conservation of energy and momentum. It is a noteworthy fact that the field equations (which, owing to previous requirements, are now unique in first approximation) satisfy the same differential identity. It therefore becomes immediately possible to insert the source into the equations in a natural manner, which is completely free from inconsistency. As soon as the source has been inserted the gravitational field can no longer vanish. As a result, the energy-momentum tensor no longer exactly satisfies the differential identity, and the laws of conservation of energy and momentum are violated. This is because we have neglected the energy contained in the gravitational field itself. In order to maintain consistency it is necessary to include this energy also in the source.

When this is done two remarkable results follow [3, 4]. First, the gravitational field then acts partly as its own source. This phenomenon, which may be described as a feedback process, leads to a characteristic and unavoidable nonlinearity of the full exact field equations. Second, it is found that standard meter sticks, which are constructed out of normal matter held together by the familiar nongravitational forces (i.e., electromagnetic, nuclear, and quantum forces), no longer maintain a constant length when they are moved from place to place in a gravitational field. Similarly, standard clocks no longer tick off equal time intervals.

The lengths and time intervals which standard rods and clocks measure are by definition constant. The variation in these quantities referred to above is only a mathematical result, based on the fallacious initial assumption that space-time is flat and that length and time intervals can be read off directly from the coordinates of a so-called Minkowskian inertial frame of special relativity. Such variations are in reality completely unobservable, for they apply equally to all rods and clocks, independently of how the latter are constructed, and hence no way exists for detecting them. Therefore, although a tensor theory constructed in the above manner is not wrong, it must be reinterpreted in such a way as to maintain consistency with the rest of physics, namely, by restoring to rods and clocks their primacy as the standards for invariant intervals. But this can only be done by abandoning the belief that space-time is flat, and the truly astonishing fact is that when this reinterpretation is carried out, the tensor theory becomes identical with Einstein's theory. It is for this reason that we may say that Einstein's theory is the most logically as well as the most intellectually compelling theory of gravity.

D. Mach's Principle

The basic postulate of the special theory of relativity states that all of the laws of nature take the same form in every inertial system. An inertial system is a rectilinear Cartesian framework of (imaginary) rods and synchronized clocks which is in uniform motion at constant velocity with respect to every other inertial system. It is necessary to assume that at least one inertial system exists. Then all the others exist as well, and it is possible to pass from one to another by performing a suitable rotation of the Cartesian framework in space and imparting to it an appropriate velocity increment. One important consequence of the basic postulate is that since no preferred absolute frame exists, the concept of absolute motion is meaningless. Only relative position and relative velocity have meaning, and only a preferred set of frames exists, namely the inertial frames, all of which have equal importance.

In the general theory of relativity, on the other hand, there are no preferred frames whatsoever. Any curvilinear coordinate system is as good as another. The laws of nature take the same form in all. Since the curvilinear coordinate systems include all possible accelerated frames (i.e., curved in time), one might suppose, by analogy with special relativity, that in general relativity there is no such thing as absolute acceleration— that only relative acceleration has meaning. This would mean that a rotating body in an otherwise completely empty universe would experience no centrifugal forces; since there is no other object for it to rotate with respect to, it would not really be rotating at all.

This idea is known as *Mach's principle*. As a motivating concept it played an important role in Einstein's development of general relativity. When expressed in words, it sounds good. Unfortunately, no one has ever succeeded in expressing it mathematically with adequate precision. When the attempt is made to go beyond the bare verbal statement, it turns out that there are as many versions of Mach's principle as there are physicists who have devoted thought to it, all incompatible and all imprecise. On only one point is there agreement; namely, general relativity does not contain it, at least without some additional assumptions which go beyond what general relativity was originally and successfully intended to be, a theory of gravity.

The reasons for this are probably simple. General relativity was built to give agreement with the Newtonian theory in the low velocity limit.

In particular, it was built to be able to handle dynamical systems involving only a finite number of mutually gravitating bodies, each confined to a limited region of space. The most obvious and natural boundary conditions for such systems are those in which space-time becomes flat at large distances, where the gravitational field tends to zero. But a flat Riemannian space-time is absolute, in the sense that accelerations with respect to it are physically detectable with no ambiguity whatsoever; for the theory of flat space-time is just the theory of special relativity, and the only absolutes which special relativity denies are those of position, orientation, and uniform motion.

Fock [5] has pointed out that general relativity is even less relativistic than special relativity, for as soon as space-time acquires bumps (i.e., curvature), it becomes absolute in the sense that it may be possible to specify position or velocity with respect to these bumps, provided they are sufficiently pronounced and distinguishable from one another. On the other hand, as soon as the bumps coalesce into regions of uniform curvature, space-time reacquires its relativistic properties. It never becomes *more* relativistic than flat space-time. All gradations between relativistic and absolute space-time are possible. In many cosmological theories a solution of Einstein's equations is adopted in which there is an absolute or cosmic time, while space itself remains completely homogeneous and isotropic, and hence relativistic.[4]

There is some belief (by no means universal) that the problem of suitable boundary conditions for the universe is intimately connected with Mach's principle. Wheeler [6] and Hönl [7] have argued that Mach's principle *is* a boundary condition and that it makes sense only if the universe is closed and finite. It would seem that this is true only in the sense that when space is finite, one feels more comfortable in asserting that the geometrical properties of space, and hence the inertial properties of bodies which occupy space, are directly determined by the distribution of matter throughout space. Unfortunately, such an assertion has not been rigorously proved and is presently suggested merely by contrast to the situation in which space is asymptotically flat, and hence infinite. In the latter case one feels not at all inclined to attribute the absolute inertial properties, which the flat regions possess, to the remote and insignificant quantity of matter contained in the effectively finite nonflat region. It must be emphasized that a decision as to whether the universe is finite or infinite has at the present time no more scientific status than an article of religious faith.

A more direct approach to Mach's principle consists in attempting to

[4] The existence of a cosmic time does not mean that one can overcome such interdictions of special relativity as the principle that no body can move faster than the speed of light.

establish an explicit relation between the inertial mass of a given body and the mass distribution of all other bodies in the universe. When there are no other bodies (or when their masses vanish), the mass of the given body should vanish so that it will exert no inertial force when accelerated. Its mass must be assumed to increase from zero as other bodies are inserted into the universe. Since, to the limits of presently available accuracy, the inertial properties of bodies are experimentally observed to be unaffected by proximity to terrestrial and planetary masses, the predominant effect must come from the distant matter of the universe, which may be regarded as having maintained an essentially constant relation to the Earth over the short time (< 200 years) during which accurate observations have been made. In this way the "fixed" stars (or "fixed" nebulae) re-emerge as the objects which endow the universe with absolute properties. A heuristic theory along these lines has been developed by Sciama [8]. Unfortunately, as has previously been remarked, no rigorous field theory embodying these ideas has ever been constructed.

E. Dicke's Ideas

A field theory which offers a partial embodiment of Machian ideas has been developed by Dicke [9]. Following a suggestion of Dirac [10] and making use of a mathematical formalism developed by Jordan [11], Dicke has introduced the idea that the value of the universal constant of gravitation is determined by the distribution of matter in the universe and not only varies from one place to another but also changes as the universe gets older. The mechanism which accomplishes this is a scalar field which is superimposed upon the tensor field of Einstein. As has already been remarked, gravity may well have a scalar component in addition to the tensor components which are necessarily present owing to the light deflection phenomenon. Experiments are not yet sufficiently accurate to decide this question. The presence of a scalar component has a certain esthetic appeal, being a natural consequence of certain unified field theories.

Dicke has been able to estimate a rough upper bound to the amount of scalar component the gravitational field can possess, without violating known observational results. With this bound imposed, he then shows that it will be extremely difficult to distinguish the scalar and tensor components in the laboratory. Some of the consequences to be expected from the presence of the scalar component are: (1) anomalous behavior of the temperatures of the Sun and Earth during past ages, necessitating corrections (e.g., by factors of order 2) to the currently estimated ages of the Solar System; (2) similar anomalous behavior of the temperatures of the stars, leading to corrections in our views about stellar and galactic evolution; (3) anomalies in the dynamical behavior of galactic clusters; (4) slight anomalies in the motions of the Earth and planets, including corrections to the perihelion precession of Mercury and variations in the Earth's axial rotation rate; and (5) corrections to Einstein's law for the deflection of light by the Sun.

The easiest of these consequences to understand is the third. The scalar field in Dicke's theory plays the role of the inverse of the gravitation constant. Like the tensor component of the gravitational field, it is produced by matter, and its magnitude is greatest where the density of matter is greatest, e.g., at the center of a galactic cluster. The presence of the matter in a galaxy therefore tends to depress the value of the gravitation constant at its center, while outside the galaxy the gravitation constant, and hence the force of gravity, is greater. This has the consequence that the cluster tends to hold together more strongly than one would predict, on the basis of Newton's or Einstein's theories, by observing the relative velocities of the galaxies in the cluster and using the local (depressed) value of the gravitation constant. The effect, although in the right direction, is unfortunately not sufficiently pronounced to provide a quantitative explanation for the actually observed peculiarities of galactic clusters [12].

Anomalies in the motions of the Earth and planets could be due to what Dicke calls "ϕ-waves," that is, oscillations in the scalar field itself, propagating with the velocity of light. The period of such oscillations could well be of the order of months or years and give rise to slight deviations from the expected planetary motions owing to the corresponding oscillations in the gravitation constant. Another possible role which ϕ-waves might play is that of providing a mechanism for detonating white dwarf stars which are close to the critical Chandrasekhar limit. A slight increase in the gravitation constant would suffice to cause such stars to undergo unstable collapse followed by a supernova explosion. It is even conceivable that such an implosion-explosion process could radiate additional ϕ-waves, which would in turn detonate further white dwarfs. This could explain the associated production of supernovae, which has been postulated as an explanation for certain very intense observed radio sources. It may be that these sources are galaxies in which, at one time, there were many stars near the critical state, about ready to become supernovae, and that by chance one went off and produced a ϕ-shock wave which set the others off.

One argument in support of these ideas is that no mechanism is known other than that of a scalar field which could radiate the requisite amounts of energy via an implosion-explosion process. This is because such a process involves predominantly an oscillating monopole, and only scalar waves can transmit monopole radiation. This is in contrast to electromagnetic waves, which transmit predominantly dipole radiation and gravitational waves, which (theoretically) transmit quadrupole radiation. The amount of ϕ-wave energy radiated by a supernova might be so great as to be detectable on Earth as a variation in the gravitation constant. Dicke has set up a program for monitoring the gravitation constant to a very high degree of accuracy over long periods of time, in the hopes of settling just such questions as these.

From the point of view of Dicke's theory, the extreme smallness of the gravitation constant is to be understood as due to the enormous quantity of mass in the universe. One must be cautious in applying this notion to the cosmological models which Dicke has constructed. Since the universe is expanding, the mean density of matter is decreasing; and one might therefore suppose that the scalar field is decreasing as well, leading to a gravitation constant which is steadily increasing with time. Dicke says that this is not so. An expanding universe involves a dynamical situation, and when the dynamical equations for the scalar field are written down, he is able to obtain a solution of them in which the field increases with time. It is to the corresponding decrease of the gravitation constant over the ages that Dicke ascribes possible anomalies in the temperatures of the Sun and stars in bygone epochs. The particular solution which Dicke obtains depends crucially on a certain choice of boundary conditions, which is potentially controversial. Therefore, it is by no means yet settled just what the theory really predicts. On the other hand, if a secularly decreasing gravitation constant is accepted, then stars would certainly have been hotter in the past when gravity was stronger. Being hotter, they would have burned faster; hence the universe (or at least our galaxy) would really be younger than is usually estimated.

F. Gravitational Waves

The relativistic version of any static field theory necessarily predicts the existence of waves. This is because of the fundamental limitation on the propagation of disturbances: no disturbance in the field (i.e., no signal) can be propagated at a velocity greater than that of light. Gravity is no exception. The main theoretical properties of gravitational waves have been known for many years. Owing to the extreme weakness of their effects, gravitational waves have never been observed experimentally. Only recently has an attempt been made to improve this situation. J. Weber of the University of Maryland has designed and built an extremely sensitive antenna for the detection of gravitational waves in the kilocycle range [13]. The fundamental principle on which the design is based is utterly simple. It goes right back to the law of geodetic deviation. As gravitational waves sweep past the antenna, they set up strains in it. Weber's antenna is a large cylinder of aluminum, and the strains induced in it are measured by piezo-electric crystals mounted on its surface. He is able to detect fractional changes in the length of his antenna equal to 10^{-17}. The antenna will soon be in operation and will monitor the gravitational waves passing through College Park, Maryland (if any) over a considerable length of time.

In principle, Weber's antenna could be run backwards, i.e., driven in such a way as to generate gravitational waves, which could then be detected by another antenna. In this way one should ultimately be able to measure the delay between emission and detection of the waves and thus verify that gravitational waves propagate with the speed of light. Unfortunately, such a measurement would have to be carried out in the wave zone to be decisive, and the sensitivity of the apparatus is still too low for this by many orders of magnitude. Moreover, there is little interest in carrying out the experiment in the near zone of the antenna, for it would then be essentially nothing but a quadrupole Cavendish experiment run with fluctuating parameters.

The strains which a gravitational wave sets up in an antenna are, according to Einstein's theory, perpendicular to the direction of propagation. Gravitational waves, like electromagnetic waves, are transverse and are usually produced in a polarized state. This makes it possible to sort them out, to a certain extent, from other strain-producing forces. Weber has analyzed seismic data on the very low frequency oscillation modes of the Earth and has obtained an upper bound on the amount of gravitational radiation falling on the Earth at wavelengths of the order of a billion miles. In this analysis the Earth itself was regarded as the antenna.

G. Current Theoretical Problems

1. ENERGY IN THE GRAVITATIONAL FIELD

In principle, the gravitational waves emitted by an antenna carry away energy which can be absorbed by another antenna. Such energy-transmitting waves are often referred to as gravitational radiation. Since gravitational radiation contains energy, it acts as a source of the gravitational field, thereby modifying itself. Owing to this nonlinear self-modification process and the effect which the gravitational field has on the metric of space-time (and on the definition of lengths and time intervals), it is impossible to give a precise statement of exactly where the energy in the gravitational radiation field is located, or how it is distributed, at any given instant. It is for this reason that many of the common engineering ideas, such as Poynting's vector, which are useful in electromagnetic theory, cannot be rigorously applied to gravity. Much of the theoretical effort in recent years has gone into finding effective substitutes for the familiar engineering concepts, so that we may begin to get a feel for the more exotic predicted properties of the gravitational field, even though experimental observations are lacking.

It would take us too far afield to describe these efforts, which are in the main highly technical. In the case of the energy problem, we shall simply state that it has been found possible to give a rigorous definition of the total energy contained in a given gravitational field, even though its precise location remains undefined. This possibility occurs only when certain boundary conditions are satisfied; namely, space-time must become flat asymptotically (i.e., at large distances). Otherwise, energy itself is undefined (e.g., it is undefined in cosmological models in which space is closed). Moreover, space-time must become flat in a certain way, that is, with a certain degree of rapidity. It is in making this latter condition precise that many of the recent developments in this area have arisen. The effort has forced us to improve our methods of classifying gravitational waves and our methods of specifying initial data, to devote greater attention to the so-called *light cone* and other characteristic surfaces which are fundamental in the description of propagation phenomena in space-time, and to reformulate the concept of conservation laws.

2. QUANTIZATION

Although general relativity has had a profound influence on the rest of physics, it has always remained largely outside the main stream of physical theory where quantum mechanics reigns. It was with the aim of removing this division in physics, and with a belief in the essential unity of nature, that the effort to bring general relativity and the quantum theory together (i.e., to quantize the gravitational field) originally began. There was no experimental motivation for the effort whatever, and it can be excused only on the grounds that it was logical.

At the beginning, the quantization effort was closely tied to the energy problem. This is because in one of the most familiar ways of looking at the quantum theory, it is the energy of a dynamical system which governs the way its quantum state develops in time. It was early recognized that the emphasis on energy singles out time for special treatment and is non-covariant. Moreover, it was not clear what one should do when energy could not be defined. Therefore, a split began to develop between those who believe that one should if necessary give up covariance itself before renouncing the energy language [14] and those, including the present author, who feel that a manifestly covariant language is both desirable and attainable [15].

The investigations of neither group are yet complete. However, at the present time the covariant approach has been pushed the farthest in the calculation of specific physical processes. This has been possible because it embodies an approximation procedure that regards the quantum theory as basically a theory of small disturbances on a classical background. In this respect it is like every other quantum field theory in existence. The proponents of the noncovariant approach argue that this is not true quantization, that something fundamentally and radically new occurs when gravitation is quantized, which can alter the basic structure of space-time itself. However, being at the disadvantage of having to tackle the most difficult end of the problem first, they have not been able to get this viewpoint off the ground. It is not inconceivable that both viewpoints are ultimately equivalent and that the covariant approach can outgrow its reliance on approximation techniques sufficiently for it to describe the fundamental modifications in space-time produced by quantization. Hints of this possibility already exist.

It is interesting to note that Einstein, had he lived to see these developments, would have said, "A plague on both your houses!" For him the quantum theory was only a way station on the road to the ultimate theory and not a permanent fact of life. He had hoped that general relativity itself would ultimately lead the way toward rescuing physics from the abhorrent indeterminism of quantum mechanics. There are few who share this view today. But even if Einstein were eventually to be proved right, this would not mean that the effort to quantize general relativity has been wasted. It has already clarified some of the difficulties encountered in quantizing other fields (notably the Yang-Mills field [16]) which bear a certain resemblance to the gravitational field and which have become of interest in recent years in connection with the problem of understanding the behavior and structure of the so-called *elementary* particles, of which there are now some thirty or forty known. Certainly, it is not too much to hope that some day all of these efforts will fall into place and yield a single, grand unified field theory.

3. TOPOLOGICAL PROBLEMS

One of the most remarkable features of the general theory of relativity is the richness of its content. We have already spoken of closed finite cosmological models, which gravity, in the guise of curvature, makes not only possible but also, under certain physical conditions of the universe, even necessary. Another possibility which is contained in the theory is that curvature may also play an equally fundamental role in the extremely small ultramicroscopic world of the elementary particles.

A closed finite universe and an open infinite universe are said to have different topology. So also in the very small it is possible to have different topological structures. It is a very interesting question to ask how such structures would behave if their curvature satisfied the laws of general relativity (i.e., the gravitational field equations).

Wheeler and his group at Princeton [17] have made the most exhaustive study of this question, although much still remains to be done. One of the topological structures to which they have devoted the greatest attention is the *wormhole*, a region of space in which gravitational lines of force converge just as if a material body were present, but in which, if one follows the lines of force, no body is encountered. Instead, one enters into a kind of hyperspacial tunnel and re-emerges in another region of space, possibly at an enormous distance from the first. Wheeler asks himself whether space-time is merely the arena in which things happen or whether it is "the all." He wonders whether the elementary particles themselves might be built out of large numbers of wormholes. He determines that it is possible to construct wormholes which have the appearance not only of mass but also of electric charge. He also finds that it is not possible to violate causality by crawling down a wormhole and reaching a distant region of space before a light signal could have got there. As long as the dynamics of the wormhole are governed by the laws of general relativity, the wormhole "throat" will either have pinched off before one can get through, or else, if one does succeed in re-emerging, an infinite amount of time will have elapsed on the outside, while only a finite amount will appear to have elapsed on the inside.

Wheeler's investigations are limited to the classical, unquantized theory. In a true theory of elementary particles, it is necessary for the wormholes to be able to annihilate one another and for wormhole pairs to be created. This cannot happen in the classical theory. There is some evidence, however, that it can happen in the quantized theory [18]. The evidence is not conclusive, since it comes from the covariant theory based on approximation techniques. However, it is the first indication that the covariant theory itself may be capable of describing the radically new properties of space-time which quantization can bring about.

In recent years topological considerations have begun to assume increasing importance, quite apart from the investigation of such fanciful objects as wormholes. If one wishes to know the structure of space-time in the large, i.e., globally, it is not enough to know merely the curvature. The topology also must be specified. It has become standard practice now to examine newly discovered solutions of Einstein's field equations as to their topological content. It is too early yet to say just what direction topological researches will ultimately take, but it is likely that theorems of topology will be among the weapons which the physicist of the future will employ with increasing frequency.

III. Experimental Tests of the Theory

A. The Classical Tests

For a theory which can predict such a wide variety of exotic phenomena, general relativity has received discouragingly little experimental confirmation. So far, the results of experiments have agreed with it within the limits of experimental error. But aside from the Eötvös experiment, only three others have been carried out. These constitute the so-called *classical tests*.

The classical tests are always quoted in a "canonical" order, the first being the frequency shift test. All theories of gravitation lead to the same prediction for this first test—even Newton's theory, provided it is supplemented with quantum mechanical arguments. This test can therefore not single out the Einstein theory from the others. However, an experimental result contrary to the prediction would demolish the Einstein theory along with all the others. The prediction is this: the frequency of a given train of monochromatic light waves will vary from place to place in a gravitational field, becoming lower in regions of high gravitational potential and higher in regions of low potential.

Until 1958, the only way that existed for testing this prediction was to observe the light emitted from the Sun and from certain white dwarf stars. According to the theory, the wavelengths of spectral lines in the emitted light should be shifted toward the red.[5] Such shifts were actually observed. Unfortunately, quantitative agreement with the theory was extremely poor. In the case of the Sun, the radial currents in the solar atmosphere (where the light originates) produce Doppler shifts which superimpose themselves upon, and obscure, the gravitational shift. Since a good quantitative theory of the solar atmosphere does not yet exist, it is impossible to say accurately how much of the total shift belongs to one effect and how much to the other. On the other hand, the predicted gravitational shift can be used to help make sense out of the radial current effect which would otherwise be totally incomprehensible.

In the case of the white dwarfs another set of problems arises. Although white dwarfs are, in principle, the most suitable stars for the observation of the red shift (since the effect is most pronounced with them), not many of them exist. Of these, only a very few (belonging to binary systems) have masses which are known with sufficient accuracy to make possible a comparison between theory and observation. The determination of the surface

---

[5] This red shift should not be confused with the cosmological red shift, which is a much more pronounced effect resulting from the high recession velocities of distant nebulae.

temperature of a white dwarf (which, together with its absolute luminosity, determines its radius) is rendered difficult by the ultraviolet absorption in the Earth's atmosphere, and it is not surprising that the agreement between theory and observation has been only qualitative. The best agreement has been achieved with the observations on 40 Eridani B [19].

In 1958, with the development of the crystal-lattice nuclear gamma emitter by R. L. Mössbauer [20], it became possible for the first time to carry out the frequency shift measurements on Earth. Mössbauer's device was put to use in 1960 by Pound and Rebka [21] to determine the shift produced over a vertical distance of 74 ft (the height of a shaft available in the Jefferson Physical Laboratory at Harvard). Using a recoilless gamma ray line having a fractional width of the order of 10^{-12}, they obtained results which agreed completely, within the limits of their experimental accuracy (~ 10 per cent), with the theoretical prediction. It is currently planned to repeat this experiment using greater vertical distances (elevator shafts in skyscrapers), and it is expected that the accuracy can be increased until the errors are reduced to 1 per cent.

The second classical test is that of the bending of light rays by the Sun. To observe the effect, one must necessarily wait for an eclipse so that the stars close to the Sun will become visible. The factors which make accuracy difficult in an eclipse observation are: (1) it is necessary to go on an expedition and to make observations with temporary apparatus set up in the field; (2) the eclipse lasts only for a brief period; (3) the weather may be bad; (4) the star field behind the Sun at the time of the eclipse is seldom the "best"; ideally the field should be as dense as possible so as to provide the best statistics; and (5) it is necessary to compare photographic plates taken during the eclipse with those taken several months later when the Sun is no longer in the same region of the sky (it is very difficult to determine accurately the scale factor(s) which connect the two sets of plates and to insure that the plates are free from the slight distortions which develop in almost any photographic emulsion).

In view of these difficulties, it is perhaps not surprising that the agreement between theory and observation is only semiquantitative, with observed results differing from the theoretical prediction and from each other by as much as 30 per cent. The agreement is good enough to rule out the scalar theory (no deflection) and Newton's theory (one-half the relativistic deflection). Thus, gravity is almost certainly a tensor phenomenon, although the observations cannot single out Einstein's theory from other possible tensor theories. Studies are currently under way to determine the feasibility of improving the data obtainable from terrestrial observations by building special coronagraph-like devices and using photoelectric methods so that it will not be necessary to wait for an eclipse.

It is only in the case of the third classical test, namely, that of planetary perihelion precession, that the observations begin to provide a real test of Einstein's theory. Here the effects depend on the nonlinearity of Einstein's field equations, and the theoretical predictions of various tensor theories differ from one another as well as from the scalar and vector theories. Here, also, the accuracy is highest. On the basis of Newton's theory, one can compute that the effect of all the other planets in the Solar System should be such as to cause a steady precession of the perihelion point of the planet Mercury, amounting to about 5552 sec of arc per century. The observed rate of precession, on the other hand, is 5595 sec per century—and this figure is derived after correcting for the general precession of the Earth's axis, which amounts to about 5026 sec per century. The difference between observation and Newton's theory is 43 sec per century, which is exactly the additional amount that Einstein's theory predicts should be present. It is remarkable that such a minute difference as this should be observable, in view of the vast amount of data which must be collected and correlated and the lengthy calculations which must be performed in order to detect it. It is a tribute to the skill with which astronomers have perfected their art and to the high degree of sophistication to which the theorists have been able to develop celestial mechanics.

The agreement between Einstein's theory and the observations of precessions of the perihelia of Venus and Earth is also very satisfactory, although in these cases the observational accuracy is poorer owing to the smaller eccentricity of the orbits.

B. Cosmological Tests

General relativity provides a theoretical framework which admits a wide variety of possible cosmological models of the universe, much wider than any other theory. Within each model there are certain requirements of compatibility which must be fulfilled, having to do with the mean density of matter and radiation, its general distribution, and its rate of expansion. The apparent ages and luminosities of the distant galaxies must also correlate with these data. The only reason for including cosmological tests as a subheading in this section is the hope that some day observations of the distant reaches of the universe will be sufficiently detailed and accurate to check the compatibility requirements, so that a definite commitment may be made either to Einstein's theory or to some other. At the present time the observational information is far too meager to permit this. Einstein's theory has merely provided the framework for two or three possible models which have some resemblance to the actual universe. Other theories have also provided viable models.

The two chief contenders for consideration as theories which are capable of describing the universe are Einstein's theory and the so-called *steady state theory* [22]. In the steady state theory the universe is regarded as being infinitely old and as having always had the same general appearance as it has today. The average mass density is kept constant, in spite of the observed expansion of the universe, by a process of continuous creation of hydrogen atoms out of nothing. The continuous creation hypothesis violates the law of conservation of energy and hence is incompatible with Einstein's theory, although certain attempts have been made to modify the latter theory so that it would embrace the steady state model as well.

In the Einsteinian models the expansion of the universe is generally regarded as having originated from a cosmic explosion at some finite time in the past, which may be taken as the zero point of time.[6] The chief difference between the various Einsteinian models has to do mainly with their mean curvature. Three cases are possible, according as the curvature is positive, zero, or negative. The curvature, however, is a very sensitive function of the mean matter density and its expansion rate, and present observations cannot yet distinguish among the three possibilities. If the mean curvature is positive, then the universe is necessarily closed and finite. It is widely believed that a zero or negative mean curvature implies the contrary; namely, the universe is infinite. It has been emphasized by Heckmann and Schücking [23] that this view is fallacious. Although the universe may be infinite under these conditions, there are also closed and finite topologies which are compatible with negative and zero curvature.

C. Antigravity

The possibility of magically nullifying the force of gravity has been a source of preoccupation for crackpots since time immemorial. Most of these people do not realize the extreme deviousness which would be required of any theory which includes antigravity among its predictions and yet which avoids incompatibility with the tremendous mass of well-established observations from the rest of physics. This is not to deny, of course, that anything short of anarchy on the part of Nature (which is in itself a strong constraint) is in principle possible. Nevertheless, no theory has ever been constructed which predicts a cheap way of overcoming gravity, without at the same time totally ignoring the rest of physics.

There are many expensive ways of overcoming gravity, some of which may even be classed as true antigravity. For example, according to Einstein's theory, gravitational radiation could in principle be used as a

[6] This does not necessarily imply that the universe did not exist prior to that instant. Pre-explosion states are perfectly conceivable.

propellant. Since gravitational waves are simply ripples on the curvature of space-time, a spaceship using this propellant could be described as getting something for nothing—achieving acceleration simply by ejecting one absolute vacuum into another. However, in view of the extreme difficulty of generating gravitational waves in sufficient quantity, it is much cheaper to stick to standard propellants.

The idea that it should be possible to build gravity shields arises from our experience with electromagnetism. There are two important factors which make it possible to build shields against electromagnetic forces, and which do not exist in the case of gravity. The first of these is the existence of charges of both signs in equal numbers, and the second is the existence of the quantum forces which hold electrons in matter and make electrical conductors as well as insulators possible. The second of these is actually the most crucial, for even if negative masses existed and the equivalence principle were violated, it would still not be possible to use quantum phenomena to build gravity shields because of the extreme weakness of gravitational forces as compared to electrical forces at the atomic level where quantum phenomena are relevant.

Note that we emphasize the violation of the equivalence principle in this connection. The common view is that in a given gravitational field positive masses would be pushed in one direction and negative masses in the other, producing a net polarization. According to the equivalence principle positive and negative masses will behave alike in a gravitational field. The only difference between the two consists in the gravitational fields which they themselves produce. Positive masses produce fields that attract, and negative masses produce fields that repel. On this basis a positive mass and a negative mass will start chasing each other, the positive kinetic energy which the one gains being compensated for by the negative kinetic energy which the other acquires. Solutions of Einstein's equations exhibiting this absurd process have actually been obtained and might be regarded as good arguments against the existence of negative masses. Another argument, which has already been mentioned, involves the breakdown of the vacuum when quantum processes are taken into account in a theory that permits negative mass and negative energy.

The most likely candidates for particles with negative mass would seem to be the so-called *antiparticles*, particularly the positron, the antiproton, and the antineutron. Experiments have been proposed, but not yet carried out, which would attempt to determine whether these particles fall up or fall down. The reason that such experiments are hard to perform is that these particles are normally produced with such high velocities that the gravitational deflection of their trajectories cannot be detected. However, Schiff [24] has pointed out that considerable indirect evidence exists that the positron at any rate satisfies the equivalence principle, i.e., falls down. The argument is that since protons and neutrons (the heaviest components of matter) are surrounded by clouds of so-called virtual particle-antiparticle pairs, an anomalous behavior of positrons (one of the components of these clouds) would imply a positive (as opposed to a null) result for the Eötvös experiment. He estimates that if a positron had negative mass, the equivalence principle would be violated to the extent of one part in 3×10^6, in comparing such disparate substances as aluminum and platinum. Such a violation seems to be ruled out experimentally.

D. Isotropy of Inertia

We have remarked earlier that Mach's principle is not contained within the general theory of relativity as usually formulated. As the work of Dicke shows, it is not necessarily incompatible with it. On the other hand, a number of physical effects which one might anticipate on the basis of certain versions of Mach's principle are clearly incompatible with general relativity. One of these is anisotropy of inertia.

In 1958 Cocconi and Salpeter [25] suggested that inertial mass might depend not only on the mean distribution of matter in the universe but also on the inhomogeneity or lack of isotropy of this distribution about any given body. The inertial reaction to a given acceleration would then vary from one direction to another. Using the most reasonable assumptions about the laws which might govern this variation, Cocconi and Salpeter concluded that the expected fractional change of inertial mass with angle should lie somewhere between 10^{-5} and 10^{-10}. They suggested that it should be possible to detect the effect of mass anisotropy, if it exists, by looking for anomalies in the atomic Zeeman effect—or better still, by looking for a diurnal variation in the line width of a Mössbauer gamma ray from a magnetically oriented nucleus bound in a crystal lattice [26].

The latter experiment was performed in 1960 by a group at the University of Illinois [27] who found that the inertial mass of the nucleons in the nucleus remained constant to better than one part in 10^{15}. Simultaneously, Hughes [28] performed a nuclear magnetic resonance experiment on Lithium 7 and searched for anomalies in the resonance line. If the nuclear inertial mass had been anisotropic, the single line would have been split into a triplet. No such effect was observed. The accuracy of this experiment was such that it was possible to assert the constancy of inertial mass to better than one part in 10^{22}. This result was later confirmed by Drever [29].

The anisotropy of inertia is thus at least twelve orders of magnitude smaller than that predicted by Cocconi and Salpeter, and general relativity is once again vindicated. It has been noted by both Dicke [30] and

Epstein [31] that the null result of the mass anisotropy experiments might actually have been foreseen if the arguments which Cocconi and Salpeter used had been pushed just a little farther. They reason that not only should the particles (nuclei, etc.) under observation exhibit the anisotropy effect but so should all fields of force within the nuclei and atoms, as well as all particles making up the apparatus and all photons used in making the observation. In short, the anisotropy should be universal and hence unobservable, another example in the classic pattern, of the conspiratorial behavior of Nature so characteristic of the relativity theory.

IV. Measurements in Deep Space

A. The Near Future

Access to interplanetary space has made it possible for the first time to consider performing a number of new experiments which are certain to have an impact on the theory of gravitation, as well as on other parts of theoretical physics. One of the earliest proposals in space research was that of using an atomic clock in a satellite to carry out a frequency shift test [32]. With the development of terrestrial measurements, using the Mössbauer effect, this experiment has been temporarily shelved. In spite of the difficulties inherent in the use of satellites for the frequency shift test, orbiting clocks may eventually give us the most accurate check on the frequency shift effects. The main difficulties concern such things as compensating for the Doppler shift arising from the motion of the satellite itself; obtaining accurate determinations of precisely where the satellite is at any instant, so that allowance can be made for the time lag in the propagation of signals; or, if time comparisons are made only cumulatively over a large number of orbits, designing an atomic clock of sufficient long-term stability. In addition, it is necessary to take into account the special relativistic time contraction effect which arises from the motion of the satellite and which superimposes itself on the gravity shift.

One of the important reasons for carrying out frequency shift measurements in satellites, as well as in various terrestrial laboratories, is to check the theoretical predictions at different altitudes and over various potential differences. Theory predicts a definite variation of the frequency shift with altitude, and it is not sufficient to obtain only one or two points on the theoretical curve to conclude that theory and experiment really agree. We shall speak of such ensembles of experiments, designed to test a theory over a wide range of parameter values, as graduated experiments. The graduated frequency shift experiments can certainly be performed in the near future. It should be pointed out also that the value of such experiments will not be confined merely to the acquisition of information bearing on gravity theory. The techniques of achieving highly accurate synchronization of signals between ground and satellite, or between one satellite and another, will be invaluable to the whole art of interplanetary navigation.

Another satellite experiment which could be performed in the near future, and which would provide a brand new test of general relativity, has been proposed by Schiff [33]. In his experiment a high-precision gyroscope would be mounted in a satellite, and its rate of precession with respect to the "fixed" stars would be monitored over a period of weeks. It might be thought that a gyroscope in space would always point in the same direction, but this is not the case. General relativity is anti-Machian in this respect— a system of axes which appears to be rotationless to an observer, in the confines of an orbiting laboratory (e.g., by dynamical measurements of Coriolis forces, etc.), will nevertheless rotate with respect to the fixed stars. The gyroscope will always keep the same orientation with respect to the local laboratory axes, but not with respect to the cosmic axes.

Three distinct effects are simultaneously at work on the gyroscope in this experiment. The first is a precession because of the gyroscope's orbital motion, with respect to the fixed stars. This is called the Thomas precession and has been well known, in the case of spinning electrons in atomic orbits, since the early days of atomic fine-structure spectroscopy. The second is a correction to the Thomas precession, because space-time is actually curved rather than flat in the vicinity of the Earth. The third effect arises from the fact that the Earth rotates and hence modifies the gravitational field in its vicinity through the addition of magnetic type dipole components. The last two effects are true general relativistic effects, and a measurement of the last effect, especially, would provide the first test of the induction phenomena predicted by Einstein's theory.

It has been proposed that spinning oriented atomic nuclei be used as the gyroscopes for Schiff's experiment. Professor W. M. Fairbank of Stanford University has also proposed carrying out the experiment on Earth, using a spinning superconducting sphere supported by a static magnetic field. In this case, it would be the terrestrial rotatory motion imparted to the laboratory that would produce the effect. Yet the effect is much more pronounced in a satellite, and in view of the fact that many of the difficulties with high-precision gyroscopes would be eliminated if gravitational contact forces were removed, it seems likely that satellite experiments will be the more successful. Moreover, the use of satellites will be essential if a set of graduated experiments is desired.

B. The Intermediate Future

1. ORBITING OBSERVATORIES

With the advent of manned laboratories in space, the character of experiments on gravity and other areas of research will undergo a drastic change. Perhaps the single most important advantage gained by human access to deep space will be the ability to see. At 2850 Å, the ozone in the upper layers of the Earth's atmosphere begins to block incoming radiation, and for all wavelengths below this down to that of the most energetic γ-rays the atmosphere is absolutely opaque. We know that a great amount of radiant energy in this ultraviolet–X-ray region is falling on the Earth. For one thing, the ozone which blocks the long wavelength end of this region is itself produced by the action of ultraviolet radiation. Aside from the fact that satellite observations have already begun to give us glimpses of this region, our present knowledge of stellar dynamics demands its presence.

The atmosphere is likewise troublesome in the infrared region. Although there are many windows in this region which permit fairly reliable extrapolation of the incoming radiation spectrum over very large ranges, there are many opaque bands. Finally, the atmosphere is turbulent, which places a limit on optical resolving power of about 0.1 sec of arc visually and 0.4 sec photographically, under the very best seeing conditions.

An astronomical observatory in space will be free of all these effects and will be able, in principle, to observe the full electromagnetic spectrum with a resolving power limited only by the size of the apparatus. It will consequently be an extremely effective instrument for increasing our knowledge of many processes taking place in the universe. It is true that many technical problems will still have to be overcome in order to utilize such an observatory most effectively, particularly in the extreme ultraviolet and infrared regions. But even in the ordinary visual region, where most of the technical problems (other than those having directly to do with the space environment) have been solved, the possibilities are staggering.

To begin with, it will be possible to observe much fainter stars than can be seen on Earth. The absence of turbulence will better allow one to separate a fixed star image from the background noise, and the noise itself will be reduced. On Earth the noise is the light of the night sky, which amounts to about 200 tenth-magnitude stars per square degree in the best directions (i.e., away from the Sun and the horizon) and which makes it impossible to detect stars fainter than about the 21st magnitude. In a spatial observatory the only noise of any significance will be the zodiacal light, which in the best directions, amounts to only 20 tenth-magnitude stars per square degree. If the observatory has a 100-in. telescope (which will certainly be feasible someday), then the theoretical resolving power at 5000 Å will be 0.04 sec of arc, and it is easy to calculate that it should be possible to detect stars which are several thousand times fainter than those detectable on Earth, corresponding to a magnitude of 31. Even in zodiacal directions 29th magnitude stars should be observable. In any case the limiting magnitude will be increased by at least 8 over an Earthbound observatory.

Now an increase of this amount corresponds to a decrease in brightness by a factor of about 1600, and this corresponds to an increase in distance by a factor of 40. Thus, even allowing for reddening and intergalactic absorption of the light coming from distant stars, by moving outside the Earth's atmosphere, one may expect to increase the linear size of the optically observable universe at least by a factor of 10. This is an enormous increase, which pushes the bounds of the observable universe out to 7 billion parsecs. (Presently the most distant observable galactic clusters are 700 million parsecs away.) At this distance the speed of galactic recession becomes greater than the speed of light,[7] so that the bounds of the "observable" universe will, in fact, *not* be observable!

It is clear that this potentiality of spatial observatories is of immense significance for cosmology and for theories of gravitation. When it becomes possible to see out to the very theoretical limits of the universe, we should be able to decide once and for all whether or not the universe possesses a large scale spatial curvature. Although it will not be possible to observe directly how this curvature changes with time, it should be possible to infer it from other plausible assumptions, and this would enable us to determine the curvature of space-time itself. This information, together with information gathered from other observations on the distribution of matter and energy in the universe, will provide the first severe test of Einstein's theory. The observed curvature may be homogeneous, or it may possess marked inhomogeneities; and it may be positive, negative or zero. The observational data will provide a real challenge to the full Einstein equations.

2. THE CLASSICAL TESTS

Many less ambitious uses of an orbiting observatory also will be of prime importance to the theory of gravitation. It will become possible, for example, to carry out graduated series of tests of the classical predictions. We have already mentioned the graduated frequency-shift tests involving atomic clocks in satellites. The deflection of light by the Sun could also

---

[7] Although local relative speeds can never exceed that of light, Einsteinian cosmology permits relative velocities greater than light at large distances. By this mechanism certain parts of the universe may be forever invisible and inaccessible to one another.

now be measured with much greater accuracy. It would no longer be necessary to wait for an eclipse; a simple coronagraph could be used. The observations could be carried out under conditions designed to give the best statistics, namely, at those times when the star field behind the Sun is the densest. Photographs could be taken as often as desired, thus further improving the statistics.

The perihelion precession tests could be carried out with much greater accuracy and over a wider range of parameters, by placing heat resistant solar satellites in highly elliptical orbits, closer to the Sun than the planet Mercury, and observing their motion for several years. Again, the orbiting observatory would be used to make the observations because of its ability to detect small objects over enormous distances. Accuracy could be further improved by the much discussed device of enclosing each solar satellite in a servo-controlled propellant-equipped shield, designed to guard it against the perturbations of solar storms and solar magnetic fields and to permit it to follow a true undisturbed gravitational trajectory.

3. Radiation and Induction Effects

Another advantage of an observatory or laboratory in deep space is its so-called gravity-free or weightless environment. In a weightless laboratory, frictional forces can be practically eliminated, and apparatus can be completely insulated against acoustic disturbances. In a frictionless environment the gradual acceleration of nearly motionless (relatively speaking) bodies over periods of days or even weeks will be easily detectable. The Cavendish experiment to determine the universal(?) gravitation constant will be performable with much greater precision than on Earth. It should also be possible to perform a graduated series of gravitational induction experiments in a controlled manner in the laboratory. For example, a massive spherical body could be set in rapid rotation with a speed limited only by its mechanical strength, and small test bodies could be allowed to fall toward it. After allowing for the slight (but unscreenable) geodetic deviation effects produced by the Sun, Moon, and planets, as well as by the laboratory itself, Einstein's theory predicts that the test body should not fall directly toward the center of the massive spinning body but should be deflected slightly to one side in the direction of rotation. It should be possible to compare these predictions quantitatively with observation.

The possibility of complete acoustic insulation implies that deep space will provide an excellent environment for antennas designed to detect gravitational radiation. One of the most vexing difficulties with these antennas on Earth is the need to insulate them from seismic disturbances, whether arising from within the Earth itself or from such things as passing trucks or even people walking. If a number of these antennas were distributed at various points throughout space, their directional sensitivity would become good enough so that it should be possible to determine accurately the direction from which detectable gravitational radiation (if any) is coming.

At the longer wavelengths, a very useful gravitational antenna may be the Moon, a world that appears to have very little seismic activity. Moreover, it has no winds or oceans to feed energy into its oscillation modes. It is not inconceivable that the Moon is such a quiet place that its background noise is less than that produced by passing gravitational waves.

C. The Distant Future

In contemplating the distant future, one can allow the imagination to run wild. Here it is necessary to take into account two factors. The first is that scientific knowledge progresses on a broad interlocking front, so that it is impossible to speculate on the future advances to be made in gravitation physics without considering simultaneously the progress which may be anticipated in other basic areas as well. The second is the factor of unpredictability. Nothing is so certain as the fact that the most startling advances to be made in gravitation physics, or in any other part of physics, will be completely unexpected. This is the reason why it is absurd to attempt to place any sort of value on one area of basic research as opposed to another.

One area which may quite possibly have great impact on gravitational research is neutrino astronomy. Increasing evidence has been accumulating that neutrinos play a dominant role in cosmology and in the final evolutionary stages of stars. Chiu [34] has calculated that in the final pre-explosion collapse of a supernova, the star radiates neutrinos at the rate of 10^{20} ergs/cm^3/sec, which is sufficient to deplete its entire energy content in a little over a day. In the shock wave produced by the subsequent explosion, as many as 10^{48} additional extremely high energy neutrinos may be produced. Where do these neutrinos go? Since the mean free path of a neutrino is about 10^{15} universes, it is evident that a very large number of neutrinos may by this time be present in the universe. Supernovae occur within the galaxy at the rate of about one per 300–350 years. Since the last one was observed in 1604, another one can be expected any century now. In spite of the fact that the mean free path of a neutrino through matter is so great, the neutrinos from such a supernova could be detected by presently available techniques.

Weinberg [35] has calculated that under certain conditions the energy contained in the wandering neutrinos may be greater than the energy contained in all the rest of the matter and radiation in the universe put together. He shows that a large number may by this time have been so red-shifted as to have become degenerate—i.e., to have attained the lowest possible energy consistent with the exclusion principle. If it were possible to detect any of the neutrinos in the resulting Fermi sea (e.g., by their effects on β-decay experiments) and to determine the energy at the top of the sea, then a very sensitive method would exist for selecting the cosmological model which corresponds most closely to the actual universe. Perhaps some day these neutrinos will be detected.

It is clear that if such a great proportion of the energy of the universe is contained in neutrinos, then neutrinos cannot be neglected in cosmological considerations. This prompts one to ask whether there may not be other similar sources of nearly undetectable energy in the universe. Among the possibilities which immediately suggest themselves are Dicke's ϕ-waves, generated by the same explosions which produce the neutrinos. ϕ-waves are even harder to detect than neutrinos, their mean free path through matter being immensely longer—of the same order of magnitude as the mean free path of gravitational waves. Gravitational waves themselves are yet another possibility.

Dyson [36] has suggested that pairs of white dwarfs may radiate immense amounts of gravitational energy. He calculates that a dwarf binary system with a period of about 100 sec and a mass equal to that of the Sun would radiate gravitational waves at a rate of about 2×10^{37} ergs/sec, which is 5000 times the Sun's optical luminosity. A pair of neutron stars (the densest form of matter) has the possibility of an even greater output of radiation. A binary neutron system, having solar mass and a separation distance of 10 km, would have a period of 5 msec (corresponding to an orbital velocity of $\frac{1}{8}$ that of light) and would radiate gravitational waves at a rate of 2×10^{52} ergs/sec. The lifetime of this system would be less than two seconds. If the initial separation were greater, the lifetime would be longer; but the end result would be the same. The loss of energy by gravitational radiation would bring the binaries together with ever increasing speed, until in the last moment of their lives they would plunge together and release a gravitational flash of unimaginable intensity.

This flash would have a frequency lying in the acoustic range and could be detected by the previously described antenna which Weber has built, at distances of the order of 100 million parsecs. The death cry of a binary neutron system, taking place in any of 10 million galaxies, could therefore be heard on Earth with presently existing equipment. Although neutron binaries are not known to exist, binary systems, in which at least one of the members is a dwarf, have been observed with periods of the order of as little as 1 hr and 21 min. It would therefore seem worthwhile to monitor the universe for events of the type which Dyson describes.

Dyson goes on to suggest that events of this type might provide evidence for the existence of extremely advanced species of life in the universe. He even suggests that at the present exponential rate of expansion of terrestrial technology and population, such an advanced state of technology, in which engineering on an astronomical scale becomes feasible, will be an eventual necessity for the human species. He describes some of the interesting technical feats which could be accomplished by a dwarf binary system under intelligent control. First, it would provide an enormous reservoir of gravitational energy which could be tapped by dropping masses into the system in such a way that they would be reejected with greatly increased kinetic energy (a restricted 3-body orbit problem) and then converting this kinetic energy directly into useful work. Second, a dwarf binary system would have the property that it could accelerate a large spaceship of normal mechanical construction, containing human passengers, to a velocity of 2000 km sec^{-1} in a very short period of time without expending any rocket propellant. The passengers would feel nothing but the ubiquitous tidal forces, and these would be hardly noticeable. Dyson suggests that a highly developed technological species might use dwarf binaries scattered around the galaxy as relay stations for long distance freight transportation.

It is clear that the possibilities for both research on and development of gravity are ultimately unlimited. The subject is even today spiced with at least one major mystery, namely, the question of the ultimate fate of neutron stars having masses in excess of three fourths of the Sun. In 1939, Oppenheimer and Volkoff [37] showed that in a cold degenerate neutron star, obeying the equation of state of a perfect Fermi gas, the very fabric of space-time would be rent asunder at the center of the star if its mass were to exceed the above amount. That means that the so-called metric tensor would become singular. Wheeler [38] has found that the critical mass would be even less if a more realistic equation of state were used. Wheeler, however, regards the presence of a metric singularity as intolerable. He states, ". . . We have come to the untamed frontier between elementary particle physics and general relativity. Of all the implications of general relativity for the structure and evolution of the universe, the question of the fate of great masses of matter is one of the most challenging. The issue cannot be escaped by appealing to stellar explosion or rotational instability, for this issue as it presents itself is one of principle, not one of observational physics."

Wheeler suggests that it is not the fabric of space-time which is crushed out of existence, but the fabric of the nucleons; that under such enormous pressures they simply dissolve into radiation—electromagnetic, gravitational, or neutrino. Whatever the final answer to this mystery may be, it seems likely that in order to discover it we shall have to make use of the full arsenal of elementary particle physics, both present and unborn, as well as the quantum theory of gravitation and, perhaps, a bit of interstellar exploration.

References

1. Forward, R. L. (1963). Guidelines to antigravity. *Am. J. Phys.* **31**, 166.
2. Brouwer, D. (1950). *Bull. Astron.* **15**, No. 3.
3. Feynman, R. P. (1957). Report of the *Conference on the Role of Gravitation in Physics,* Chapel Hill, North Carolina.
4. Thirring, W. (1959). Lorentz-invariante Gravitationstheorien. *Fortschr. Physik* **7**, 79.
5. Fock, V. (1959). "The theory of space time and gravitation." Pergamon Press, New York.
6. Wheeler, J. A. (1963). Report of the *Conference on Relativistic Theories of Gravitation,* Jabłonna, Poland, 25–31 July 1962.
7. Hönl, H. (1962). *In* "Physikertagung Wien," E. Crucke, Ed., Physik Verlag, Mosbach/Baden.
8. Sciama, D. W. (1959). "The unity of the universe." Doubleday, New York.
9. Brans, C., and Dicke, R. H. (1961). Mach's principle and a relativistic theory of gravitation. *Phys. Rev.* **124**, 925.
10. Dirac, P. A. M. (1938). A new basis for cosmology. *Proc. Roy. Soc. (London)* **A165**, 199.
11. Jordan, P. (1955). "Schwerkraft und Weltall." F. Vieweg, Braunschweig.
12. "Conference on the instability of systems of galaxies," *Astron. J.* **66**, 533 (1961).
13. Weber, J. (1960). Detection and generation of gravitation waves. *Phys. Rev.* **117**, 306.
14. Dirac, P. A. M. (1958). The theory of gravitation in Hamiltonian form. *Proc. Roy. Soc. (London)* **A246**, 333; (1959), Fixation of coordinates in the Hamiltonian theory of gravity. *Phys. Rev.* **114**, 924; (1963), The Physicist's picture of nature. *The Scientific American* **208**, No. 5. 45.
15. DeWitt, B. S. (1962). The quantization of geometry. *In* "Gravitation, an introduction to current research" (L. Witten, ed.). Wiley, New York, 266.
16. Yang, C. N., and Mills, R. L. (1954). Conservation of isotopic spin and isotopic gauge invariance. *Phys. Rev.* **96**, 191.
17. Misner, C. W., and Wheeler, J. A. (1957). Gravitation, electromagnetism, unquantized charge, and mass as properties of curved empty space. *Ann. Phys.* **2**, 525. Fletcher, J. G. (1962). Geometrodynamics, *In* "Gravitation, an introduction to current research" (L. Witten, ed.). Wiley, New York, 412.
18. DeWitt, B. S. (1962). Quantization of fields with infinite-dimensional invariance groups. III Generalized Schwinger-Feynman Theory. *J. Math. Phys.* **3**, 1073.
19. Popper, D. M. (1954). Red shift in the spectrum of 40 ERIDANI B. *Astrophys. J.* **120**, 316.
20. Mössbauer, R. L. (1958). Kernresonanzfluoreszenz von Gammastrahlung in Ir^{191}. *Z. Physik* **151**, 124.
21. Pound, R. V., and Rebka, G. A., Jr. (1960). Apparent weight of photons; attempts to detect resonance scattering in Zn^{67}; the effect of zero-point vibrations. *Phys. Rev. Letters* **4**, 337, 397.
22. Bondi, H. (1960). "Cosmology," 2nd ed. Cambridge.
23. Heckmann, O. and Schücking, E. (1962). Relativistic Cosmology. *In* "Gravitation, An introduction to current research" (L. Witten, ed.). Wiley, New York.

24. Schiff, L. I. (1959). Gravitational properties of antimatter. *Proc. Natl. Acad. Sci. U.S.* **45**, 69.
25. Cocconi, G., and Salpeter, E. E. (1958). A search for anisotropy of inertia. *Nuovo Cimento* **10**, 646.
26. Cocconi, G., and Salpeter, E. E. (1960). Upper limit for the anisotropy of inertia from the Mössbauer effect. *Phys. Rev. Letters* **4**, 176.
27. Sherwin, C. W., Frauenfelder, H., Garwin, E. L., Luscher, E., Margulies, S., and Peacock, R. N. (1960). Search for the anisotropy of inertia using the Mössbauer effect. *Phys. Rev. Letters* **4**, 399.
28. Hughes, V. W., Robinson, H. G., and Beltram-Lopez, V. (1960). Upper limit for the anisotropy of inertial mass from nuclear resonance experiments. *Phys. Rev. Letters* **4**, 342.
29. Drever, R. W. P. (1961). A search for anisotropy of inertial mass using a free precession technique. *Phil. Mag.* **6**, 683.
30. Dicke, R. H. (1961). Experimental tests of Mach's principle. *Phys. Rev. Letters* **7**, 359.
31. Epstein, S. T. (1960). On anisotropy of inertia. *Nuovo Cimento* **16**, 587.
32. Singer, S. F. (1956). Application of an artificial satellite to the measurement of the general relativistic "Red Shift." *Phys. Rev.* **104**, 11.
33. Schiff, L. I. (1960). Possible new experimental test of general relativity theory. *Phys. Rev. Letters* **4**, 215.
34. Chiu, H. Y. (1962). Neutrino astronomy. *Intern. Sci. and Technol.* August.
35. Weinberg, S. (1962). Universal neutrino degeneracy. *Phys. Rev.* **128**, 1457.
36. Dyson, F. J. (1961). Gravitational machines. Gravity Research Foundation essay, New Boston, New Hampshire.
37. Oppenheimer, J. R., and Volkoff, G. M. (1939). On massive neutron cores. *Phys. Rev.* **55**, 374.
38. Wheeler, J. A. (1962). "The degenerate star, a relativistic catastrophe." Lecture delivered at the Institute for Space Studies, National Aeronautics and Space Administration, New York.

Integration of Payload and Stages of Space Carrier Vehicles

Geoffrey K. C. Pardoe

Space Division
Hawker Siddeley Dynamics Limited
London, England

I. Introduction

However successful the design and efficient the operation of the individual stages of a multistage space carrier vehicle, the overall mission success depends fundamentally on the integration of these stages with each other.

In the design of a space carrier (launch) vehicle, there are theoretical methods that can be used to determine the performance required of the individual stages to achieve the overall mission; this basic relationship required between stages will be briefly analyzed here. However, within the boundaries and constraints imposed by the theoretical treatment, a successful design will depend ultimately on factors that cannot be expressed mathematically—factors such as the engineering solution to the interface problem between stages and to the interaction problems of various systems within the stages, both in the pre-flight checkout and the in-flight performance. These problems are solved by experience and are descriptive rather than mathematical; therefore emphasis will be given to them in this form here. This chapter will attempt to summarize the past achievements and outline the new problems to be solved to achieve the exciting missions of the future.

A review of the space carrier vehicles to date exposes the inelegance of matching stages, which has arisen from the sheer necessity of combining stages separately developed as ballistic missiles or space vehicles, and it is only in more recent years that the hybrid vehicles of the past have begun to be replaced with stages properly designed with relation to each other. The trends which are now being established in the "building block concept" for multistage vehicles will be examined to see what effects they will have on the future designs.

The basic idea of "building blocks" should provide the maximum flexibility in the composition of multistage rocket vehicles for the tasks in hand, with the minimum of new development programs for rocket stages which sometimes differ only slightly in performance. It is therefore necessary to examine in the course of this discussion what effect different payloads (probes or satellites) have on individual rocket design and staging. Again, particular examples will be taken to illustrate the various alternatives.

Inevitably, many of the data and examples given in this review are based on American equipment. To scientists and engineers of the world, it is of immense value that much information is available on American achievements to date and that their plans for the future are known in considerable detail. The successes of Russian missions certainly speak for themselves, but from an engineering viewpoint there is a sense of great frustration from the lack of evidence as to how these missions have been accomplished. Therefore, only occasional reference can be made to Russian activity in this particular problem area, and in general American data will be predominant.

There is one European field of work of particular interest: the European Launcher Development Organization. In America the integration problems are difficult: the different stages of rocket are constructed by different contractors, then assembled perhaps by yet another contractor, and finally fired from the launching pad by still another organization. In Europe the problem is intensified since a three-stage carrier vehicle is being developed, each stage being constructed by a different *country* belonging to ELDO.

Within the nations concerned, various companies are contributing systems and subsystems. Moreover, the launching range, the tracking and telemetry, and test satellites are all the responsibility of yet *other* countries. It is not really necessary to underline what an immense problem of administration and technical liaison is involved, and how important it is for the methods of integrating payloads and rocket stages to be well understood and properly planned. The European organization is still very young and, in the case of the second and third rocket stages, little evidence has yet been made available about how the problems are being overcome. The first rocket stage, the British Blue Streak, is in an advanced stage of development. Evidence of this and some of the other European activities will be used in discussing the unique problem of stage integration.

II. Dynamic Requirements

A. Basic Staging Theory for Various Missions

As a prelude to discussing the problems of integrating stages and payloads, it is useful to review briefly the basic staging requirements. For an analysis of the general dynamics of space flight the reader is referred to Ref. [1]. Ideally, for maximum efficiency, while the rocket engines of a space carrier vehicle burn and propellant is so consumed, the attendant structure weight should reduce accordingly until, at the instant that the propellants are completely consumed, there is no structure weight and only the payload! This is obviously impossible to achieve, and the ideal can only be approached in a series of discontinuous steps. A "step" consists of breaking down the overall vehicle from the theoretical size necessary to give the total characteristic velocity condition, into a series of practical stages, each stage consisting of a self-contained propulsion unit. There is a half-stage variation of this which will be discussed, whereby engines only are shed, without their associated tank unit. In general, however, when each stage reaches the all-burnt condition, the dead weight of the remaining tankage, propulsion system and its control systems are shed, or staged, leaving the next stage to burn, thus imparting its increment of velocity to the overall vehicle.

The question then arises as to the best distribution of thrust and size, and number of stages for the appropriate mission. At one extreme, for simplicity, the performance of a single stage must be considered in order to see whether the added reliability and simplicity of only one complete propulsion system outweighs the disadvantages of inefficiency by carrying into orbit the excessive deadweight of the vehicle. In practice, no single-stage rocket has ever achieved orbital condition; the closest approach to it is the one-and-a-half stage Atlas in which only the booster engines are shed soon after lift-off.

With contemporary limitations on propulsion system efficiency and on structural efficiency, more than one stage must be involved.

The problem is clearly open to mathematical analysis as follows: As an initial assumption, take equal specific impulse of the propellants of each stage. Then let

M_n = total mass at beginning of nth stage burning

RM_n = dead weight of nth stage (i.e., it is assumed that the dead weight of nth stage is proportional to the weight of the stages above it)

r = typical stage number

M_p = payload mass ($= M_{n+1}$)

Δv_n = velocity increment of each stage

I = impulse

The total velocity increment is then given by

$$\Delta v = gI \sum_{r=1}^{n} \log \frac{M_r}{RM_r + M_{r+1}} \tag{1}$$

Differentiating partially, one obtains

$$\frac{M_{r+1}}{M_r} = \frac{M_r}{M_{r-1}} \tag{2}$$

Mass ratio for rth stage $= \dfrac{M_r}{RM_r + M_{r+1}}$

$$= \frac{1}{R + (M_{r+1}/M_r)} \tag{3}$$

Mass ratio for $(r-1)$th stage $= \dfrac{M_{r-1}}{RM_{r-1} + M_r}$

$$= \frac{1}{R + (M_r/M_{r-1})}$$

$$= \frac{1}{R + (M_{r+1}/M_r)} \qquad \text{[using Eq. (2)]} \tag{4}$$

Therefore, for maximum total vehicle velocity, the mass ratios of each stage are equal and the velocity increment added by each stage is equal.

In general, each stage will not have the same specific impulse or structural fraction (i.e., dead weight as fraction of fully fueled weight); thus the above generalized analysis is immediately open to modification. Such is the nature of the problem that the overall optimization is no longer possible by simple analytical treatment when all the practical constraints are taken into account, and vehicle optimization studies can be carried out in practice only by lengthy procedures using extensive digital computer facilities. Firm ground rules, therefore, can clearly not be given here, but for the simple case to indicate the nature of the problem, it is thus possible to define the ideal situation for a particular mission with regard to the staging configuration of an overall vehicle. In practice there are various constraints to achieving this ideal, which will now be considered.

B. Engineering Constraints on Staging Theory

There are many engineering constraints which influence the final configuration of a rocket that emerges from the contractor's factory to meet the mission requirement. In the early days of orbital flight, missions were achieved in the most expedient way possible, with little emphasis placed on engineering elegance. For example, the early payloads in the Explorer series were as little as 0.05 per cent of the total lift-off weight of the carrier vehicle on the pad and this was typical of what could be achieved in a low orbit using the vehicles and engines existing at the time. One of the main engineering factors influencing the final configuration is the range of engines, or even the range of developed rocket stages, available to do the job.

The first capability of space flight evolved largely out of the ballistic missiles currently being produced—namely, Jupiter, Thor, and Atlas. While many engineers associated with the design of an IRBM or ICBM had in mind that they might eventually have a space capability, no major space-oriented design criteria were provided in the choice of configuration to attain the original military ballistic requirement. Therefore, it was only a question of accepting each vehicle for what it was at the time, and determining what upper stages could best be amalgamated with it to produce a space capability.

The payload requirements in the first generation of ballistic missiles could be achieved largely with a single-stage vehicle (in the case of Thor and Jupiter) or a one-and-a-half-stage vehicle (in the case of Atlas). In the interests of military reliability, the minimum number of stages to do the task paid dividends. Clearly, the fixed design of the first stage posed a considerable constraint on the ability to optimize its integration into a multistage carrier vehicle.

Even in the case of designing a complete carrier vehicle specifically for a task, a further influence on the vehicle designer is the range of rocket engines available to him. In theory, of course, a rocket engine could be produced precisely to meet most of a specific requirement, but in practice the rocket engine specialists have usually anticipated events to some extent, possessing an engine which has gone through its preliminary evaluation stages. For the vehicle designer to ask for a new engine would mean perhaps delaying the program. This problem faces the designer not only of the first stage, but also of any stage of a multistage carrier vehicle, and can be seen to lead to a deviation from the optimum staging configuration.

A multistage carrier should be optimized against a specific mission; that is, the staging ratios and configuration performance of the vehicle will be different for a payload to be placed in a synchronous orbit than for one placed in a low elliptical orbit. There is no theoretical reason why this optimization should not be done, but in practice it is quite impracticable to design a specific vehicle for a specific orbit. The problem has to be solved in one or more of the stages, whereby propellant is left out in exchange for increased payload; alternatively, if a small payload is required, the vehicle tanks are completely filled. It is possible, of course, to make a complete change of the uppermost stage or even occasionally the next stage; this will be part of the building block concept to be dealt with later. It can be seen that such constraints will lead to further deviation from the optimum.

Another factor that must be taken into account is the fact that the complete vehicle at lift-off from the pad requires a positive acceleration, usually in the order of 0.3 g. Although the lift-off thrust will exceed the vehicle weight by about 30 per cent, for other stages it is no longer necessary to maintain a positive thrust/weight ratio, and values of thrust/weight ratio between 0.7 and 1 are frequently used in the upper stages. Reference to the analysis in Section II, A shows that this practical requirement of different thrust/weight ratios in the first stage and upper stages leads to a further deviation from the optimum staging conditions.

It is not only in the vehicle design that constraints are imposed on the configuration; the nature of a space vehicle is such that the ground support equipment at the launching range has a great influence in all stages of its development program and its design. The way in which the range equipment influences the design can be seen in relation to the first- and second-stage predicted impact areas. All stages except the last propulsion stage (which injects the payload into orbit) re-enter the atmosphere at some point. In the case of the upper stages, their size is such that, with the high speed they experience in the suborbital atmospheric flight, they would probably burn up, thus posing no ground impact problem. However, the first (and sometimes the second) stage is not so easily destroyed, so it must be allowed to impact in suitably uninhabited areas.

In the case of the American Cape Kennedy range, most launching directions are over the sea and so impact presents no great problem. In the British/Australian range at Woomera the first-stage impact is quite likely to be on land. The first planned impact area for the Blue Streak first stage was in the Gulf of Carpentaria, some thousand miles down range. Some time after the program had begun it was realized, from optimization studies, that a greater performance could be achieved by using a much shorter impact range for the first stage. However, the required 600 miles optimum was unsuitable for safety reasons, and thus a further compromise had to be made by using an impact range of 500 miles from the launching pad. This is certainly closer to the optimum than the original one for most of the missions, but is not the best and again influences the staging efficiency. To give some idea of the extent of this problem, a decrease of first-stage impact range from 600 to 500 miles from the launching pad has the effect of reducing a final payload into 300 miles circular orbit from 2800 to 2600 lb. The second stage case is less severe in the ELDO vehicle being launched from Australia, but the problem exists, and the same principle would apply in influencing staging efficiency.

Yet another feature associated with the ground support equipment concerns the case of radio controlled multistage vehicles; it is the location of the down-range guidance and control stations. The stations must clearly have within sight at the time of sending guidance signals, the vehicle they are controlling into orbit. Unfortunately these stations cannot always be placed in the ideal position in association with the staging requirement. The practical locations of these guidance stations in their control of individual stages give rise to further deviation from optimum staging.

All the above features, together with other engineering problems, are reflected to a very significant degree in the efficiency of the staging ratios that can be achieved as well as each individual stage design. Each stage is further exposed to individual constraints, such as the construction methods that can be used in association with whatever propulsion system is currently available. The combination of these constraints sometimes leads to a certain amount of canceling out of detrimental effects, but the overall effect is undoubtedly to reduce the payload that the combined vehicle can inject into orbit.

C. Payload Dynamic Requirements

1. Basic Performance Specifications

To this point the discussion has centered on the optimization of various stages to provide the appropriate injection velocity to the payload capsule. The requirements so far have been expressed in terms of velocity vector, but when the end product, namely the payload, is considered, further requirements arise from the actual separation of the payload from the last burning stage.

The separation of a satellite from the final stage takes place under two different conditions: the last propulsion stage may be stabilized by having had a spin rate imparted to it (as an alternative to an active form of steering control); or, alternatively, in the case of a larger, more refined last propulsion stage, the steering control may be exercised by auxiliary jets or vanes in the main efflux and the separation of the capsule would therefore take place under virtually nonspinning conditions. In either case the payload designer has to be assured that his payload will finish separating from this stage within a known set of limits. In the case of the spinning assembly, the payload generally must be de-spun to some rotational value after separation; but in a few cases the payload will have been designed to operate at the spin rate of the final stage. In addition to the rotational effects, the payload must be separated from the final stage with certain linear accuracy.

Final injection accuracy is typically within 10 to 200 ft/sec. Separation is often achieved (as will be discussed later) with springs or posi-grade jets, and care must be taken to ensure that the resulting velocity increment relative to the final stage falls within the appropriate limits. The size of these limits depends to a large degree on the nature of the payload itself, as to whether the final accuracy of injection into orbit relies primarily on the last main rocket propulsion stage, or whether there is to be a post-injection correction by means of a position control system in the satellite. The latter system would certainly be involved for a satellite that had to keep stationary in relation to the Earth or to an adjacent satellite. In this case the opportunity exists for the final main propulsion stage injection limits to be relatively wide, subject to the propellant capacity of the satellite's station-keeping system.

It should be borne in mind that the tighter the accuracy of injection of the payload at separation, the less the demand on the station-keeping equipment in the satellite and vice versa. This particular feature can be optimized, taking into account the overall satellite propulsive system requirements, for attitude and station-keeping systems in the satellite, to see if the propellant is best contained within that system or within the previous propulsion stage. There is often an advantage in leaving the payload position control equipment to carry out the fine velocity correction, since by their nature the jets are small in size, sometimes with thrusts of only a few pounds—or even a small fraction of a pound—for the task of maintaining the satellite in its orbit. These low thrusts are very suitable for the fine control, whereas the last propulsion stage may well have a thrust of several hundreds or thousands of pounds.

For a satellite which has post-injection correction (by virtue of the position system), prior injection velocities of from 50 to 300 ft/sec would be typical, 300 ft/sec being that order of velocity accruing from a simple, spin-stabilized solid rocket final propulsion stage; 50 ft/sec could be expected from a fairly refined type of guided liquid rocket end stage. If no post-injection correction is available, then secondary jets must be fitted to the final main propulsion stage and in this case accuracies of from 0.1 to 10 ft/sec should be achievable. The nature of the payload and mission requirement clearly has a considerable bearing on the type of separation system called for.

A similar problem exists in the pitch and yaw axes of the satellite, in that the separation mechanism of the satellite from the end stage must keep the pitch and yaw rates within predetermined limits. This is required so that the subsequent attitude control system in the satellite can correct these rates to the prescribed values appropriate to the mission. This demands considerable care in the alignment of the separation devices, be they springs, gas jets, or rockets, plus a careful calibration or balancing of these devices to maintain the limits. Relative separation velocities of from 1 to 2 ft/sec are typical.

The requirements are particularly tight in the case of an Earth-oriented satellite using, for example, a gravity gradient stabilization system as its primary means of control, or a space stabilized satellite with a minimum control system. The tolerance is widest, and indeed may not exist at all, for a satellite equipped with an omnidirectional antenna, instrumented for its mission in a manner that enables it to tumble in a random motion during

INCORRECT EJECTION ANGLE

CORRECT SPEED BUT INCORRECT ANGLE AT LAUNCH RESULTING IN SATELLITE PERIOD-ICALLY DIPPING INTO ATMOSPHERE GIVING LIMITED NUMBER OF CYCLES.

EXCESSIVE SPEED

INITIAL SPEED TOO GREAT (E.G. 5 M.P.S.) ORBIT BECOMES SWEEPING ELLIPSE

IDEAL CIRCULAR ORBIT

PERFECT MATHEMATICAL ORBIT (IMPROBABLE) HEIGHT 300 MILES SPEED 4·737 M.P.S. OR 17,054 M.P.H.

INSUFFICIENT SPEED

CORRECT ANGLE BUT INSUFFICIENT SPEED. LACKING THE NECESSARY RESOLVED CENTRIFUGAL FORCE SATELLITE DROPS BACK TO EARTH.

FIG. 1. Effect of injection errors on typical satellite orbit.

its orbit. The main requirement for pitch, yaw, and roll separation characteristics, then, would be to ensure a clean separation with no fouling between the satellite and its attachment lugs and the end stage. Clearly depending on circumstances, the pitch, roll, and yaw rates can vary over a very wide limit. It is not, therefore, sensible to quote typical values, other than to say that for the general Earth-oriented, attitude-stabilized satellite, pitch and yaw separation rates of from 10 to 20 deg/sec in a random direction on a satellite would be easy to deal with.

It is, of course, fundamental to the design of the overall vehicle that the vector velocity of the satellite at injection into orbit should be within prescribed limits for the success of a mission. Figure 1 gives a general indication of the effect of different errors of injection on the subsequent trajectory of the satellite; such errors could accrue from malperformance of any of the stages of the launcher, but the onus for accuracy of injection mainly rests on the final main stage, discussed above. There is a similar effect on the azimuth plane, where the plane of the orbit can be incorrect and, depending on mission requirement, could prove embarrassing. This would be particularly true in the case of injecting a satellite into a synchronous orbit above the equator, where a bad injection accuracy of several degrees in the azimuth plane could cause some difficulty in the tracking of the satellite in association with fixed ground antennae since it would otherwise be stationary so far as longitude is concerned.

It will be seen later that all these considerations have some influence on the method of integration of the payload to the satellite and the stages of the carrier vehicle.

2. INFLUENCE ON OVERALL VEHICLE DESIGN

The payload carried at the front end of a multistage launch vehicle will experience a variety of vibration modes. The vibration from the rocket engine, which is the most important source of vibration, will be considerably attenuated at the payload location; however, the vibration spectrum there can still be severe, and the payload designer must take due account of the probable vibration levels. There are limits to what can be tolerated —particularly with some of the sensitive electronic equipment and gyros. Therefore, with regard to the survival or performance in these items, the dynamic payload limitations may well influence to some degree the method of payload installation and the vehicle design. There is little the vehicle designer can do to reduce the amount of vibration originating from a combustion chamber, but the environment of the equipment can be eased by providing antivibration mounting of the payload or the equipment within the payload. Typical vibration levels that a satellite would experience on the upper end of a launch vehicle such as Thor-Delta would be

2 g rms at lower frequencies (less than 100 cycles/sec) and as much as 12 g rms at higher frequencies (between 700 cycle/sec and 200 cycle/sec). These vibrations could be both longitudinal and transverse, and have also a random Gaussian distribution of 0.1 g^2/cycle/sec superimposed on the range 20 to 2000 cycle/sec.

FIG. 2. Thor Delta three-stage carrier showing increased diameter fairing over satellite payload. (Courtesy of NASA.)

The satellite experiences a severe temperature environment, resulting from the dynamic heating created by the ascent through the atmosphere. To deal with this and the dynamic pressure a fairing is always placed over the satellite during this part of the flight, comprising a further influence on the satellite design. The equipment, which may need to be erected from the satellite when in orbit, now has to be suitably folded and enclosed within the fairing envelope during the ascent. There is a moderate amount of flexibility in the extent of this fairing and various satellites; for example, Tiros and other American satellites have been launched within a bulbous fairing of considerably greater diameter than the final upper stage as seen in Fig. 2. The design of this fairing can have a considerable influence on the efficiency of the overall vehicle. It should be kept in mind that it is a requirement that the fairing be jettisoned as soon as possible after the dense atmosphere is left behind. The earlier the jettisoning of this fairing the less energy is required to lift its dead weight into altitude; consequently, the fairing must not only protect the vehicle from the dynamic pressure on the ascent and kinetic heating, but must be extremely reliable from the point of view of jettisoning, and yet be of minimum weight. It is clearly a significant feature of the overall design, as premature ejection would mean excessive pressure and temperature impingement on the satellite. Retarded ejection would mean an inability to reach orbital conditions specified; or, in the extreme case, inability of the satellite to separate from the final stage and erect its antennae, solar cell paddles, etc. Since the weight of the fairing can be in the same order of weight as the payload it covers, efficiency of design in this area can be significant in achieving a successful mission.

III. First Generation Multistage Space Carrier Vehicle Evolution

A. Single-Stage Carrier Vehicles

Perhaps the most important, and certainly the most dramatic, propulsion stage of a satellite launcher is the first stage. Not only does it have the greatest thrust of any part of the vehicle, by definition, but closely allied with this is the problem of lift-off and integration with the launching pad. When discussing the integration of stages of a carrier vehicle it is most appropriate to visualize the first stage in many ways as a "second" stage, the first stage being the actual launching pad itself and its release equipment—albeit fixed and (it is hoped) firmly anchored to the ground.

Many of the problems faced in subsequent stage integration and separation are faced to a maximum degree at this first primary separation, namely the clean mechanical lift-off of the carrier from the pad and the clean separation of the power supply lines, fueling lines, etc., which are

FIG. 3. Atlas carrier at lift-off showing main launcher assembly and connections to pad. (Courtesy of NASA.)

vital in the final preparation and countdown phases of the total vehicle. The extent of these connections can be seen in Fig. 3, which shows the vehicle at lift-off from the pad.

It is interesting to compare the mechanical release gear arrangement of the American Atlas and the British Blue Streak carriers, shown in Figs. 4 and 5. The Atlas scheme uses an open framework assembly with two main

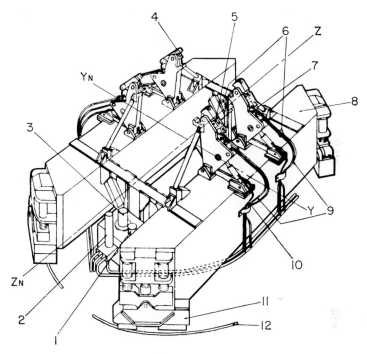

Fig. 4. Diagram of launcher assembly for Blue Streak carrier vehicle. 1, Vent filter; 2, accumulator; 3, pump unit; 4, top jaw movable; 5, release jaw to toggle linkage; 6, fulcrum pins; 7, release mechanism; 8, box section beams; 9, hydraulic pipe lines to and from release jacks; 10, hydraulic pipe line to erection jack; 11, retractable bogie unit; 12, outer track. (Courtesy of Hawker Siddeley Dynamics Ltd.)

release arms diametrically opposed across the propulsion bay; these take the main thrust loads and the vehicle weight, with two lighter, balancing connections at 90 deg to these. At lift-off all four arms pivot in the manner evident from the photograph, allowing a clear passage for the propulsion bay accelerating vertically from the pad. On the other hand the British Blue Streak vehicle (part of the European Launcher Development Organization's three-stage carrier) has a more compact release gear comprising four arms in two pairs, each picking up on opposite ends of two main beams

Fig. 5. Blue Streak attachment to launcher assembly showing two of the four hydraulically operated arms. (Courtesy of Hawker Siddeley Dynamics Ltd.)

which pass through the center of the propulsion bays. Four lugs, located at the connection points are enclosed by conformal jaws at the upper end of the main release gear arms. At the instant of lift-off these conformal jaws are opened by hydraulic jacks and the four arms are pivoted back by further hydraulic actuators, again leaving a clear zone for the vehicle to lift off.

If we bear in mind that the typical vertical acceleration of a rocket at lift-off is about 0.3 g, we see that the vehicle spends a relatively long time in rising the 10 or 15 ft necessary to clear these arms, and a considerable study of the separation dynamics is necessary for the design of this release gear.

A further feature associated with the dynamic motion of the carrier at lift-off is the method of carrying services to the front end of the vehicle or to the upper stages. Again using Atlas as an example, the services are carried by a "gooseneck," shown in Fig. 6. In the case of the Blue Streak

FIG. 6. Atlas rocket on launching pad showing the connection methods for services to the front and rear end of the vehicle. (Courtesy of NASA.)

vehicle a mast located adjacent to the release gear carries the front end services. In the case of the Atlas pad the top section of the gooseneck, carrying the electrical services, swings out of the way on separation of the electrical plugs, whereas with the Blue Streak (Fig. 7) the mast is fixed and only the connections are swung clear. In both cases the dynamics of the vehicle in the first few seconds of flight is an important feature to avoid mechanical interference.

A further feature of the separation of this first stage from the launcher pad has to do with the fluid disconnect couplings. Propellants need to be pumped in the tanks and other fluid services (for example, liquid helium, liquid nitrogen, hydraulic fluid) also require fluid couplings. Some of these services (particularly topping up with liquid oxygen) have to be carried out until the last moment before lift-off, and this means that the disconnection of these couplings must be done remotely. A common way of achieving this is for the vehicle to fly up and "lift out" from the couplings; this is the method adopted for Atlas and Blue Streak. Bearing in mind the size of the lateral loads which can be experienced by the vehicle because of engine gimballing during lift-off, some considerable flexibility must be allowed in these couplings to avoid their damage by fracture during the few inches of movement necessary to effect separation. Considerable design and development work on special rigs needs to be carried out in support of this subsystem.

A similar fly-off technique is often used in the case of the electrical connections to the rocket. For the same reason, a certain amount of flexibility must be incorporated in the connection so that it is not damaged during the inch or so movement required to separate the pins. Furthermore, precautions must be taken on the electrical connections to ensure that the particularly hot, dirty, wet environment (associated with the high induced flow and rocket jet impingement and reflection present at the time on the launching pad at lift-off) does not interfere with the exposed electrical connections. In guided missiles and space technology, electrical plugs are notorious as the source of trouble and failure, and a considerable amount of development work must go into producing successful plugs for the connection of the vehicle to the site. These plugs can carry many hundreds of channels of information and command signals.

In discussing vehicle integration it should be borne in mind that vehicles have now been launched with lift-off thrusts of 1,500,000 lb, with first stages measuring 22 ft across the base; thus sheer size and power are significant factors. Vehicles will be launched during this decade at thrusts of 7,500,000 lb and even perhaps 12,000,000 lb.

FIG. 7. Blue Streak on static test stand showing the cable mast adjacent to the rocket carrying services to the front end. (Courtesy of Hawker Siddeley Dynamics Ltd.)

B. One-and-a-Half-Stage Carrier Vehicles

The difficulties of achieving orbital flight with a single-stage carrier vehicle can be relieved to some degree by adopting the use of a one-and-a-half-stage configuration. This is a fairly unusual configuration. The only real example of this in the present generation of large carrier vehicles to date is the Atlas. In a one-and-a-half-stage vehicle the booster propulsion units

FIG. 8. Atlas rocket at launch showing general configuration of propulsion bay and vernier motors. (Courtesy of NASA.)

themselves are staged or separated; but this does not include their tankage since it is left with the remaining propulsion unit. This means that there is a single main tank structure, comprising liquid oxygen and kerosene tanks; the propellants in these tanks initially feed three engines, the two booster units and the central sustainer unit. The two outside units have a thrust of over 150,000 lb; the center unit has a thrust of some 55,000 lb (all sea level values). The two booster units are set diametrically opposite on each side of the central sustainer. These are assembled into a propulsion bay of 10 ft diameter containing the appropriate control system equipment associated with the booster chamber swiveling, power distribution, instrumentation, etc. Lift-off of the vehicle is achieved using all three of the engines; roll control and fine velocity correction for cutoff is achieved on the Atlas by use of two small vernier motors mounted some 15 ft forward from the base of the vehicle on the outside of the main tankage forward of the staging joint. The configuration can be seen by reference to the photograph in Fig. 8. The booster engines burn for just over 2 min in a typical mission and then each engine is shut down prior to staging.

The staging apparatus is rather unique, and comprises (Fig. 9) two rails parallel to the longitude axes, supported a short distance below the skin of the propulsion bay. Mating with these rails are rollers, attached to the propulsion bay, so that when the staging command is given the front end of the propulsion bay is unclamped from the main structure and the continued acceleration on the main structure (derived from the continued burning of the sustainer engine) accelerates the main structure away from the propulsion bay. The latter then rides down the rails which prevent interference between the equipment of the propulsion bay and the sustainer engine; the bay then falls cleanly away after the end of the rail has been traversed. It is a considerable credit to the ingenuity and engineering skill of the Atlas designers that the system works reliably and cleanly, since the tolerances are quite tight between the complex equipment within the propulsion bay and the sustainer engine assembly.

Further complication comes from the fire wall which is placed across the base of the whole Atlas vehicle. The purpose of this wall became apparent after the first few failures of Atlas rockets, when it was found that recirculating airflow at the rear end carried high temperature gases and flame into the propulsion bay, with subsequent fire occurrence. The fire wall at the rear end prevented this recirculation of air, steering movement of the combustion chambers being allowed by flexible asbestos collars between the chambers and the main fire wall. At staging it is necessary to ensure that there is no holdup to the division of motors at the fire wall throat level.

A further complexity associated with staging of only the rocket engines

Fɪɢ. 9. Exposed sustainer pack of Atlas missile around which booster propulsion bay slides on rails, shown in bottom left and top right. (Courtesy of NASA.)

themselves is the need to disconnect and seal the low pressure propellant supply lines between the tanks and the engines. This has led to the development of special quick-sealing, rapid shut-off couplings for the external low pressure delivery pipes to the bay. The axis of these couplings is parallel to the longitudinal axis of the missile. The question of fluid couplings and their disconnection problems will be dealt with in detail in Sections VI, A and VI, B.

The effect of discarding the dead weight of the engines and their control system is to increase the overall efficiency of the vehicle. With this combination of one-and-a-half stages, the Americans have successfully injected the whole series of one-man Mercury space capsules into low circular orbits between 100 and 165 miles in altitude. This is a payload into orbit of about 3000 lb with a vehicle lift-off weight of some 280,000 lb, representing payload percentage of about 1 per cent, which is no mean achievement in contemporary rocket performance.

C. Two-Stage Carrier Vehicles

This category of vehicles has been more common in space activities to date. Examples of two-stage vehicles used to achieve orbital flight are Thor Agena A, B, and D; Atlas Agena A and B, etc. It would seem also that much of the Russian space activity has been achieved by using two-stage rocket vehicles, if we are to accept the implications from the published information concerning the manned space flight work to date in which two phases of acceleration seem to have been experienced by the cosmonauts. Other examples of the two-stage vehicle are the Atlas Centaur and Titan II.

The two-stage configurations used to date have by no means had optimum staging ratios; in each case the Agena B and the Agena D second stages have been used with both the single-engined Thor and the three-engined Atlas. (Of course to be rigorous in this analysis, with Atlas the number of stages should strictly be termed $2\frac{1}{2}$.)

With the two-stage vehicle the problems are different from those of the one-and-a-half stage case previously considered, since here the tankage stays with the engines it has been feeding with the propellants. The interface problem is therefore less severe, and comprises mainly electrical signaling and monitoring connections between the stages, together with mechanical attachment devices. Electrical connections pass information along the length of the vehicle and, in some absolutely necessary cases, power supplies. These lines can also be used to provide signals for the next stage light-up of the engines during flight. With the second stage of a two-stage vehicle, a variety of geometric configurations may be utilized.

In the two extremes so far, one (Atlas Centaur) has a second stage with the same diameter as the first stage; but, on the other hand, a considerable reduction of diameter has been used as in the case of the Thor Agenas. The latest of the Able series, Able Star, is manufactured by Aerojet-General Corp., and contains the Aerojet A.J. 10-104 engine, burning inhibited red fuming nitric acid and UDMH, producing a thrust of 7870 lb. This stage is used on top of the Thor rocket produced by Douglas Aircraft

FIG. 10. Agena upper stage rocket. (Courtesy of NASA.)

Corp. It is powered by the R/D.DM-21, 165,000-lb thrust engine, burning liquid oxygen and kerosene. The Thor Epsilon (Thor Able Star) is capable of putting 900 lb in a low circular orbit, and has been used in such programs as Transit and Courier.

The Agena stage (Fig. 10) produced by Lockheed Missiles and Space Co. uses the Bell Aerospace Corp. 8096 engine burning inhibited red fuming

Fig. 11. Thor Agena two-stage rocket combination. (Courtesy of NASA.)

nitric acid and UDMH, producing a thrust of 16,000 lb. This, in combination with Atlas, has been used with the Samos, Midas, Mariner, and Ranger projects, and it is capable of orbiting 5000 lb in low circular orbit or putting 750 lb to escape velocity. Thor Agena B and D (Fig. 11) have been responsible for many orbital missions, particularly the Discoverer satellite program. They can place 1600 lb into low circular orbits. The Agena stage

Fig. 12. Atlas Centaur two-stage rocket configuration. (Courtesy of NASA.)

has the particular facility of relighting in flight, since the IRFNA/UDMH propellants are suitable for this purpose. It is also able to orient itself in space after a coasting period, and has been particularly useful in this role in the recovery of Discoverer capsules from orbit. It will be noticed that both the Agena and Able Star stages use liquid rockets.

One of the most interesting two-stage vehicles being developed by the Americans is Atlas Centaur. Centaur is the first of the new generation of high energy upper stages using liquid oxygen and liquid hydrogen propellants. The combination of Atlas and Centaur will be able to orbit 9000-lb payloads or provide escape velocity for 2300-lb payloads. It is scheduled to launch the Mariner B and Surveyor Probes. The Centaur vehicle is produced by General Dynamics/Astronautics and has two Pratt and Whitney RI 10-A.3 engines, each of 15,000-lb thrust developed on liquid oxygen and liquid hydrogen propellants. Centaur is 10 ft in diameter, the same as the main tankage of Atlas, which, in its role as a Centaur launcher, has been altered to maintain the full 10-ft diameter forward to the adaptor section between the two stages. The normal Atlas, it will be recalled, has a tapered front tankage section. The Atlas Centaur configuration is as shown in Fig. 12.

The largest two-stage carriers are Saturn I and Saturn I-B, both in advanced stages of development, and both powered by first stages with eight 188,000-lb thrust H-1 engines for a total stage thrust of over 1,500,000 lb. The Saturn I's second-stage incorporates six Pratt and Whitney RL 10-A3 engines for a total thrust of 90,000 lb. Its low orbital payload capability is 20,000 lb. Saturn I-B substitutes a single J-2 engine for the six smaller engines, increasing the payload capability to 32,000 lb. The first stages of these carriers are built by the Chrysler Corporation.

D. Three-Stage Carrier Vehicles

The three-stage carrier vehicle provides a more elegant solution to the orbital problem with the contemporary range of engineering techniques available to the rocket designer. Thor Delta is a typical three-stage vehicle, the single-engined Thor rocket with a thrust of 165,000 lb carrying a second-stage with a single AGC AJ 10-118 rocket engine, burning white fuming nitric acid and UDMH, with a thrust of 7700 lb. Atop is the third stage, the Allegany Ballistics Laboratory 248 rocket, which is powered by a solid rocket producing a thrust of 3100 lb. This combination is able to orbit 750 lb into low circular orbit or 140 lb to escape. Thor Delta has been used in such programs as Explorer, Syncom, Telstar, Tiros, and Relay. Again the interface problem is similar to the two-stage vehicle in

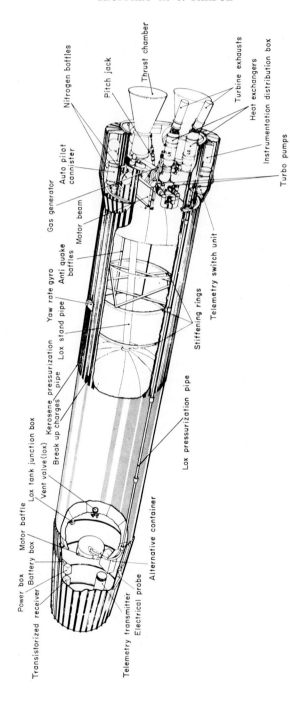

Fig. 13. Cutaway drawing of Blue Streak first-stage (Courtesy of Hawker Siddeley Dynamics Ltd.)

that no fluid connections are involved between the stages, but merely mechanical and electrical connections.

Moving to the European field, we find that two other interesting three-stage vehicles are in the course of development. The more important of these is the European Launcher Development Organization's carrier [2] based on the British Blue Streak first stage [3].

Blue Streak is a single-stage rocket powered by two engines, designated RZ12, each producing a thrust of 150,000 lb and burning liquid oxygen and kerosene [4]. The vehicle is of a similar design to that of Atlas, in that it relies on thin, pressurized stainless steel tanks for the liquid oxygen and kerosene; the two are separated by a single-thickness diaphragm across the width of the 10-ft diameter. The engines are mounted side by side in the propulsion bay, the diameter of which is 9 ft (compared with the 10 ft of Atlas). The general arrangement of the vehicle is similar to that shown in Fig. 13 [5]. The engines burn at nominally constant thrust and are not throttled, although they experience the usual thrust increment as the vehicle progressively enters hard vacuum conditions.

The second stage of the ELDO launcher is being developed by the French, and consists of a two-meter diameter tank section containing nitrogen tetroxide and unsymmetrical dimethyl hydrazine propellants; these supply four engines producing a total thrust of 70,000 lb. The complete stage weight is 25,000 lb.

The third stage of the ELDO vehicle is being produced by Germany and is also of two-meter diameter, but the thrust is considerably lower, being produced by two engines each with a thrust of 2300 lb. The third-stage propellants comprise nitrogen tetroxide and a mixture of UDMH and N_2H_4.

The combination of these three stages, shown in Fig. 14, should be able to put payloads weighing some 2300 lb into a 300-mile circular orbit. This vehicle is unique in that each stage is being produced not only by a different company, but by a different country; and in certain cases the vehicle itself is being produced by a consortium of different aerospace contractors. This in the case of France and Germany has led to the organization outlined in Section IX. The interface problems between stages are certainly accentuated, not only in the technical sense but in the administrative sense. This, too, will be examined in the subsequent sections.

The other European three-stage vehicle of interest is the French Diamant, now in advanced development. The first version of this vehicle (Fig. 15) has a liquid propellant first stage of 4.59-ft diameter with a thrust of about 60,000 lb. Propellants are nitric acid and turpentine—rather an unusual combination. In the first version of the vehicle the second and third stages are solid rockets, the combination being capable of injecting a pay-

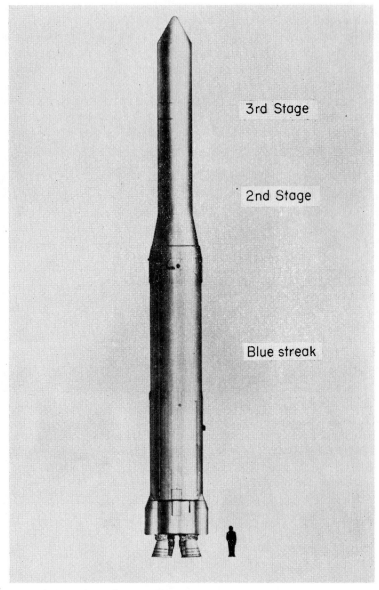

Fig. 14. European Launcher Development Organization three-stage satellite launching vehicle. (Courtesy of Hawker Siddeley Dynamics Ltd.)

Fig. 15. French Diamant three-stage space carrier vehicle. (Courtesy of Sereb)

load of between 110 and 180 lb into an orbit height between 230 and 800 miles (perigee and apogee). The French have several different versions of Diamant planned, progressively replacing the second and third solid rocket stages by liquid rockets, first with medium-energy propellants, then with liquid oxygen and liquid hydrogen, and finally liquid hydrogen and fluorine. The last combination should be capable of injecting a payload of

some 800 lb into the same elliptic orbit. The Diamant is being developed
by the SEREB consortium in France, to be described later.

The largest three-stage carrier, and the largest carrier of any kind
known to be in development in the world, is the Saturn V. Its Boeing-
built first, or S-1C, stage provides a total thrust of 7,500,000 lb from five
liquid oxygen-kerosene F-1 engines. The second, S-2, stage is built by
North American Aviation and contains five J-2 engines burning liquid
oxygen and liquid hydrogen and rated at 200,000 lb each. The third stage
is the Douglas-built S-4B, powered by a single Rocketdyne J-2 engine.
The carrier is some 350 ft high and is being developed under the direction
of NASA's George C. Marshall Space Flight Center to launch the Apollo
spaceship to the Moon. It has an orbital capability of 220,000 lb and an
escape capability of 90,000 lb.

E. Four-Stage Carrier Vehicles

The only four-stage vehicle to have been used to any extent is the
American Scout, including the Blue Scout version. This is an all-solid
rocket capable of putting payloads weighing about 150 lb into 300-mile
circular orbits, or somewhat lower payloads into higher orbits by the
addition of another stage or two. In the case of the Hughes proposal for
a 52-lb synchronous satellite system, six stages were proposed.

The basic Scout first stage is Algol II, built by Aerojet, whose XM 68
engine produces a thrust of over 120,000 lb. The second stage is a Thiokol-
built Castor whose XM 3320 or XM 75 engine produces 55,000 lb of thrust.
The third stage is Antares II built by the Allegany Ballistics Laboratory,
with engines designated X254 or X259 producing thrusts, respectively, of
13,600 and 19,000 lb. The fourth stage in the basic Scout vehicle is the
Altair II, being made again by the Allegany Ballistics Laboratory with
engines designated X248 or X258 with respective thrusts of 3100 and
5000 lb.

A particularly interesting four-stage configuration is the Titan III,
consisting of a modified Titan ballistic missile with extra booster stages
and an upper stage added. The first-stage boosters consist of two solid
rockets strapped one on each side of the main vehicle. Each of these rockets
is developed by the United Technology Corporation, and is a five-segment
10-ft diameter rocket producing a thrust of 1,000,000 lb (Fig. 16). The
vehicle (in effect, the second and third stages) which these solid rocket
motors boost into flight consists of the basic Titan II, the first stage of
which is powered by the Aerojet AGC XLR/87-AJ 5 rocket engine, burning N_2O_4/N_2H_4 with a thrust of 430,000 lb. The second stage of the

FIG. 16. Model of Titan III space carrier vehicle. (Courtesy of NASA.)

Martin vehicle contains a single AGC XLR 91-AJ 5 engine, also burning
N_2O_4/N_2H_4 and UDMH at a thrust of 100,000 lb.

The fourth stage of the Titan III is placed atop the Titan II core;
it is powered by two engines burning N_2O_4/N_2H_4 and UDMH, with a
combined thrust of 16,000 lb. Titan III is capable of placing 28,500 lb in a
low circular orbit, or 8000 lb to escape. At present, the carrier is scheduled
for the MOL manned orbital laboratory project. It is of interest to compare the 28,500-lb payload capability of this rocket with its 2 million-lb
lift-off thrust, and the payload of 20,000 lb of the Saturn I with its 1.5
million lb lift-off thrust. The former shows a slight improvement over the
latter.

It would be inappropriate to attempt to give an exhaustive consideration of all possible four-stage vehicles at this part of the chapter, as
various three-stage vehicles can readily have a four-stage adaptation.
Indeed, although the ELDO vehicle, considered in Section III, D above,
is basically a three-stage vehicle consideration is being given to the use of
an apogee motor, either as a small solid or liquid motor, which would give
an added velocity increment for injecting the satellite into higher orbits;
a particular application of this would be for injection into a synchronous
orbit. The use of this apogee motor considerably increases the high altitude
payload capability, and is an interesting demonstration of the building
block concept of upper stages, which will now be examined.

IV. The Building Block Concept for Upper Stages

A. Individual Stages

By noting the increasing diversity of vehicles discussed in the preceding
section, as the number of stages increases it can be seen that there exists
considerable flexibility in choosing upper stages in association with the
smaller number of first stages. It is very sensible in any major program to
plan the development of a family of rocket motors and a family of rocket
stages so that various choices can be made among these combinations to
produce the best result for a given mission. This logical process is frequently referred to as the building block concept. Some idea of a few typical
vehicles and upper stages that have already been used, or are planned to
be used, is given in Fig. 17. It is worth discussing the individual stages
which are now available or should be produced to ensure full flexibility in
this matter.

The general size of stages can be classified roughly as follows. Frequently the ratio of thrust between adjacent stages of a vehicle is between
4:1 and 6:1. The first generation of American carriers (namely the Atlas

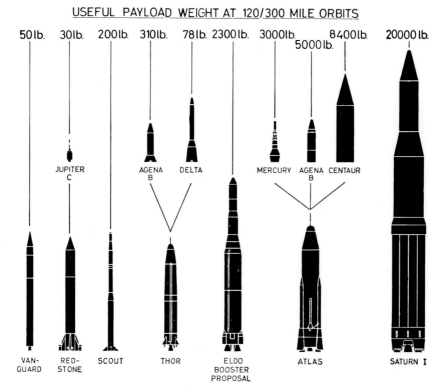

FIG. 17. Typical carrier vehicles with various upper-stage combinations.

type) has a thrust of some 300,000 to 400,000 lb. The next generation (namely Saturn I-IB and Titan III) has a thrust between 1,500,000 and 2,000,000 lb. Saturn V has a lift-off thrust of 7,500,000 lb. Taking the separate stage thrust ratios of the latter, we see that they are (1) 7,500,000 to (2) 1,000,000 to (3) 200,000.

So far as the Centaur stage is concerned, it was primarily designed for use with the Atlas and the engine thrust was to some extent governed by the size of upper stage that the Atlas could conveniently handle—one example of the engineering constraints referred to in Section II. One feature which affects the suitability of various stages to be used in appropriate places in a building block concept is of course the structure design and the configuration. In general, a vehicle using liquid hydrogen (in the high energy propulsion category) will be much larger than medium energy vehicles because of the low density of hydrogen. So far as an upper stage is concerned, the fact that it is smaller than a first stage means that there is a reasonable amount of tolerance in the absolute size that can be accommodated; but if, for example, a 7,500,000-lb stage used liquid hydrogen and liquid oxygen as propellants, then its size would be massive compared even with the present stage which uses liquid oxygen and kerosene. It is consequently not always best, when designing a stage, to use automatically the propellants of highest efficiency.

There emerges a pattern in which first stages tend to be of medium energy (solid or liquid), with upper stages of high energy becoming progressive'y more exotic as the number of the stage increases. This is reasonable from an overall engineering design point of view as well as analytically. The greater advantage in using high energy propellant efficiently is gained, in the use of higher stage numbers, so far as the choice of specific impulse is concerned.

Thus far the discussion has been limited to chemical propellants; future multistage rocket vehicles will incorporate higher energy propulsion systems based on nuclear and electric drives. The maximum specific impulses which can be expected from a reasonable combination of chemicals is in the order of 370 sec for sea level conditions, rising to above 480 sec for vacuum conditions. This is the peak for fluorine/hydrogen, although it is worth noting that ozone/hydrogen in theory will give slightly higher values. Nuclear propulsion systems, however, have the potentiality of considerably greater specific impulses—in the range of 750 sec and more. Various types of electrical drive—for example an ion rocket—have specific impulses of 2000 or 3000 and by suitable choice of parameters specific impulse as high as 5000, or even more, can be obtained, although at the expense of using extremely low thrusts. This last aspect is a feature of these electric drive engines where thrusts are measured in a few millipounds although they may be sustained for a considerable time—hours, weeks, or even months, depending on the type of electric supply system incorporated in the overall design.

The nuclear rocket falls halfway between the chemical and electric engine in its capability. The Rover nuclear rocket engine project has a design thrust of some 55,000 lb. This is clearly a thrust more suitable to upper-stage than to lower-stage propulsion.

A further feature of the choice of a nuclear rocket for the first stage of a vehicle is that most nuclear rockets conceived at present consist of a reactor used to heat and so to expand a working fluid (for example hydrogen) by passing the hydrogen through the reactor core. The heated gas is then expanded through a normal nozzle as in a chemical combustion chamber, thus producing a momentum change and subsequent thrust. This is relatively straightforward but poses the fundamental problem of an exhaust efflux which is extremely radioactive. If this engine were used in the first stage, the flame bucket and the whole area of the launching pad would become extremely radioactive as the engine ignited and lifted off. Not only would this constitute an extreme hazard to personnel, but it is impractical at this stage of development to consider a single-shot capability of a launching pad from the logistic point of view.

It is theoretically possible to create a clean efflux from a nuclear rocket, free of radioactivity; when this becomes a reality (assuming that the specific impulse and efficiency of the rocket engine can be maintained), nuclear propulsion for first stages will be possible. It should be borne in mind that the thrust levels must be increased accordingly. A further overriding feature may well arise: while the reactor for a nuclear drive is not in the same form as a nuclear weapon, it can constitute an explosive hazard. Rocket launching pads, the associated blockhouse, and the general range facilities must always be designed for the worst possible case, and this would impose extremely severe restrictions on the design of the whole complex against the event of the ultimate nuclear failure. It is therefore unwise to activate the nuclear stage on the launching pad. This provides an even further reason for the incorporation of nuclear propulsion in upper stages.

As at present conceived the Rover nuclear rocket engine being developed by America may be flown in RIFT (Reactor in Flight Test) program, and may be carried aloft as an upper stage on the Saturn I vehicle. The nuclear rocket therefore falls into a fairly clear category as a second-stage propulsion unit for carriers developed in the late 1960's, and as a third-stage propulsion unit for somewhat more advanced vehicles.

With electric propulsion systems, the choice of natural stage is a more clear-cut matter. In general, electric propulsion will nearly always be incorporated in the uppermost stage of a multistage launching vehicle. The reason is fairly straightforward, since the low thrust and long burning time of such an engine dictate that the propulsion unit must operate on the lowest possible mass—which of course is the payload. An exception may be made in the case of an interplanetary mission where it may be necessary to retain some higher thrust chemical capability with a propulsion stage associated with the later injection of the payload into the planetary orbit and the return journey. In this case it is conceivable to use electric drive to accelerate the combination.

B. Integrated Vehicles

We have had some examples in Sections III,C, III,D, and III,E above of the combination of various stages into integrated vehicles. The general trend is to use a high number of stages if the mission requires a low payload into a deep space trajectory, namely lunar or interplanetary missions, and a lower number of stages if the mission requirement is for a high mass pay-load into a low circular orbit of a few hundred miles, for example.

A further concept in the integrated vehicle category is the Nova project (Fig. 18). Nova, if it evolves as a development carrier, would follow

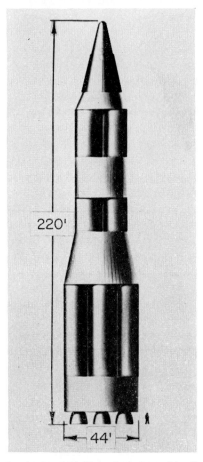

FIG. 18. One example of advanced Nova rocket combination, 6-engine first stage. Mission capability: 300 N.M. orbit, 290,000 lb; 24-hr orbit, 60,000 lb; lunar probe, 100,000 lb.

Saturn V and would have a minimum of 12,000,000 lb lift-off thrust using eight F-1 rocket engines. A typical second stage, designated N-2, might use four AGC M-1 engines burning liquid oxygen and liquid hydrogen producing a second-stage thrust of approximately 5,000,000 lb. The smallest of the variations possible for the third stage, N-3, might contain one J-2 rocket burning liquid oxygen and liquid hydrogen, with a thrust of 200,000 lb. Larger Nova designs have been proposed, with up to 30,000,000 lb of take-off thrust. A variety of propulsion systems have been examined, including very advanced designs permitting single stage-to-orbit techniques with 2,000,000 lb payloads.

C. Limitations of Techniques

The building block concept cannot be extended indefinitely. Reference to the brief analysis of the dynamics in Section II reminds the reader of the ratio requirements between the stages, which leads directly to a physical limitation of the number of stages that it is reasonable to visualize at this point in the space era. It should be recalled that each time a stage is added not only is there an increase in the engineering problems with the connections between the stages, but there is a corresponding decrease in reliability by virtue of the added connections and elements which can go wrong. The Saturn program is probably typical of the limitations of the technique in current technology with the three main propulsion stages of the V program plus the three stage-modules of its Apollo payload. Considerably more mission flexibility, even within the fixed combinations of rocket stages, can be achieved by the technique of different propellant loading, mentioned earlier.

One result of applying different propellant loading techniques to different injection profiles (according to the varying requirements of different missions) would be a corresponding difference in coasting times of various rocket stages. Thus a further factor that must be considered in the design and choice of stages in a building block plan is that a low-temperature, high-energy, volatile propellant boils off to a considerable degree when left to coast in a transfer or parking orbit for several hours in direct sunlight. Various techniques are used to minimize this propellant boil-off, ranging from insulating the whole stage to insulating just one part of it and arranging that that part should be attitude controlled to face the Sun during the coasting period. (The latter technique has obvious complexities and lower reliabilities associated with the requirement for continuous control.) Again, as a general trend, the higher the orbit, and so the deeper into space, then the longer the coasting times required between, or in, the various stages.

Each stage must be made with a performance as flexible as possible, since each new size of stage and type of propellant bring the attendant high development costs and new test facilities associated with such work. A properly designed pattern of stages, therefore, should not only cater to a combination of different major missions but also provide sufficient flexibility in each stage to provide performance overlap to deal with minor mission changes.

V. Payload Installation on the Final Stage

A. Attachment Methods

The dynamic motion of the satellite or space probe upon separation from the final propulsion stage of the carrier requires low angular and lateral disturbance, with its relative motion to the final stage being confined within certain limits. The latter are usually determined by the extent that the satellite or probe can correct itself in position and attitude after the separation phase is completed. In general, the attachment of payload to the final propulsion stage is by a rigid connection, antivibration mounting being provided to components, if necessary, within the payload itself. The attachment interface (or junction) with the final stage does not always necessarily coincide with the separation interface. When connection and separation are common the attachment is often carried out by a set of clamp rings which are opened up by gas pressure, or released by explosive bolts (Fig. 19). A variety of devices have been designed to facilitate a clean separation, and the majority use variations of this particular theme. Either gas pressure is generated in a common chamber which then distributes a radial pressure to remove each clamp ring, or an explosive bolt is used to sever a spring-loaded cable attachment, which in turn flies off and removes the clamp connections.

After the mechanical connection is severed, two basic methods of separating the satellite from the final stage are then used. One involves a compressed spring which automatically expands when the mechanical attachment is severed; in the other, some form of gas pressure is used either to force a piston out of a cylinder, or in small thrust jets located on the spacecraft. Separation from the final propulsion stage is often delayed by up to 40 or 60 sec after nominal cutoff of the last stage to ensure against final spasmodic burning (sometimes referred to as "chuffing") of the propulsion system after main cutoff accelerating the propulsion stage, so that it catches up with the spacecraft and collides with it. Unless a specific velocity increment is required, a typical separation velocity between the spacecraft and the final propulsion stage is one or two feet per second, easily achievable by the spring. The rocket jet system of separation has greater flexibility in providing a velocity increment.

A further feature of the spacecraft is the shroud or fairing over it during boost. The fairing is there for two reasons. The first is to protect the payload

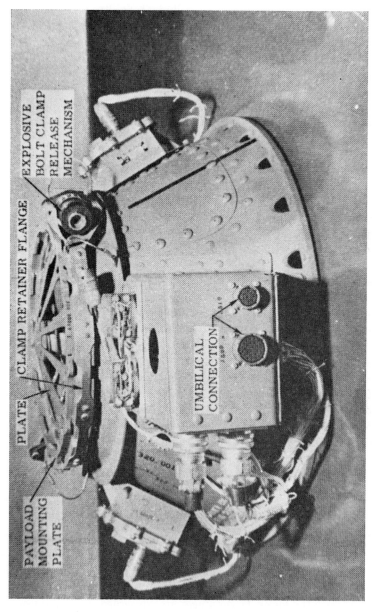

FIG. 19. Typical mechanical interface junction between stages of the Chance Vought Corp. Scout rocket showing explosive bolt mechanism (Courtesy of NASA, Langley Research Center).

from kinetic heating and dynamic pressure on the ascent; as an indication
of the intensity of this, the shroud can easily be heated to temperatures of
500 or 600°F during the passage through the thinning atmosphere. The
second use of the shroud is to provide a rigid support for an incidence meter
placed at the head of a vehicle to give it refined guidance data during its
first moments of flight. Sometimes the fairing supports other instrumentation used during the development or operational use of the vehicle.

The attachment of the fairing is usually made to the last main propulsion stage and not to the spacecraft payload itself. In some cases where
the last propulsion stage is very small the shroud attachment can then
even be to the penultimate stage. This particular method was used on a
Thor Able carrier wherein the payload was attached to the third-stage
support structure by means of two, semicircular, segmented tension strap
clamps which were held together by explosive bolts. A spring cartridge on
the forward end of the engine provided a separation velocity between the
third stage and payload of approximately $1\frac{1}{2}$ ft/sec. Separation of the
payload occurred upon release of the bolts by a signal from the timer.
The third stage, which was an NX248A7 solid rocket propulsion engine,
was attached to the second stage by means of four hinge-mounted petals.
This system was designed to enable spin stabilization of the third-stage
rocket and the payload. The spin table was attached at the forward end
of the guidance compartment. A spin rate of about $2\frac{1}{2}$ to 3 rpm was imparted to the spin table and upper stages by two rockets mounted on the
spin table. A similar principle applies to other satellites which are spun
prior to separation from the final propulsion stage.

The fairing or shroud is seldom in one piece, usually being in two, three,
or four petals, which are split open by explosive bolts and driven out by
springs or by gas jets located between the segments.

In all cases where mechanical equipment is separated in orbit, careful
dynamic analysis must be made to ensure against chance of subsequent
collision between the parts, even during the ascent phase.

B. Influence on Vehicle Mechanical Design

The influence of the satellite on the vehicle design is almost exclusively
limited to the upper stages. If the payload is particularly sensitive this may
demand vibration limits, and so could influence the design of the upper
stages, where the attenuation is low between its propulsion unit and the
payload. So far as the lower, larger stages are concerned, there are usually
bases for their selection connected with the overall performance required
to achieve the mission. It is therefore unlikely that the satellite design, as
such, would influence the main design of the carrier, but its mission would

of course have considerable influence on the design of the upper stages. An example would be the provision of coasting phases.

A difficult situation can occasionally arise in which the payload is of a size which is larger than the diameter of the final propulsion stage. This situation has led to one or two American carriers having a payload fairing sometimes nearly double the diameter of the last propulsion stage. This, in turn, has led to many extra test programs to evaluate the aerodynamic characteristics of the combination, with particular regard to kinetic heating during the ascent phase through the atmosphere. In one or two cases this bulge at the front end has proved extremely expensive to perfect (in terms of time and money in the overall program) and in general is best avoided if possible. Therefore, at the beginning of a vehicle development program, it is wise to choose an upper-stage configuration which, if possible, has the largest possible diameter, consistent with performance, and so will not involve a re-entrant angle behind the payload fairing. Indeed, while it is vital to minimize the *weight* of a satellite, care in determining its *size* is often given too little priority, a fact which becomes apparent only in the later phases of installation design of the payload. In choosing his upper stages, the designer would be well advised to exercise as much imagination as possible as to the variations of satellite, probe, or spaceship configurations he may be called upon to accommodate in later missions. Flexibility again is the keynote to success.

VI. General Engineering Problems on Stage and Payload Integration

A. Mechanical Interface Problems

There are certain features of mechanical engineering common to the mating of all stages of a multistage carrier vehicle. There are two mechanical problems involved in the connection between two stages: first, to retain a rigid mechanical connection while the stages are combined; and second, to provide a reliable clean separation system on staging. The expression "interface" applies to the whole general area in which mechanical (and electrical) problems are encountered and solved.

It is common practice to retain the separation interface design and development under the control of one manufacturer. The desirability of this is obvious with regard to compatibility. In the case of the Atlas in one of its roles of being mated with an upper stage, the interface between the two manufactured stages was a bolt attachment to the adaptor section at the front end of Atlas. There were two basic reasons for this: (1) the staging interface had to be maintained to tighter tolerances than the bolted interface; and (2) since the adaptor belonged to the stage above it, it could be used in stage check-out prior to mating.

The most obvious place for mechanical interface is not always the best; in the case again of Atlas, the point of transition between the unpressurized conical tank section extension and the forward bulkhead might be considered reasonable, primarily because it was the end of the liquid oxygen tank and was already an interface for the weapons system. A more reasonable interface, however, may well be *forward* of this point, about half way up the adaptor between the Atlas and the upper stage. Such a change would permit all installations on the forward end of the tank (for example pneumatic lines, boil-off valves, electrical and telemetry lines) to be made with the adaptor in place.

A further problem is also associated with the physical problem of mating the stages. Men working in the servicing tower are likely to drop nuts, bolts, spanners and so on; these objects falling on to the exposed front dome of a thin-walled propellant tank can cause considerable damage, not only immediately but potentially, particularly if the front dome is concave inwards which is usual, thus involving a cusp between the outer unpressurized adaptor section to the front end of the tank and the tank itself. Objects falling into that difficult area obviously cause problems. If, however, the transition can be made far enough forward, level with the most forward of the objects associated with the front end of the rear stage, then all of the necessary electrical and other connections between the stages can be brought to a conveniently constructed mating joint; and, if necessary, a protecting diaphragm can be placed right across the mating section.

So far as vehicle staging is concerned, clamps have been used extensively in the past, but more recently the use of prima-cord has become popular. This device will unzip the primary structure with high reliability and considerable precision. Latches have been used in many instances to separate stages, and in fact the propulsion booster bay of the Atlas missile separates from the main vehicle by means of latches. These have proved very reliable, and a typical system is to have a helium storage bottle with pyrotechnic squibs which open valves between the bottle and pneumatic solenoids which then operate the latches (together with the associated wiring and pneumatic lines). Prima-cord, however, provides a ring main which is completely disconnected on initiation, and so offers a simpler system. Not only can connections be severed in this way, but, by its explosive nature, it can also be used to sever normal load-carrying structure, thus eliminating the equivalent of mechanical cutouts. Complete duplication is advised with prima-cord to achieve the highest reliability.

So far as mechanical precision between the actual mating faces is concerned, it is common for the contractor responsible for one side to provide the interface tooling fixture to be used by the other contractor for his side of the interface. This will ensure not only that manufacturing errors will be eliminated, but any unnoted engineering changes will become apparent as a result of the use of a common jig design. Obviously, too, it provides great confidence that the sections really will meet. This aspect is particularly important where perhaps the contractors are not even in the same country, which is the case with the ELDO vehicle. While at the time of this writing no detailed policy had been published on this matter, it is clear that where dimensional systems and language differ between the contractors building each side of the interface, extreme care must clearly be taken to ensure that the general principle above is followed. In the case of the ELDO program a further difficulty occurs, as it is quite possible that the separate stages of this three-stage vehicle will not in fact ever meet until they all reach the launching pad in Australia. The problems that can arise, therefore, in the absence of the best liaison between designers in the early stages do not have to be emphasized.

The problem of the mechanical interface is not limited to tolerances and loads, but also affects the choice of materials. The materials chosen on each side of the joint must be compatible, and careful control must be exercised to ensure that no galvanic interface situation is created, particularly where the sides of the joint are being made by different contractors. The usual engineering problems obviously also apply, such as the choice of coefficients of expansion of the materials. Also the choice of the materials must allow for any stresses which may be set up because of differential movement of the material on each side of the interface. The problem is particularly sensitive, since not only must loads be transmitted during the combined vehicle operation, but friction must not be allowed to interfere with the staging operation.

One of the major aspects is the aerodynamic loading of the total configuration; and this, too, must be carefully integrated in the design. Moreover, in thin-walled vehicles, buffet conditions must be considered. Even though an adaptor between the stages may be satisfactory under normal circumstances, from the point of view of buffet, it may be that when it is deformed through a bolt interface in combination with the expansion of a pressurized tank on the other side of the interface, it will no longer resist buffet. Failures produced can be fairly subtle, in that the buffet loads may be transmitted across the interface and cause a failure there, even though they do not cause a direct mechanical failure where the buffet occurs.

Tests must be conducted by elaborate wind tunnel programs on both rigid and flexible models, to determine the aerodynamic spectral density, and the response of the configuration to the loading produced thereby.

For this purpose dynamically similar models are also required. Wind tunnel tests of course help considerably in determining the problems to be solved, but are not conclusive in themselves since the nature of the interface loading is most complex in real size and in real flight; in general, over-design in this area is a reasonable precaution. It should also be borne in mind that there may be major changes in the stages above the stage which is being connected to the first stage as a result of different mission needs. These variations will influence the attachment problem.

Extensive testing of the mechanical interface is carried out before the carrier ever takes to the air. Specimens of the forward end of the stage are built and are tested in conjunction with the adaptor and the rear end (at least) of the next stage. Both static and dynamic tests are carried out which must be planned early in the development program. Problems that may show up in such tests could require changes to be made in tank shapes, when pressurized tanks are involved in either side of the junction, and considerable induced stresses become apparent with pressurized tank configurations. The use of cryogenics on one side or the other of the interface involves material effects.

With an interface the normal problems of structural design are all present, with the additional concern of divided responsibility for the actual connection between the stages. The proving, therefore, of the mechanical integrity of the interface and its separation characteristics involves extensive tests with regard to pressure, cryogenics, heat, vibration, and static loading, all culminating in in-flight test. Reference has been made to the case involving not only the transmission of loading, but also the need for spinning up the next stage, in relation to the preceding one, before staging takes place. This requires a special test rig of its own, to ensure that the appropriate spinning rates are achieved as part of the separation sequence.

Finally, mention has been made of the fact that there is a special mechanical interface problem between the carrier and the launch pad. Similar requirements arise for mechanical rigidity before lift-off and yet a clean separation when the vehicle is released. Not only are similar design problems encountered, but special test rigs on a large scale have to be constructed to check asymmetrical loading as well as symmetrical loading, since, of course, the vehicle is controlling itself by swiveling the combustion chambers during the lift-off process. In addition to normal temperature effects, in this case there is direct impingement, too, of the rocket efflux on parts of the structure, and while in general the physical separation will have been made *before* the efflux impinges on the mechanical links, all such effects must be taken into account in these critical first few seconds of flight.

B. Electrical Interface Problems

The electrical interface between stages may involve any of the systems normally employed on the complete vehicle, namely the guidance, control, telemetry, range safety, and electrical power. In general, electric power is confined to a particular stage and so each stage generally has its own system. However, there are cases where power must be taken across an interface which subsequently has to be broken. The reluctance to involve this is apparent, since it is one thing to take signals of a monitoring type across a breaking joint, but another to carry any high voltage or current.

With the electrical interface, the same criteria apply as with the mechanical interface, namely design integrity while the faces are joined (involving compatibility, reliability, etc.) and clean separation when required. It is necessary to specify very closely *all* of the electric features of the connections between the stages, and to ensure that all contractors who may be involved with the systems directly connected between the stages, and any other systems which may be secondarily affected, are completely familiar, and in agreement, with the interface plan.

With regard to the system design, the extensive use of relays is involved to isolate power supplies as much as possible in the various stages and signal lines. While this can be achieved by the use of relays, the very use of these devices involves further difficulties with regard to reliability. Problems will be encountered, and sometimes extra signal conditioning or further stages of amplification must be provided in telemetry circuits to minimize this effect. Various types of circuits may be made across the joint, in which activation of a process may be achieved by either energizing or de-energizing the circuits. Some of these circuits could include the self-destruction system fitted in each stage. In the case of malfunctioning of the appropriate initiation circuit used when the stage begins to go outside of the range safety boundary, care must be taken to choose the type of circuit that fails safe, in the wider sense of the word.

The relay problem has been mentioned, and particular care must be taken in both the design and test of these to ensure proper sequencing of events, particularly during a staging maneuver. Functions slightly out of sequence, even by a few milliseconds, can cause complete failure (if, for example, range safety destruction systems are not properly de-energized before a fail-safe break occurs). Sometimes the receivers, which take the range safety signal from the controller at the pad, are located in one stage. The signals are transmitted through the interface to other stages.

Another electrical problem is found in the case in which the stage ahead of the one about to burn out must have its propulsion system ignited before staging. This requires a degree of precision in the sequencing operations, and often destruction of the vehicle could occur if the second-stage motors were ignited and burnt at full thrust for an excessive period of time before the stage behind cleared away.

Discussions with designers of multistage rockets reveal that they must face various interesting problems in logic. In general, there are two or three systems which can be the source of specific commands—perhaps the guidance of the first stage or the control of the first stage or the program device of the second stage—and any one of these devices could initiate a variety of commands. Electrical connections, therefore, between the various stages must allow signals to be suitably passed from any of these systems, but the priority of this information must be properly discriminated. Typical signals involved are vernier cutoff, the requirement to disarm premature separation destruct systems, retro-rocket firing to ensure the separation of the stage, the initiation of the second-stage programmer device, the activation of the separation circuits themselves, and finally the actual separation. The task of staging is sometimes so complex that some redundancy should be built into a good system to ensure that any single failure would not cause complete failure of the whole mission. To do that while still keeping events in the right sequence to avoid (for example) vernier propellant depletion is an interesting problem that does not, by any means, necessarily have a unique solution.

Within the vehicle control system another interesting interface feature arises: In a long, multistage vehicle such as the ELDO three-stage carrier, the first-stage control system uses rate gyros as part of the dynamic monitoring loop. Analysis and test of the optimum location for these rate gyros show that some must be located at vibration nodes and/or antinodes, which are in fact on the second or third stages. This means, therefore, that the necessary feedback signals from these items of equipment must be connected through the interfaces to the second and third stages and yet combined into the electronic autopilot circuitry of the first stage.

With regard to the guidance system, in some carriers commands are received by receivers located in the upper stages, and then transmitted through the interfaces to the first stage. The value of such an arrangement is apparent if guidance is to be exercised during the whole of the flight, since locating the receivers and transponders associated with the guidance system in the third stage avoids duplication or triplication in each of the lower stages. However, it does require that a high quality electronic connection be established throughout the length of the vehicle to ensure integrity and compatibility.

An overriding feature of all electrical interface design is that the number of connections between stages should be kept to an absolute minimum.

In the same way that special test rigs had to be created for evaluating mechanical problems, so must these be built for the electrical problem. It is sometimes necessary to create a piece of test equipment which is referred to as a stage simulator. This equipment receives and transmits the same electrical signals (excluding radio frequency) as the flight equipment does. In some cases this test gear is built by the stage contractors and supplied to the other contractors as required; in other cases it is built by the contractor planning its use. In either event such a piece of equipment has two important criteria: First, it must be design controlled and updated as carefully as the flight equipment itself if the value of the equipment is to be maintained; second, to ensure proper simulation, it must use actual flight hardware such as relays, harnesses, etc. Impedance matching and component time of activation are other important features to be resolved by the use of such equipment.

As well as direct electrical signals, potential radio frequency interference must be controlled between stages in the same manner as within an individual stage. This means, therefore, that in checking r.f. interference, due allowance must be made for *all* electrical systems used in *all* stages, and not only that of a single stage in a single test. Not only must this work be done in the contractors' establishments, but also the final testing of the combined vehicle on the pad must make allowance for such induced effects.

Actual electrical connections across the interface are achieved by plugs, sockets, and (in the case of the carrier to the pad) umbilical connections. Extensive tests on these to ensure their quality of performance in the appropriate conditions must be carried out. Relays, plugs, and sockets are among the greatest causes of failure in guided missiles and space rockets, and extensive environmental testing must be done to ensure successful operation of these in the actual program.

C. Propulsion System and Propellant Compatibility

Brief mention should be made of the implications of different stages of rocket vehicles using different propellants. In theory a stage is self-contained, and the choice of propellant may have no direct bearing on the stage above or below it. In practice, the vehicles are fully fueled before lift-off from the pad and if there is a possibility of hypergolic action between the propellants of two stages, then the fact that some propellant may be spilt and come into contact with similar spillage from a lower or upper stage must be taken into account. It is not only the question of pad safety that is involved here; but, particularly in the case of cryogenic propellants, the mechanical distortions and loading which can occur as a result of the extremely low temperatures are facts to be taken into account in the design of the interfaces.

In the case of the Atlas, the point has been made that at staging the low pressure liquid oxygen and kerosene lines are broken between the main tank and the propulsion bay. This leads to special problems involved in maintaining proper sealing at, and subsequent to, the instant of separation of these main lines.

The use of different propellants in different stages involves a variety of handling arrangements during the preparation and the loading of the vehicles. There is also a logistic problem introduced by the use of different propellants at different stages of the carrier; this is reflected in the storage and loading systems of these propellants on the site as well as in the design of the vehicle itself. The vehicular loading conditions may dictate a certain sequence of events in loading propellants. In the case of Blue Streak for example, the rear kerosene tank has been structurally stiffened to ensure that, even if the internal pressurization of this tank should fail after the liquid oxygen tank above it has been filled on the pad, there will not be a complete failure of the vehicle itself, because the structure is sufficiently strong to take the combined weight of the overhead structure including the oxidant tank when filled.

D. In-Flight Systems

Some features of the control and guidance systems have been mentioned. In the case of instrumentation and telemetry an interesting problem exists in deciding the allocation of monitoring points between various stages and the attendant transmission of data between stages by different telemetry senders.

At one extreme it can be argued that telemetry senders should be placed in the final stage, and all the information from the preceding stage activity should be sent through the interfaces up to the final stage, thus saving the weight of additional transmitters. However, against this is the fact that the lower stages sometimes require a larger number of channels of information; this means that the transmitter (if it is to be common in the upper stage) must be fairly large. Unfortunately, a large weight in an upper stage gives a greater penalization to the overall vehicle performance than the same weight in a lower stage. This throws the balance perhaps back toward diversifying the transmitters, thus saving the overall payload. The relative merits of this problem can be decided only by taking the particular set of conditions and determining the emphasis that a particular vehicle may wish to have placed on either its performance or reliability.

The European ELDO carrier is being developed with a minimum number of test firings. Consequently, it is important to ensure that every flight is extremely well instrumented and monitored. There may, therefore, be reason here to sacrifice some small payload at the expense of higher reliability, and hence it could be visualized that each stage might carry separate transmitters; yet the final stage would also be capable of transmitting some information of the first-stage performance.

E. Pre-Launch Check-out of Overall Systems

No discussion of the general engineering problems of stage and payload integration would be complete without emphasizing the importance of pre-flight checking of the combined carrier as well as the individual stages. The quality of early planning has had considerable influence here on the smoothness with which a combined carrier vehicle is ultimately prepared on the pad and subsequently successfully launched. Unless a plan is determined early in the program, a ridiculous situation may occur in which a tardy contractor comes along with a set of check-out equipment well adapted to his particular stage, but quite incompatible with the other stages and with the launch complex arrangements. Fortunately, such a position is rather elementary by present standards and should never occur. Nevertheless, even the obvious is sometimes missed, and the point is made here that a clear policy on the type and use of check-out equipment of various stages should be made early in the program.

This point is particularly true when it comes to the integration of payloads themselves. A three-stage vehicle (for example) may be kept in a similar form for a variety of missions, but each mission requires a different payload. In addition the payload could contain a variety of different experiments, each perhaps constructed by a different contractor and being done for a different group of scientists. Therefore, not only is it important to have check-out equipment of each propulsion stage compatible, but the payload check-out equipment must be compatible too. In some cases portable check-out equipment is made available for the payload, and this is taken up to the payload when it has been installed on the vehicle. The payload is then checked out and not dealt with again during the final check-out procedure. There is sometimes a case here in which final check-out of the payload is done by r.f. transmission between the payload and a telemetry receiver on the launch complex, thus avoiding the need to pass information down through physical connections through the interfaces of the vehicles and on to the pad.

F. General Considerations

The overall examination of the requirements of integration suggests that a proper system approach should require that, within the overall design of the vehicle combination, each stage should be able to operate with the maximum possible deviation from its nominal performance, consistent with the capabilities of the other stages, to correct not only its own tolerances, but those arising from other stages to accomplish the desired mission. Each stage is capable of adding an increment of velocity to the stage above it; if each stage controls this increment of velocity which it adds, then such control should provide reasonable limitations in the velocity at which the next takes over. Other things (such as system complexity) being equal, such an approach will maximize flight reliability. A further extension of this approach, in the specific case of total performance, makes one examine the possibility of burning all but the final controllable stage to individual stage propellant depletion. The orderly combination of performance tolerances with such an approach will result in greater *total* assured performance than will be the case if each stage must have precise burn-out velocity requirements, thereby wasting some propellant left in each stage. If this technique results in a significant increase in system complexity, the relative merits of performance requirements against potential reliability effects must be carefully weighed to determine the best method.

As a final comment it may be worth remembering that a chain is as strong as its weakest link. So it can be said of a space carrier vehicle, for however capable each stage is of performing on its own, unless it is started in the right manner following on the work of the preceding stage, and unless it hands over properly to the subsequent stage, it might as well not be there. Such is the value of integration efficiency.

VII. Extraterrestrial Multistage Integration

A. Earth Orbit Assembly

To date, the problem of integrating various stages of a carrier vehicle has been confined to the launching pad prior to lift-off toward an orbital condition. This will not be the case in the future, when orbital assembly techniques are used. The principle of orbital rendezvous is well understood, even though (at the date of writing) it has not yet been carried out in practice.

Rendezvous in orbit around the Earth may take place for reasons of logistic supply (in the case of the manned space station). Or it may provide the capability of undertaking a lunar or interplanetary mission by using a combination of relatively small rocket vehicles rather than requiring a development program for a massive multistage vehicle such as Nova to achieve a mission by the direct method. The use of orbital assembly permits an earlier capability and a more economical one. For example, it has been reported that, if a two-rocket, Earth orbit, rendezvous assembly technique is used for the lunar manned mission, the cost of the ultimate achievement will be only some two thirds that of the direct approach using Nova. However, in terms of overall energy expended, the balance is probably favorable in the case of the direct shot.

Whatever the reasons, the need for orbital rendezvous and assembly exists, and the associated techniques must be developed. The implications of integrating payloads in the space environment should, therefore, be considered briefly.

The first obvious problem is the actual motion of offering up one stage to the other—without the convenience and comfort of a normal launching pad servicing tower! This will be the culminating moment of the rendezvous procedure when the two vehicles have been launched independently into a similar orbit, and when the final corrections to the coincident orbit have been made by small rocket jets either by automatic or manual means. It is quite possible that the final phase of *docking*—the act of mating the parts—will be carried out under visual control by a man in one craft or in the other. The mechanical problems, therefore, start with the two space craft a few feet apart, slowly converging.

Two techniques may be followed: first, the actual mating operation may be conducted remotely by control from within the spacecraft; or second, the mating operation may be achieved by the astronauts' leaving their craft, in the safety of their individual pressurized life support suits and equipped with small individual propulsion packs enabling them to operate in the vicinity. In the first of the two techniques, the provision of a tapered adaptor section in either of the stage interfaces will clearly facilitate the final offering up of one stage to the other, and the actual connection could well be the reverse procedure of the separation technique previously described in the more orthodox case (excluding, of course, the explosive methods which tend to be nonreversible).

Even after the assembly in orbit the combined craft may well then be injected into a transfer orbit, and so after propulsion when the stage which has been mated on to the main craft has achieved its task, it may have to be separated again. In this event it is unlikely that the same interface used in the original mating operation would be used for the separation sequence, since both present somewhat different problems and engineering solutions. However, let it be assumed that the attachment interface is being made good; either remotely operated clamp rings have come into operation in the automatic case or in the manual case good old-fashioned nuts and bolts are inserted between two flanges fixing the sections together. We can imagine the feelings of the astronaut about to insert his bolt only to find that the same jig has not been used to drill the holes on one side of the interface as on the other. If ever there existed a case for good preplanning in design, this is it. It seems, however, that it would be fairly safe to assume that this particular operation will be straightforward, and the astronaut may return safely to his capsule to conduct the next stage of the mission.

If subsequent staging is required between the two sections which have previously docked in orbit, then the same principles previously described for the injection cases would be brought into operation.

The first American equipment having a rendezvous and docking in space is expected to occur in 1965 with the two-manned Gemini capsule and an unmanned Agena satellite. It is not intended that this particular combination will then perform an ignition and further transfer into another orbit, although undoubtedly part of the test schedule will include a sequence to examine how this will be done when the need arises.

Several concepts for manned space stations have suggested that they be orbited in segments. One of the earlier suggestions was that each segment should comprise the empty tank section of the main vehicle which has achieved the orbital altitude. However, some form of compartment will be injected into the appropriate orbit and a further compartment (possibly of identical design) will then be offered up to it. In this particular case it seems unlikely that an automatic docking procedure will be followed, since by that time sufficient experience will have been gained for astronauts to move freely outside their spacecraft in the orbit environment. The offering up of the two sections or more could then be carried out in a fairly orthodox manner, seals being included within the interfaces to ensure that the total station will be fully sealed when the parts have been assembled. It is realized that no problem really exists with the size or mass of the objects to be handled by astronauts in orbit because of the weightless condition there. The effect of size will be merely to slow down the speed at which large objects can be offered up to each other. It is this feature that enables such advanced concepts as massive radio telescopes measuring several thousands of yards in maximum dimension to be proposed.

It may be expected that the relatively orthodox engineering techniques of assembling stages of carrier vehicles on Earth will be applied fairly readily in the assembly of stages in the orbital environment.

B. Cis-lunar and Interplanetary Assembly

As an alternative to the technique of assembling stages in orbit around the Earth, the possibility exists of assembling them in orbit around the Moon, on their way to a planet, or even around a planet itself. It is not necessary, or even appropriate, in this chapter to examine the relative performance features to be derived from each of these different techniques. It is enough to say that certain advantages can be shown to exist in total energy requirement by adopting some of these techniques, depending on the mission objectives. In the case of the American plan to land their first astronauts on the Moon, the procedure to be adopted consists of a multi-module spaceship being sent into a lunar orbit. The Lunar Excursion Module (LEM), carrying two of the three crew members, will be detached from the main two-module craft for the descent to the lunar surface. After initial exploration has taken place the module will subsequently lift off from the lunar surface and enter into a lunar orbit coincident, ultimately, with that of the main Apollo spaceship. The LEM will next rendezvous and dock with the Apollo. The crew will then transfer into the main spaceship, which will return to Earth, leaving the LEM behind.

No significantly different problems of principle should arise in this lunar assembly operation from those described previously in the general case of orbital assembly around the Earth. It would perhaps be extending the moral of good preplanning a little too far to point out again the need for compatibility between the interface connections of the two parts to be assembled in the vicinity of the Moon.

Perhaps one feature should be mentioned here. Whereas in the Earth assembly condition it is unlikely that each part to be connected will remain for long in the space environment before being joined, this is by no means the same case in the category of lunar orbit assembly or intermediate planetary orbit assembly, where the equipment may well have to be exposed to the space environment for days or even weeks. This would lead to attendant temperature, vacuum, and even micrometeorite effects, all of which must be taken into account in designing the interface to ensure that stage connections will be smoothly achieved.

So far as the size of the equipment is concerned, the mission itself will dictate the extent of propulsion required to accelerate the mission spacecraft to its required velocity. There are no hard and fast rules that can be indicated at this point without considering a series of detailed missions, which would be inappropriate.

There is the further question of integrating stages of rockets on the

Moon or on the planets. This will undoubtedly bring along its own particular type of problem, and in all of these remote assembly conditions special tools, jigs, and equipment must be provided to enable the operation to be achieved. The mundane task of bolting stages together in the servicing tower on Earth will certainly take on a different character when transferred to orbital, lunar, or planetary conditions.

VIII. Launching Site and Ground Support Facilities

A. Launching Pad

The design of a launching pad is influenced by more than the size of a multistage carrier vehicle; it is also influenced by the specific needs of each of the stages, and particularly of the payload. For example, in the case of the Mercury manned satellite missions, a clean room was built in the servicing tower around the platform level of the entrance to the Mercury capsule. This was to ensure that after the space craft had been properly checked the astronaut would be able to enter it without any danger of carrying along foreign bodies and also to ensure that no dirt or other particles entered the satellite. There are other similar precautions which have to be taken with unmanned satellites and space probes with the object of maintaining their standard of functioning between final preparation and lift-off.

To date in America, the assembly of multistage vehicles has been done in a fairly orthodox manner, each stage being progressively lifted by a winch and lowered to the stage below it with access platforms available to facilitate the actual mating process. Already servicing towers over 300 ft high have been produced (as in the case of the Saturn I carrier) and the problem gets progressively worse as vehicle sizes increase.

Perhaps the present 320-ft tower for Saturn I will be the last and largest *movable* version of this arrangement, since the intention for later vehicles is to assemble the various stages in a separate building remote from the launch pad. The stages will be assembled onto a launcher base mounted on a special caterpillar-tracked crawler vehicle. Titan III will be assembled this way, and Saturn V will be constructed and erected by such a technique in a building that will stand about 500 ft high to accommodate the complete vehicle. On completion of the preparatory and check-out requirements (excluding engine firings) the carrier vehicle will be moved, complete in its vertical position on the crawler, along a specially prepared road to the launching pad itself, where the first stage will be attached via the launcher base on the crawler to the pad. The vehicle will then be ready for launch. Such a process should enable greater utilization of the pad, and the preparation of the vehicle in better controlled conditions than those at present experienced. It is significant that the Russian Vostok 3 and Vostok 4 manned satellites were launched from the same pad within 24 hr of each other. This would suggest that in 1962 the Soviets were already using this remote erection and assembly technique, the vehicle then being taken along to the pad attached, and launched within a matter of hours.

B. Ground Support

The immediate ground support equipment in the vicinity of the launch pad increases in proportion to the number of stages which are incorporated in the multistage vehicle. The complexity of the payload has considerable bearing on this ground support equipment, and particularly if the payload consists of a manned capsule; then all the apparatus for the controlled transfer of the astronaut from his preparation area to his actual pad must be included. Involved, too, is the multiplicity of the contractors' preparation areas for each of the stages and in the case of the ELDO vehicle, different nationalities are concerned.

The transportation and handling of the various stages present different degrees of difficulty. Until the advent of Saturn I transportation was difficult, but fairly orthodox, special trailers being prepared which could be taken onto a standard highway.

For the Saturn I the transportation of the first stage presented serious problems because of its sheer size (60 ft long by 22 ft in breadth). To transport the bay from Huntsville, Alabama, to the Cape Kennedy, Florida, launching pad special barges had to be constructed to take the carrier down the Mississippi River; then, sea-going barges had to take it across the Gulf of Mexico and so round to Cape Kennedy where it was off-loaded at a special dock.

This is indicative of the magnitude of the problem involved in transporting such massive stages, today. Yet even larger stages are now being developed. Transportation by road becomes quite impractical and even transportation by water will pose severe problems if the constructor's plant is not adjacent to a waterway. Therefore, large space carriers vehicle stages of the future may have to be manufactured *in situ* at the launch pad, and even upper stages could experience the same order of difficulty as the present Saturn I first stage.

The sheer size of carrier vehicles creates particular mechanical engineering problems for the handling of the stages. It is recalled that in many instances the vehicle is made of a fairly thin skin, the stability of which is obtained by internal pressurization. The Blue Streak uses a special handling frame which has hydraulic tensioning gear to stretch the tank, as well as pressurization to provide some stability during transportation. It is of interest, too, that transportation loads influence some of the design criteria for the carriers themselves and with their payloads. The designer envisioning his overall vehicle must certainly be aware of these aspects when laying out his plans.

IX. Organization of International Multistage Carrier Vehicle and Payload Programs

A. International Satellite Programs

Reference has been made throughout the technical discussion to several different carrier vehicle and satellite programs of an international nature. These will briefly be reviewed here.

First, the United States, through the National Aeronautics and Space Administration, has offered facilities to other countries to launch satellites, which may be wholly or partly constructed by these other countries. This offer was taken up early by the British and, as a result of their collaboration, the Ariel satellite was launched by the Americans in July 1962. The satellite structure, power supplies, etc., were constructed in America, but the experiments were designed and built in England and sent to America for integration at Goddard Space Flight Center. This is a particularly complex form of collaboration, in which part of the payload is built on one side of the Atlantic and part on the other. Not only had the components and the equipment to be built to perform correctly in their own right, but clearly the overall specification of the performance within the confines of the total system had to be extremely carefully defined. Ariel was a satellite weighing 150 lb and of a fairly orthodox design with four solar cell paddles. The Thor Delta carrier, which put it into orbit, was orthodox, and all aspects worked successfully; the preparation and launching were conducted with the aid of British scientists and engineers involved with the program. The third in this NASA United Kingdom series, designated U.K.3, will have the complete satellite built in the United Kingdom as well as its five experiments. The satellite is scheduled for launch by the Scout carrier, which originally had been intended for the whole of the series.

Another successful example of this form of cooperation was the Canadian Alouette satellite. It weighed 320 lb and contained equipment to enable the examination of the topside characteristics of the ionosphere looking from above. The satellite itself was constructed by De Havilland Aircraft Company of Canada, who also were responsible for the unique extendible "STEM" antenna units. The remainder of the electronic and power supplies was constructed within Canada and incorporated through the Department of Defense. The satellite was launched on a Thor Delta provided by NASA. It again performed extremely well, thus demonstrating another achievement in the field of international collaboration. The principle is now being extended and the French and Italians are building satellites which also will be launched by American rockets.

In all of these examples suitable ground support equipment had to be provided and integrated with the overall vehicle, and care had to be taken to ensure compatibility in the performance of the separate pieces of equipment.

B. European Carrier Vehicle Development Programs

In 1962 the European Launcher Development Organization (ELDO) was formed to develop a three-stage carrier vehicle based on the British Blue Streak, which was described earlier in Section III,D. The organization to achieve this requires Britain to develop the first stage, France the second stage, and Germany the third stage. The test satellite will be developed by Italy, the down-range tracking equipment will be developed by the Belgians, the telemetry system by the Dutch, and the development trials range at Woomera will be provided and operated by the Australians.

One of the tasks the ELDO vehicle will undertake is to launch a variety of research satellites sponsored by the European Space Research Organization (ESRO). ESRO's task is limited to the development and production of satellites and not upper stages, although associated with the satellites may well be propulsion systems for their position keeping and attitude stabilization which could, therefore, constitute (within the looser meaning of the term) a further propulsion stage. In ESRO there are ten countries involved, the preceding seven of ELDO (excluding Australia) and Sweden, Switzerland, Spain, and Austria. In this case the satellites will not be produced completely in ten different countries, but will be developed at the European Space Research Technical Center (ESTEC) located at Delft in Holland. Data processing—both that before launching of rockets and that of all the scientific information obtained afterwards—will be carried out at Darmstadt in Germany. In Italy there will be an advanced projects laboratory, charged with the task of examining such things as electric propulsion and attitude stabilization systems for incorporation in the satellite. The headquarters of ESRO and the headquarters of ELDO are both in Paris.

The preceding organizations have a major task ahead of them to coordinate all the features of a multistage carrier vehicle and its satellite payloads. If they are successful it will be a unique operation and a great achievement in terms of engineering co-ordination.

X. General Comments

Various aspects of integrating stages of space vehicles have been dealt with in general terms. One of the more interesting points that has emerged from the discussion is that while it is possible to optimize to a considerable degree, by analytical means, devices as large, complex, and expensive as rockets, the final solution of multistage carrier vehicles is only a broad compromise with the best that could be achieved.

One of the most important features of planning for the future is the choice of stages to ensure that the building block concept can be carried out as efficiently as possible, and that the widest flexibility is provided in individual stages as well as in their various combinations. Since reliability is of such paramount importance in space missions, the interface problems must be kept uppermost in the minds of the designers, and test facilities and schedules must be set up to ensure that missions do not fail because of lack of attention to the problems of integrating the individual stages. Integrated design of monitoring and check-out equipment to deal with the various stages will become increasingly important; and it is obvious, too, that in the future very severe mechanical problems will arise from the sheer size of assembling and integrating massive rockets. Each of the stages will undoubtedly be the subject of continuously striving for low structural weight, and in the future we may see an even greater diversity of stages when nuclear propulsion and electric propulsion are incorporated in the upper levels.

The inconsistencies and incompatibilities of today's rockets will seem small in comparison with the problems to be faced in the next ten years and beyond. Even present day assembly techniques will seem rather simple. However, it would appear that the day of the single-stage launch vehicle is nowhere in sight. Perhaps the only possibility for this will be in some development of the aerospace plane concept, in which hybrid propulsion systems within the same airframe may enable a vehicle to take off from an airfield and accelerate into orbit. Even here, present conception already suggests a two-stage vehicle would be more appropriate, though both would be manned and recoverable. So it seems there will still be the problem of stage integration, which has become already a vital feature of the conquest of space.

Acknowledgments

The author would like to acknowledge the assistance given to him by Mr. P. L. V. Hickman at the Hawker Siddeley Dynamics Ltd. and to Mr. P. E. Culbertson at

General Dynamics/Astronautics for contributing various comments in the course of preparing this material. Finally, the author would like to thank Hawker Siddeley Dynamics Ltd. for permission to publish this work.

References

1. Thompson, W. T. (1961). "Introduction to Space Dynamics." Wiley, New York.
2. Europe's Heavy Launcher (17 August 1961), *Flight*, pp. 220–226.
3. Samson, D., ed. (1963). "Development of the Blue Streak Satellite Launcher." Pergamon, London.
4. Cleaver, A. V. (February, 1962). The Blue Streak rocket propulsion system. *J. Brit. Interplanet. Soc.*, pp. 259–269.
5. Pardoe, G. K. C. (1963). Blue Streak as first stage of a satellite launcher. *Proc. 12th Intern. Astronaut. Congr.* Academic, New York (Vol. 2).

Navigational Instrumentation for Space Flight

Saul Moskowitz and Paul Weinschel

Navigation and Guidance Staff
Space Division
Kollsman Instrument Corporation
Elmhurst, New York

I. Introduction

We are today at the threshold of an historical transition in manned space flight. Man's essentially passive function in space travel is about to be replaced by his active participation in this operation. Spacecraft *guidance* is about to be superseded by *space navigation*.

This chapter[1] presents an analysis of the instrumentation requirements for space navigation. Space navigation implies the transformation of primary observed data into position and trajectory information. The requirements exist, therefore, for performance of the requisite primary measurements and the calculations required to provide meaningful navigation information. The exact form of these operations is largely dependent upon the mission. (The problem of circumlunar flight, for example, encompasses certain unique aspects which are not encountered during Earth-orbital flight.) In all cases the limiting constraints imposed on any space navigation system are the characteristics of the available observables. A property of a terrestrial, celestial, solar, or orbiting body must be measurable before it can provide useful information. These observables must be defined before specific instrumentation or data reduction techniques can be set forth.

The fundamental data (observables) available for space navigation are not dependent upon the particular navigation scheme employed. The selection and utilization of the data, however, do vary to a large extent with the specific concept. It is possible to employ any one of a number of implicit or explicit navigation schemes for a given mission, whether the mission be Earth-orbital, rendezvous, circumlunar, or deep space flight. The choice between implicit or explicit techniques is dependent upon on-board computational capacities, the available observables, and the operational requirements of the mission.

The capabilities and limitations of the human operator in the performance of the space navigation function have a significant effect upon the selected instrumentation. It is now a widely accepted premise that complete automaticity in the navigation function is not an ultimate design objective. Even where automatic computation is employed, the primary observations must (at a minimum) be initiated by the operator and may possibly be performed in their entirety by him. For back-up and redundant navigation functions (both observation and computation) man is an essential element in all feasible approaches.

The major missions and their navigational requirements are discussed in greater detail within the following sections, followed by descriptions of equipment configurations which satisfy the specific problems. These instrumentation schemes are based upon proven engineering concepts, rather

---

[1] A brief summary of the concepts advanced herein was originally published by the authors under the title "Instrumentation for Space Navigation" in the *IEEE Transactions on Aerospace and Navigational Electronics*, September, 1963.

than ideas requiring significant technological breakthroughs for their realization.

II. The Missions

A. Classification

All of the space missions scheduled for the predictable future fall into the following three basic groupings:

(1) Earth-orbital flight
(2) Lunar flight (circumnavigation or landing)
(3) Deep space flight (circumnavigation of Mars or Venus)

These classifications categorize the total mission. To establish the characteristics and requirements of space navigation instrumentation it must be recognized, however, that each of these missions has aspects or phases common to all. These common phases then dictate the type, speed, and accuracy of the required observations and computations.

The phase classifications indicated below prove particularly useful in the definition of the system requirements and specification of the responsive system characteristics:

(1) Boost to injection
(2) Early midcourse flight
(3) Late midcourse flight
(4) Near-body (orbital) flight
(5) Variation of orbital trajectory
(6) Rendezvous with target in orbit
(7) Earth re-entry and landing

The interrelationship of these methods of classification is strictly in accordance with the significant visual effects during space flight shown in Table I.

Under normal conditions, guidance during the boost-to-injection phase will be accomplished by means of conventional, automatic techniques. It is quite feasible, however, that during emergency situations the astronaut could actually "fly" the carrier vehicle. Tests supporting this contention have already been performed [1].

B. The Mission Phases

The definitions of the fundamental phases of space flight are summarized below:

TABLE I. DEFINING CHARACTERISTICS OF MISSION PHASES

Phase of flight / Mission	Earth-orbital flight	Lunar-flight (circumnavigation or landing)	Deep space flight (circumnavigation of Mars or Venus)
(1) Boost-to-injection	Parent body in immediate proximity. Mission begins or aborts. Conventional guidance (gyroscopically stabilized platform with accelerometers) employed.		
(2) Early midcourse flight	Not applicable.	Not applicable.	Characterized by target body at a great distance from spacecraft and still (navigationally) a point.
(3) Late midcourse flight	Not applicable.	Target body of finite extent and appears as some stage of a crescent. Landmarks on target body observable.	
(4) Near-body flight	Characterized by large angular extension (up to almost 180°) of reference body. Trajectory parameters must be determined for orbital change, (Earth) re-entry, circumnavigation, and initiation of return orbit (Moon, Mars, Venus).		
(5) Variation of orbital trajectory	Requires data on present and desired trajectory. Possibly necessitates tracking another body in orbit to obtain its motion.	Applicable if lunar landing is to be performed by spacecraft.	Not applicable.
(6) Rendezvous with target in orbit	Target must be acquired and tracked for closure guidance.	Target must be acquired and tracked for closure guidance. Applicable to those lunar missions involving rendezvous.	Not applicable unless complete spacecraft is to be assembled in Earth-orbit.
(7) Earth re-entry and landing	Final phase of space flight requires determination of trajectory relative to Earth and location of landing zone. Data obtained from terrestrial and stellar sightings.		

1. Boost to Injection

The initial portion of a lift-off trajectory during which the astronaut does not directly control the spacecraft or orbital vehicle, except under conditions of extreme emergency, is the boost-to-injection phase. The parent body is in immediate proximity. The mission begins or aborts.

2. Early Midcourse Flight

This phase is characterized by the observational limitation that the target celestial body is a great distance from the spacecraft and therefore is still essentially a point. (Note, however, that the parent body may exhibit a finite disk which is of some use to the navigation function.)

3. Late Midcourse Flight

The target celestial body is of finite angular extent and appears as some phase of a crescent. Landmarks may then be observable and utilized for accurate reference sightings.

4. Near-Body Flight

This phase comprises Earth-orbital, lunar, and planetary circumnavigation, lunar and planetary landing, boost phase parking orbit, and de-boost point for Earth re-entry flights; it is characterized by large angular extension (up to almost 180 deg) of the reference body.

5. Variation of Orbital Trajectory

A number of space missions require that the astronaut alter his trajectory to meet changing situations, initiate rendezvous operations with targets in either Earth orbits or lunar orbits, or prepare for re-entry and landing at a safe base. A special case of variation of orbital trajectory requires that the orbit of the spacecraft be matched with the orbit of a target vehicle.

6. Rendezvous with Target in Orbit

The normal operations required for refueling, the construction or repair of orbiting space stations, or return to the primary vehicle after a lunar landing all embody a rendezvous phase of space flight. It is necessary to consider the acquisition and tracking of three classes of rendezvous targets:

(1) Cooperative targets
(2) Passive noncooperative targets
(3) Evasive noncooperative targets

7. EARTH RE-ENTRY AND LANDING

Self-contained navigation systems must include the capability of returning spacecraft to a safe Earth base. Thus, in addition to considering the aerodynamic characteristics of the spacecraft, it is necessary to obtain data relating to its present and predicted motion. Attitude control is of great importance and must, at some point, be based upon the determination of the local vertical. (Different classes of lifting vehicles require different solutions to this aspect of the overall problem.)

The basic missions in terms of phases 1 through 4 (defined in Table I) are depicted in Fig. 1. Phases 5, 6, and 7 are encountered in conjunction with near-body flight.

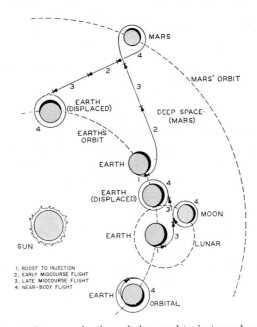

FIG. 1. Space navigation missions and trajectory phases.

C. The Mission in Terms of Its Phases

1. ORBITAL FLIGHT

This phase consists of the boost-to-injection phase, near-body (orbital) phase, possibly orbital change and rendezvous, re-entry, and landing. During the near-body phase, the navigation techniques must provide

information which may be employed to attain the desired orbit, initiate
orbital changes, and determine the initial conditions for Earth re-entry.
Such operations are characterized by extreme proximity to the reference
body. The angular extension of the Earth approaches 180 deg as a limit.
Significant observation of other celestial bodies is periodic throughout the
orbit. Characteristic of all phase 4 operations, rapidity of determination
of navigation parameters is most important. During orbital flight, rendezvous operations may be required. It is then necessary to acquire and track
the target vehicle.

2. Lunar Flight

This phase may consist of the boost-to-injection phase, late midcourse,
near-body (circumlunar), late midcourse, terminal flight, re-entry, and
landing resulting in phase 1, 3, 4, and 7 operations. Lunar landing also
includes lunar orbital, landing, lunar boost, and rendezvous phases if the
excursion module concept is employed. Since orbital change may also be
required, the overall mission comprises phase 1, 3, 4, 5, 6, and 7 operations.
The circumlunar (or lunar orbital) phases are similar to Earth-orbital flight.
Upon leaving the vicinity of the Moon, the navigation system must permit
a second phase of late midcourse guidance. This is followed by near-body
flight, re-entry and landing phases.

3. Deep Space Flight

Sequentially, this phase consists of the boost-to-injection, early midcourse, late midcourse, circumnavigation, early midcourse, late midcourse,
terminal flight, and re-entry and landing, resulting in phase 1, 2, 3, 4, and 7
operations. Additionally, there exists the possibility of rendezvous in Earth-
orbit to take on additional propellant.

III. The Observables

A. Definition of Terms

The observables of space flight, while often quite varied and numerous,
are not all equally useful or practical for a particular space mission.
Stadimetric ranging to the Sun, for example, would prove of little value
for circumlunar navigation. Measurement of optical doppler shifts is an
excellent theoretical source of velocity information; however, the practical
limitations of current spacecraft instrumentation preclude the effective
employment of such data. To permit simplicity (and resulting reliability)
of the space navigation instrumentation it is necessary, therefore, to select

the observables to be exploited in the navigation system with extreme care. For completeness, most of the available observables are discussed herein, with emphasis on those most suited to practical instrumentation.

The description of the observables is logically considered from the chronological aspect of the space mission. Accordingly, the ensuing discussions are again subdivided by phases:

(1) Near-body flight
(2) Early midcourse flight
(3) Late midcourse flight
(4) Rendezvous operations

The observables of the orbital change or re-entry and landing operations (as shown in Table I) are implicitly included within the above categories.

B. Observables of Specific Missions

1. NEAR-BODY FLIGHT

Parking orbits, Earth-orbital flight, and Earth re-entry may be considered representative of *near-body flight*. The techniques for Earth-referenced flight are essentially identical to those referenced to the Moon, Mars, or Venus. The differences within the other requirements incur no loss of generality in this discussion.

Figure 2 illustrates the phenomena which can be classed as meaningful

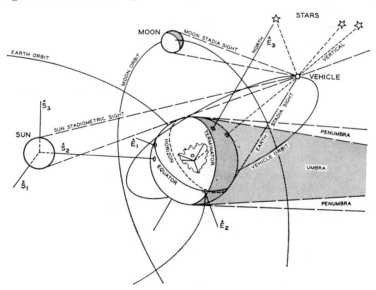

FIG. 2. Observables of near-body flight.

observables. Star sights, local body sights, stadimetric sights, occultation sights, shadow areas, recognizable features, and direct velocity measurements are indicated. Not all of these observables will be available over the entire orbit. Also, some will prove much more useful than others.

The following combinations of readings provide sufficient information for the computation of position and attitude:

(a) Two-star, Sun and Moon intercept
(b) Two-star, Earth and Moon intercept
(c) Two-star, planet and Sun intercept
(d) Star, Earth stadimetric, and Moon stadimetric intercept locus
(e) Star, Earth stadimetric, and Sun stadimetric intercept locus
(f) Two-star, Earth vertical, and Earth stadimetric intercept
(g) Earth, Moon, Sun stadimetric, and line of position intercept locus
(h) Combinations of the above readings with the Earth or Moon terminator line
(i) Combinations of the above readings with the umbra line
(j) Two-star and heliographic subpoint recognition
(k) Two-star and map recognition of subpoint on the Earth
(l) Two-star, abberation velocity magnitude and direction
(m) Two-star and Earth or Moon occultation measurement
(n) Two-star, stadimetric ranging, and orbital phase determination
(o) Two-star and Earth stadimetric angular rate of change measurements
(p) Three local body lines of intercept locus

The above measurements and the alternate possible combinations fall into three main groupings:

(1) Line of position intercepts (a, b, c, h, j, m, p)
(2) Range angle intercepts (e, f, g)
(3) Coordinate system and vertical intercepts (j, k)

Cases (l), (n), and (o), not categorized above, are combinations wherein the plane of motion is directly measured. All of the above readings are obtainable from an appropriate space sextant.

Systems utilizing data available during limited portions of the flight may not be desirable. Further, measurements with respect to the nearest body are inherently more accurate. In addition, systems utilizing measurements encompassing several of the variables for material reduction of the computation effort are preferred. Within the above restrictions, measurements between selected navigation stars and either the local vertical or horizon (groups 1 and 2 above) fall within this category.

Instrument and computational limitations restrict the choice of feasible

sets of observations. Depending upon the uncertainty in the initial conditions for near-body flight, requirements may be formulated in terms of
either construction of orbital constants or comparison of sightings referenced to pre-flight computed trajectory data. The significance of each
approach is discussed in Section IV.

2. Early Midcourse Flight

Early midcourse flight, within the previously established definitions, is
encountered only in deep-space missions. During this phase of the trajectory the target body is still essentially a point (for instrumentation
purposes). Consequently, no useful range information (i.e., by means of
stadimetry) is available. All position information must be derived by
various methods of triangulation (explicit or implicit).

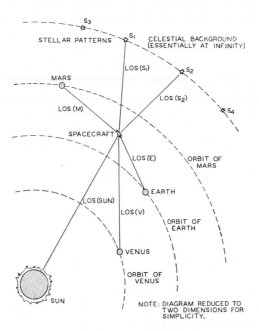

Fig. 3. Observables of early midcourse flight.

The fundamental observations are essentially line-of-sight angles (i.e.,
the angular relation of finite distant bodies to the stellar background, as
measured from the vehicle). The two-dimensional simplification shown in
Fig. 3 illustrates a typical situation. From the spacecraft the angular
subtense between the target planet (Mars) and two stars in the celestial

background may be measured. These are the included angles between the lines of sight LOS(M) and LOS(S₁), LOS(M) and LOS(S₂). Similarly, other pairs of angles to various bodies in the Solar System may be measured. In Fig. 3, Venus, the Sun, and Earth are shown as representative bodies. Under certain conditions Jupiter or Saturn could also prove useful. The position of the spacecraft, as determined from these measurements, is characterized by large absolute errors, but not necessarily large angular errors. Therefore, as the spacecraft approaches its target the instrumentation phenomenon of "accuracy convergence" occurs.

No practical method for direct measurement of meaningful velocity data exists at this time. The Sun is the only body in the Solar System generating radiation with well-known frequency lines and of sufficient intensity for practical measurements. This permits the measurement of the spacecraft's radial velocity with respect to the Sun. In a transfer orbit, however, the spacecraft velocity relative to the Sun is quite small, so that the error in the relative velocity measurement is large. Spaceborne instrumentation capable of attaining even this limited accuracy is rather delicate and is predicted to be of questionable utility within the space environment. No practical method is known at this time for measurements of that component of velocity tangent to the orbit of the vehicle. In general, it appears that all required measurements will be those essential to either explicit (direct) or implicit determination of position.

3. Late Midcourse Flight

The major observable characteristic of late midcourse flight is that the target body is of finite angular extension. Ranging measurements (by means of stadimetric and/or landmark sighting) now become possible. Further, trajectory fixing by such sophisticated methods as occultation correlation may now be considered.

The entire midcourse flight in a lunar mission falls within the "late midcourse" classification. Similarly, when the distance between the spacecraft and its planetary target approaches the magnitude of Earth-lunar ranges, the above techniques are applicable. In addition to the applicability of sextant measurements (as during early midcourse flight) a number of potentially more accurate techniques exist for acquisition of the data employed for position and trajectory fixing. The more prominent of these measurements are depicted in Fig. 4. Since current knowledge of the Moon is greater than man's knowledge of any other body in the Solar System, the ensuing discussion of late midcourse observations emphasizes reference to this body. These observables, of course, have their counterparts with respect to Mars or Venus.

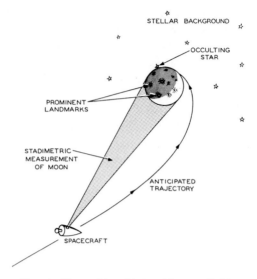

FIG. 4. Observables of late midcourse flight.

a. Stadimetry

Measurement of the apparent angular extension of a celestial body, combined with knowledge of its diameter, provides information for the computation of range to that body. If a spacecraft is sufficiently close to the body (Earth-lunar distances), useful data for purposes of position fixing and navigation may be derived by this method. The mathematics of such an operation are simple. The geometric relationships of stadimetry are depicted in Fig. 5, while the magnitudes of these measurements for Earth/lunar flight are shown in Fig. 6.

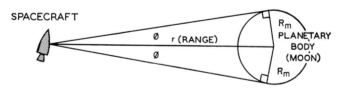

FIG. 5. Stadimetric measurements of planetary and lunar bodies. $r = R_m/\sin \phi$: r = range from spacecraft to center of body; R_m = known radius of body; 2ϕ = measured angular extension of body.

b. Occultation

By the accurate observation of the time and location of an occultation of a star (by the Moon), an observer in cislunar space can determine a fine line of position. By combining this information with stadimetric information, the observer should be able to fix his point of observation along this line, and thus know his position in the space between the Earth and Moon.

The occultation observation should include the actual time within one-tenth of a second, and the point of immersion or emersion relative to lunar coordinates. The observer must refer these data to the lunar ephemerides.

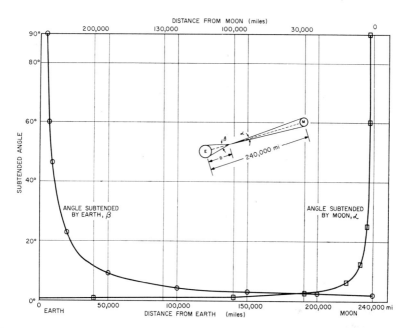

Fig. 6. Angle of subtense of Earth and Moon as a function of distance between them.

As the Moon moves in its Earth-perturbed orbit around the Sun, it carries a "shadow cylinder" for every star in the universe (in reality, cones with negligible apex angles). The diameter of each of these cylinders is 2160 mi. Any motion of observer or Moon which changes the direction of the Moon-observer axis (with respect to celestial coordinates) will cause the observer to enter new shadow cylinders and to emerge from others.

The observer may perceive the relevant phenomena under the following conditions:

(1) Immersion of a star along a path into the dark limb of the Moon.
(2) Emersion of a star along a path from the dark limb.
(3) (4) Emersion and immersion with respect to the bright limb. (At the bright limb, however, there is a penalty of three magnitudes of brightness relative to the dark limb.)

The rate of observable star occultations, i.e., the number of stars disappearing and reappearing per unit of time, is a function of:

(1) The subtense solid angle of the Moon (or Earth); see Fig. 6.
(2) The angular rate of the Moon-observer axis with respect to the celestial coordinates.
(3) The density of the observable star population in the direction of the Moon (or Earth).

For example: consider the frequency of the possible observations of immersions of 8th magnitude stars from the vicinity of the Earth. At the Earth-Moon distance, the Moon subtends an angle of approximately 0.5 deg. The Moon, in its eastward motion about the Earth, moves about $\frac{1}{2}$ sec of arc per second, or 1 deg per 2 hr. Therefore, it sweeps out 1 sq deg solid angle in 4 hr. Thus, one star of the 8th magnitude (or dimmer) will be occulted every 4 hr. Of course, the stars are not evenly distributed and this calculation applies for a region of average density along the ecliptic. If, for example, the Moon was passing through the Beehive in Cancer, the frequency would be greater.

This sample calculation of frequency of occultation by the Moon observed from the vicinity of the Earth is almost the worst case. As the observer approaches the Moon, the angle subtense of the occulting body increases inversely as the distance. Therefore, halfway to the Moon the occultation frequency would be on the order of four times the above, or once every hour.

c. Earth and Moon Surface Features

As seen through a spaceborne instrument, the surfaces of the Earth and Moon can present a set of mapped features which are sufficiently distinct to serve as reference points for star-to-Earth or star-to-Moon sextant sightings. Because these bodies are essentially spherical, the features most useful as targets are those on the nearest 10 per cent of the surface. In the case of the Earth, the atmosphere significantly degrades the accuracy of any sight to Earth landmarks in the same way that celestial references are degraded for astronomers on Earth. Further, the Earth targets are subject to the obscuration of cloud layers.

Lunar observations are favored by the best of "good seeing." The lunar targets for accurate visual reference are also abundant. A rough estimate of the number of craters larger than some diameter d (if d is less than 200 km) on the visible hemisphere of the Moon is given by the formula (d in kilometers)

$$N = \frac{300,000}{d^2}$$

Craters which are convenient as targets, insofar as both recognizability and subtense angle are concerned, can be considered to subtend about one arc minute as seen by the viewer. Considering a magnification of 30, and actual subtense of 2 arc sec, and assuming an Earth-Moon viewing distance, the craters of interest may be estimated to have diameters of 2 to 4 km. By formula then, some 50,000 craters are available.

d. Use of Earth Airglow

The phenomenon of airglow may be utilized as a visual reference for Earth diameter measurement. From viewing distances of 100 mi and beyond, the visual constituents of the airglow will be seen as an outer shell of the Earth's atmosphere for both the day and night portions of the Earth.

The most noticeable and persistent component is the 5577 Å (green) line of atomic oxygen, which emanates from a layer of the ionosphere in the region centered at 97 km.

Although the brightness of the green airglow component is near the limit of vision for an Earthbound observer, the viewer from a space vehicle enjoys several advantages. First, he looks edgewise at the shell rather than through it. Second, he is not hampered by the attenuating effect of the Earth's atmosphere; and finally, he sees the glow in contrast to the blackness of outer space. The existence of a well-defined sharp airglow edge has been visually confirmed during manned space flight [2].

4. RENDEZVOUS OPERATIONS

The constraints imposed by the "observable" data limit any feasible rendezvous instrumentation scheme. These constraints are based upon the specific characteristics of:

(1) The target vehicle
(2) The celestial background
(3) Solar and planetary variables
(4) The (nonstellar) field background

The target vehicle must be visible, or made visible to the observer. Therefore, it must reflect solar illumination, provide its own illumination source, or reflect directed narrow beam illumination from the interception spacecraft (i.e., laser generated beams of radiation in the visible portion of the electromagnetic spectrum). Operational considerations determine characteristics to be designed into the target vehicle. If rendezvous maneuvers must be effected during passage through the Earth's shadow, the first approach (above), that of diffuse reflection of solar illumination, is not applicable. It may be necessary, therefore, to provide system compatibility with various alternate modes of target illumination.

Measurement of angular change of the line of sight of the target relative to the intercepting spacecraft must be made with reference to a fixed frame or coordinate base. The celestial (or stellar) background provides this reference.

An instrumentation system employing the stars in this manner meets the requirements of accepted rendezvous techniques. The angular field of view required for the tracking instrumentation must be determined so as to assure acquisition of the target and the reference stellar background under all attitude situations.

The solar and planetary effects upon sighting are not negligible. Even without an atmosphere, light scattering within an instrument cannot be absolutely reduced to zero. Thus, there is a limit on the ability to "see" a target or the celestial background. This limit is established to a large extent by the position of the Sun relative to the given line of sight.

The intensity of the background field is also a factor. In the Earth's atmosphere, for example, background intensity represents the most severe constraint, simply because of high illumination (due to scattering) of the daytime sky. In outer space, the problem is not as severe, but is still significant.

IV. Concepts of Space Navigation

A number of approaches have been advanced as solutions to specific space navigation problems. By the broad nature of their respective definitions, these approaches fall into either the "implicit" or "explicit" classes of navigation. This section defines both of these categories.

A. Explicit and Implicit Navigation

Most conventional, nonspace navigation schemes are explicit in formulation. That is, the navigational measurements (observables) are reduced to positional, and velocity (vector) information. The outputs of such navigation systems are present position, and possibly predicted future position or commands as required to attain a desired future position. The primary advantage of explicit navigation lies in the fact that the navigator is provided with direct data of vehicle position and course. Explicit navigation, however, also has the disadvantage of necessitating an extraordinary amount of computation for the more complex missions, such as deep space flight.

Implicit navigation denotes the fact that true present or future position is not determined. Rather, certain abstract conditions are met with the prior knowledge that fulfillment of these conditions assures adherence to the desired course or trajectory. (Most ballistic weapons systems employ implicit guidance concepts.)

The implementation of implicit navigation is dependent upon a well-defined reference trajectory, from which only small deviations are permitted. The validity of this concept, in terms of on-board computation, vanishes when the spacecraft exceeds these small deviations. A crossover point exists, at which the instrumentation for explicit navigation actually becomes simpler than that for implicit navigation. The choice between the two approaches must be made, therefore, on the dual basis of mission and mode of operation.

The navigation concepts for four illustrative mission phases are discussed herein. These are Earth-orbital near-body flight, early and late midcourse flight, and rendezvous. Navigation by explicit concepts illustrates Earth-orbital flight, while both phases of midcourse flight may be solved by the same implicit technique. A modified proportional navigation ratio approach is employed for solution to the rendezvous problem.

B. Earth-Orbital Flight

Completely unrestricted Earth-orbital flight is now technologically feasible. For navigation purposes it is necessary, therefore, to consider all possible orbits, with no *a priori* preference given any one orbit. This major requirement consequently suggests an explicit scheme of navigation.

For practical reasons, the following discussion embodies the assumption that the only gravitational source to be considered is the Earth. The perturbations introduced by the Moon, Sun, and other planets remain negligible relative to the overall accuracies obtained from the measurements of the relevant observables.

Earth-orbital flight, accordingly, occurs in the gravitational field of a single body resulting in the classical two-body problem. A complete solution to this navigation problem calls for the determination of the six parameters describing the conic section in three-dimensional space.

A workable solution is facilitated by the observation that the six parameters may be divided into two groups: two parameters describing the size and shape of the orbit and the four remaining parameters describing the orbit orientation and time of passage over perigee. The significance of this grouping is discussed below.

The geometric parameters are of great significance since they define the safety of the orbit, in terms of collision with the Earth's atmosphere. The initial decision to abort a mission for safety reasons must be made on the basis of the values of geometric quantity; the perigee radius. An orbit is safe only if the perigee radius, d_c (distance of closest approach), is greater than some predetermined value. The other geometric parameter is ϵ, the eccentricity of the conic section. It is then possible to calculate the period of orbit and orbital angle (true anomaly) as a function of time.

The reason for the above grouping of orbital parameters is that three measurements of range to the center of the Earth (and knowledge of the times of these measurements) are sufficient to determine ϵ and d_c. Therefore, the number of measurements and the amount of calculation necessary to make the abort decision is significantly smaller than would be required if one scheme for the simultaneous determination of all six parameters were employed. A number of approaches put forth in the recent literature advocate the simultaneous determination of the six orbital parameters [3]. Such simultaneous solutions force the navigational system to possess the capacity for extensive digital computation. The advantage gained by reducing the required computation to the point where manual techniques would prove entirely satisfactory cannot be overemphasized. In the succeeding material it will become evident that the entire space navigation problem can be formulated in a manner which need not rely upon the perfect performance of automatic equipment.

After the safety of the orbit has been determined (from the geometric parameters) a set of star sightings proves adequate for completion of the total problem; that is, the calculation of present and predicted future position anywhere along the orbit.

1. Geometric Parameters

The determination of the geometric parameters of the orbit must be considered in terms of the observables of the mission. To achieve the simplest possible instrumentation, it is necessary to transform the classical equations of motion into quite a different form. Instead of measuring the time rate of change of the orbital angle (true anomaly) through continuous tracking of the Earth's center against a fixed inertial reference, or other similarly difficult operations, it is desirable to employ nothing more than a small set of discrete stadimetric ranging measurements. The approach described herein is only one of a number based upon the same rule; the reformulation of the classical equations of motion to utilize only simple measurements of the most significant observable—the angular diameter of the Earth.

Start with Eq. (3-16) of [4] in its squared form

$$\left(\frac{dr}{dt}\right)^2 = \frac{2}{m}\left(E - V - \frac{l^2}{2\,\mu r^2}\right)$$ (1)

where

m = the reduced mass of the vehicle
E = the total energy of the vehicle
l = the total angular momentum of the vehicle
V = the potential energy of the vehicle

It is possible to show that for an orbit which is a conic section

$$r = \frac{A}{1 + \epsilon \cos \psi} \tag{2}$$

where ψ is the true anomaly, and A a parameter, and

$$\epsilon^2 = 1 + \frac{2El^2}{m^3k^2} \tag{3}$$

where $k = GM$, the product of the gravitational constant and the Earth's mass. From Eq. (2) at the perigee of the orbit, $\psi = 0$ and $r \equiv d_c$ so that

$$A = d_c(1 + \epsilon) \tag{4}$$

If we combine Eqs. (3) and (4) with (1)

$$\left(\frac{dr}{dt}\right)^2 = \frac{k}{r^2}\left[\left(\frac{\epsilon - 1}{d_c}\right)r^2 + 2r - d_c(1 + \epsilon)\right] \tag{5}$$

Equation (5) expresses range rate in terms of the unknowns ϵ, d_c and range r as desired. (Note that all time derivatives of range will be replaced by their differences to yield approximate equations which are valid for orbits of low eccentricity.)

It is also required that the second derivative of range with respect to time be obtained. This quantity is computed by observing that

$$\frac{d}{dr}\left[\left(\frac{dr}{dt}\right)^2\right] = 2\left(\frac{dr}{dt}\right)\frac{d}{dr}\left(\frac{dr}{dt}\right) = 2\left(\frac{d^2r}{dt^2}\right)$$

Therefore, d^2r/dt^2 is obtained from the differentiation of Eq. (5) with respect to r

$$\frac{d^2r}{dt^2} = \frac{k}{r^3}[d_c(1 + \epsilon) - r] \tag{6}$$

To solve Eqs. (5) and (6) for ϵ and d_c it is necessary to express the derivative of r with respect to t in terms of the given observables. Let r_1, r_2, and r_3 be the three measured values of range and let Δt be the time interval between measurements. Approximately

$$\left.\frac{dr}{dt}\right|_{t=t_2} = \frac{r_3 - r_1}{2\Delta t} \tag{7}$$

$$\left.\frac{d^2r}{dt^2}\right|_{t=t_2} = \frac{r_1 + r_3 - 2r_2}{(\Delta t)^2} \tag{8}$$

then from Eq. (6)

$$A = r + \frac{r^3}{k} \frac{d^2 r}{dt^2} \tag{9}$$

and from Eqs. (4) and (5)

$$\epsilon^2 = \frac{A}{k} \left(\frac{dr}{dt}\right)^2 + \left(\frac{r - A}{r}\right)^2 \tag{10}$$

If Eqs. (7) through (10) are combined, the following set of equations is obtained for ϵ and d_c in terms of observables of the Earth-orbital mission

$$A = r_2 + \frac{(r_2)^3}{k} \frac{(r_1 + r_3 - 2r_2)}{(\Delta t)^2} \tag{11}$$

$$\epsilon^2 = \frac{A}{k} \left(\frac{r_3 - r_1}{2\Delta t}\right)^2 + \left(\frac{r_2 - A}{r_2}\right) \tag{12}$$

$$d_c = \frac{A}{1 + \epsilon} \tag{13}$$

Other methods not dependent upon the above approximations are also available for determination of the orbital parameters.

2. Prediction of Future Position

The determination of ϵ and d_c requires a set of timed ranging measurements. Solution for the other parameters requires the use of celestially derived data, which is essentially an extension of conventional celestial navigation techniques.

First, consider the form of the basic data or observables. For purposes of computation, the coaltitude of the reference navigation star (the angle between the spacecraft's local zenith and the star measured as an arc along the celestial sphere) proves to be the most useful measurement. For space flight this angle is not so easily obtained, since no gravity-derived vertical reference is available. It is necessary, therefore, either to establish a vertical optically and effect simultaneous star sightings or to measure the altitude of the star above the apparent horizon. This quantity, designated γ in Fig. 7, is not the true star altitude because of the depressed horizon. It is required, therefore, to subtract the sum of γ and ϕ (the stadimetric half angle) from 180 deg to obtain α, the star coaltitude. If the spacecraft is in an orbit of low eccentricity, ϕ can be measured just prior to the star sighting and still provide sufficient accuracy. Otherwise, it is possible to calculate ϕ by use of equations previously derived

$$\phi = \arcsin \frac{R_e}{r} \tag{14}$$

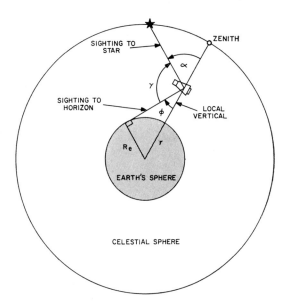

Fɪɢ. 7. Determination of star coaltitude.

Combining Eqs. (2) and (4), we calculate r of Eq. (14) from

$$r = \frac{d_c(1 + \epsilon)}{1 + \epsilon \cos \psi} \tag{15}$$

The true anomaly, ψ, is a function of time and the specific orbit. It is possible to store ψ in parametric form and simply read off the appropriate values as required.

Prediction of future position is equivalent to the determination of celestial latitude and longitude of the spacecraft's zenith as a function of time. This is accomplished by measuring coaltitudes to stars whose right ascension and declination are known. The times at which these measurements are made must be noted. Figure 8 presents the geometry of the future position problem. In this diagram P_1, P_2, P_3, and P_4 represent the positions of the zenith on the celestial sphere at the times the altitudes to the reference navigation stars are measured. The projection of the spacecraft orbit (the intersection of the plane of the orbit and the celestial sphere) on the celestial sphere is the great circle through the four zenith points designated the "track of zeniths." Three stars A, B, and C are used. The coaltitudes of A, α_1, and α_3, are measured at the points P_1 and P_4. The coaltitudes to B and C, α_2 and α_4, are measured from points P_2 and P_4; V represents the future position of the vehicle at time t.

The arc lengths ψ_1^2, ψ_2^3, ψ_3^4, and ψ_4^v may be computed or obtained from the stored parametric representation of ψ as a function of time and the particular orbit. The arcs ϕ_A, ϕ_B, and ϕ_C are the right ascensions of the stars A, B, and C. The arc lengths NA, NB, and NC are the codeclinations of A, B, and C. The quantities to be determined are ϕ_v and N_v, the celestial longitude and the coaltitude of V.

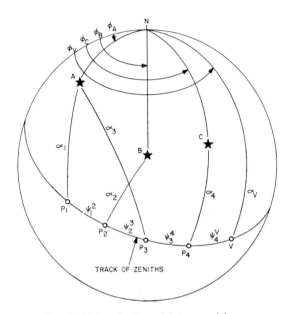

FIG. 8. Determination of future position.

To avoid ambiguities, it is necessary to employ three stars which do not lie along the same celestial great circle. (One situation can still result in an ambiguity, but it is easily avoidable.)

C. Interplanetary and Lunar Midcourse Flight

1. SELECTION OF IMPLICIT TECHNIQUES

Operational considerations presently suggest the utilization of implicit navigation schemes for the midcourse phases of lunar and deep space flight. Here the position and velocity of injection establish the spacecraft trajectory. Errors in position and velocity cause deviations from the desired trajectory. These must be corrected at a "best" time. The desired reference trajectory is fully defined as a function of time. If the spacecraft sensing equipment were perfect, any deviations from the reference trajectory could be detected shortly after injection. Then, if the required corrective maneuver could be executed perfectly, no other measurements or corrections would be needed. This is obviously not the case. Both the measurement of the deviation and the application of the corrective maneuver will include errors. Therefore, more than one correction will almost always be required.

At all points on the standard trajectory the desired position of the spacecraft is known. Angular measurements from this reference position to selected celestial bodies may be precomputed. These angular measurements taken within the actual trajectory will therefore provide information of the spacecraft's position deviation from the reference trajectory. A minimum of three successive measurements is required to determine this deviation. If more measurements are available, the resulting redundant information may be utilized by statistical techniques, leading to a more accurate estimate of deviation.

When measurements are made at more than one point on the reference trajectory, sufficient information exists for the determination of the velocity deviation. A velocity correction may then be computed and applied if deemed necessary. This scheme exhibits a tendency, however, for the actual trajectory to oscillate about the reference trajectory, resulting in a nonoptimum utilization of propellant. Overcorrection may frequently occur because of measurement errors.

To minimize propellant consumption, an optimum navigation technique must be utilized. This guidance scheme would employ perturbation techniques, translating angular measurements to position deviations from a reference trajectory. From a least squares adjustment of current position deviation and the expected deviation (based upon the past history of the actual trajectory) a best estimate of the miss distance at the target can be acquired at each measurement point. The estimated variance of the miss distance serves as an indicator of the statistical uncertainty of available knowledge and is used as an integral part of the navigation logic employed to eliminate excessive corrections.

This implicit approach to space navigation can be realized at various levels of sophistication. For purposes of illustration, the concept postulated by the Massachusetts Institute of Technology Instrumentation Laboratories for an unmanned interplanetary reconnaissance probe is considered first [5]. Note that this is a relatively simple scheme and is applicable only to very restricted cases of spaceflight. It does provide a good starting point, however, for the consideration of the more sophisticated approach described in Section IV, C, 3.

2. A Simple Approach

An Earth-Venus transfer trajectory is chosen for reference purposes. The six necessary observables selected for navigation purposes are:

(1) Angle between Sun and Earth $= A_1$
(2) Angle between α-Centauri and Earth $= A_2$
(3) Angle between Arcturus and Earth $= A_3$
(4) Angular subtend of Earth $= A_4$
(5) Angle between Sun and Arcturus $= A_5$
(6) Angle between Sun and Mars $= A_6$

As the spacecraft approaches the target planet, the latter's angular subtense becomes finite (for instrumentation purposes) and may be employed in place of Eq. (4) above. The illustrative calculations are for the second set of observations taken during the flight.

The differences between the measured values of the A_i and the precomputed stored values are designated δA_i. These differences are the basic inputs to the navigation scheme. The fundamental position deviation vector at the time of the second set of observations

$$\boldsymbol{\delta r}_2 = (\delta r_{20},\ \delta r_{21},\ \delta r_{22}) \tag{16}$$

is a function of the δA_i multiplied by their respective precomputed influence coefficients.

$$\left.\begin{aligned}
\delta r_{20} &= C_{01}\delta A_1 + C_{02}\delta A_2 + C_{03}\delta A_3 + C_{04}\delta A_4 + C_{05}\delta A_5 + C_{06}\delta A_6 \\
\delta r_{21} &= C_{11}\delta A_1 + C_{12}\delta A_2 + C_{13}\delta A_3 + C_{14}\delta A_4 + C_{15}\delta A_5 + C_{16}\delta A_6 \\
\delta r_{22} &= C_{21}\delta A_1 + C_{22}\delta A_2 + C_{23}\delta A_3 + C_{24}\delta A_4 + C_{25}\delta A_5 + C_{26}\delta A_6
\end{aligned}\right\} \tag{17}$$

At the time of the first set of observations a position deviation vector

$$\boldsymbol{\delta r}_1 = (\delta r_{10},\ \delta r_{11},\ \delta r_{12}) \tag{18}$$

is computed. Both these vector quantities are employed to compute the velocity correction $\Delta \mathbf{V}_2$ necessary to achieve the desired trajectory.

$$\begin{aligned}
\Delta V_{20} &= d_{00}\delta r_{10} + d_{01}\delta r_{11} + d_{02}\delta r_{12} + d_{03}\delta r_{20} + d_{04}\delta r_{21} + d_{05}\delta r_{22} \\
\Delta V_{21} &= d_{10}\delta r_{10} + d_{11}\delta r_{11} + d_{12}\delta r_{12} + d_{13}\delta r_{20} + d_{14}\delta r_{21} + d_{15}\delta r_{22} \\
\Delta V_{22} &= d_{20}\delta r_{10} + d_{21}\delta r_{11} + d_{22}\delta r_{12} + d_{23}\delta r_{20} + d_{24}\delta r_{21} + d_{25}\delta r_{22}
\end{aligned} \tag{19}$$

The above equations are simple enough to be solved by pencil and paper as well as through automatic computational techniques. As the implicit navigation concepts become more sophisticated, however, the required computations become increasingly more complex, finally necessitating the employment of a digital computer of significant capacity.

3. Introduction of Statistics

The previously described navigation scheme was based upon the original work contained in Ref. [5]. This approach called for the utilization of redundant measurements and their adjustment by least squares techniques. In this manner a probable best estimate of position deviation is obtained. Thus, the accuracy of the determination of the position deviation at each navigation fix is increased but there is no assurance that the accuracy of the navigational solution is improved with each successive fix. It is possible, however, to increase this accuracy by means of the statistical data adjustment concept developed in Ref. [6].

At each measurement point an estimate of the miss distance is made from a "maximum-likelihood" adjustment of measured position deviation and predicted position deviation. The estimated variance of the miss distance serves as an indication of the statistical uncertainty of available knowledge. This quantity is employed as a decision expression indicating the velocity correction to be applied.

Present study indicates that the assigned number of decision points for a representative lunar trajectory should not exceed forty. A decision point is a point in time at which either a measurement or a correction is to be made. It also appears that the number of correction decision points should not exceed three. As with the simple implicit navigation concept considered previously, the sequenced schedule of measurements is predetermined for the entire flight. Without an applied velocity correction, the statistical value of position and velocity deviation information improves with the number of measurements. For each preselected measurement point the associated covariance matrix is predetermined. With an application of corrective thrust, additional uncertainties are introduced into the system. However, noting the three-correction restriction, it is quite feasible to supply tables for the second and third correction points that give the predetermined constants as a function of ΔV_1 withinmeas urement accuracies. These tables would also indicate revised measurement schedules if necessary.

Again, the differences between the measured values of the A_i and the precomputed stored values are designated δA_i. At the nth decision point

$$\left.\begin{aligned}
\delta r_{n,1} &= a_{n,1,1}\delta A_k + a_{n,1,2}\delta A_{k+1} + a_{n,1,3}\delta A_{k+2} \\
\delta r_{n,2} &= a_{n,2,1}\delta A_k + a_{n,2,2}\delta A_{k+1} + a_{n,2,3}\delta A_{k+2} \\
\delta r_{n,3} &= a_{n,3,1}\delta A_k + a_{n,3,2}\delta A_{k+1} + a_{n,3,3}\delta A_{k+2}
\end{aligned}\right\} \tag{20}$$

If $\delta \mathbf{r}_A$ is the currently expected position deviation at intercept, then the discrepancy vector $\boldsymbol{\epsilon}_n$ is

$$
\begin{aligned}
\epsilon_{n,1} &= \delta r_{A,1} + b_{n,1,1}\delta r_{n-1,1} + b_{n,1,2}\delta r_{n-1,2} + b_{n,1,3}\delta r_{n-1,3} \\
&\quad + C_{n,1,1}\delta r_{n,1} + C_{n,1,2}\delta r_{n,2} + C_{n,1,3}\delta r_{n,3} \\
\epsilon_{n,2} &= \delta r_{A,2} + b_{n,2,1}\delta r_{n-1,1} + b_{n,2,2}\delta r_{n-1,2} + b_{n,2,3}\delta r_{n-2,3} \\
&\quad + C_{n,2,1}\delta r_{n,1} + C_{n,2,2}\delta r_{n,2} + C_{n,2,3}\delta r_{n,3} \\
\epsilon_{n,3} &= \delta r_{A,3} + b_{n,3,1}\delta r_{n-1,1} + b_{n,3,2}\delta r_{n-1,2} + b_{n,3,3}\delta r_{n-3,3} \\
&\quad + C_{n,3,1}\delta r_{n,1} + C_{n,3,2}\delta r_{n,2} + C_{n,3,3}\delta r_{n,3}
\end{aligned} \quad (21)
$$

The components of the three correction vectors (residuals) γ_1, γ_2, and γ_3 are

$$
\begin{aligned}
\gamma_{n,1} &= d_{n,1,1}\epsilon_{n,1} + d_{n,1,2}\epsilon_{n,2} + d_{n,1,3}\epsilon_{n,3} \\
\gamma_{n,2} &= d_{n,2,1}\epsilon_{n,1} + d_{n,2,2}\epsilon_{n,2} + d_{n,2,3}\epsilon_{n,3} \\
\gamma_{n,3} &= d_{n,3,1}\epsilon_{n,1} + d_{n,3,2}\epsilon_{n,2} + d_{n,3,3}\epsilon_{n,3} \\
\gamma_{n,4} &= d_{n,4,1}\epsilon_{n,1} + d_{n,4,2}\epsilon_{n,2} + d_{n,4,3}\epsilon_{n,3} \\
\gamma_{n,5} &= d_{n,5,1}\epsilon_{n,1} + d_{n,5,2}\epsilon_{n,2} + d_{n,5,3}\epsilon_{n,3} \\
\gamma_{n,6} &= d_{n,6,1}\epsilon_{n,1} + d_{n,6,2}\epsilon_{n,2} + d_{n,6,3}\epsilon_{n,3} \\
\gamma_{n,7} &= d_{n,7,1}\epsilon_{n,1} + d_{n,7,2}\epsilon_{n,2} + d_{n,7,3}\epsilon_{n,3} \\
\gamma_{n,8} &= d_{n,8,1}\epsilon_{n,1} + d_{n,8,2}\epsilon_{n,2} + d_{n,8,3}\epsilon_{n,3} \\
\gamma_{n,9} &= d_{n,9,1}\epsilon_{n,1} + d_{n,9,2}\epsilon_{n,2} + d_{n,9,3}\epsilon_{n,3}
\end{aligned} \quad (22)
$$

(Note that the component indices 1, 2, 3 apply to γ_1, 4, 5, 6 apply to γ_2, and 7, 8, 9 to γ_3.) Application of these correction vectors yields improved estimates of δr_A, δr_{n-1}, and δr_n. (Note that the original values of the components of δr_A, δr_{n-1}, and δr_n are designated by the superscript °.)

$$
\begin{aligned}
\delta r_{A,1} &= \delta r^{\circ}_{A,1} + \gamma_{n,1} \\
\delta r_{A,2} &= \delta r^{\circ}_{A,2} + \gamma_{n,2} \\
\delta r_{A,3} &= \delta r^{\circ}_{A,3} + \gamma_{n,3} \\
\delta r_{n-1,1} &= \delta r^{\circ}_{n-1,1} + \gamma_{n,4} \\
\delta r_{n-1,2} &= \delta r^{\circ}_{n-1,2} + \gamma_{n,5} \\
\delta r_{n-1,3} &= \delta r^{\circ}_{n-1,3} + \gamma_{n,6} \\
\delta r_{n,1} &= \delta r^{\circ}_{n,1} + \gamma_{n,7} \\
\delta r_{n,2} &= \delta r^{\circ}_{n,2} + \gamma_{n,8} \\
\delta r_{n,3} &= \delta r^{\circ}_{n,3} + \gamma_{n,9}
\end{aligned} \quad (23)
$$

The miss distance at time of intercept, estimated at the nth decision point, is

$$
\begin{aligned}
\delta b_{n,1} &= e_{n,1,1}\delta r_{A,1} + e_{n,1,2}\delta r_{A,2} + e_{n,1,3}\delta r_{A,3} \\
\delta b_{n,2} &= e_{n,2,1}\delta r_{A,1} + e_{n,2,2}\delta r_{A,2} + e_{n,2,3}\delta r_{A,3} \\
\delta b_{n,3} &= e_{n,3,1}\delta r_{A,1} + e_{n,3,2}\delta r_{A,2} + e_{n,3,3}\delta r_{A,3}
\end{aligned} \quad (24)
$$

$$
\delta b_n^2 = \delta b_{n,1}^2 + \delta b_{n,2}^2 + \delta b_{n,3}^2 \quad (25)
$$

The velocity error at the nth decision point is then

$$
\left.\begin{aligned}
\delta V_{n,1} &= f_{n,1,1}\delta r_{n-1,1} + f_{n,1,2}\delta r_{n-1,2} + f_{n,1,3}\delta r_{n-1,3} \\
&\quad + g_{n,1,1}\delta r_{n,1} + g_{n,1,2}\delta r_{n,2} + g_{n,1,3}\delta r_{n,3} \\
\delta V_{n,2} &= f_{n,2,1}\delta r_{n-1,1} + f_{n,2,2}\delta r_{n-1,2} + f_{n,2,3}\delta r_{n-1,3} \\
&\quad + g_{n,2,1}\delta r_{n,1} + g_{n,2,2}\delta r_{n,2} + g_{n,2,3}\delta r_{n,3} \\
\delta V_{n,3} &= f_{n,3,1}\delta r_{n-1,1} + f_{n,3,2}\delta r_{n-1,2} + f_{n,3,3}\delta r_{n-1,3} \\
&\quad + g_{n,3,1}\delta r_{n,1} + g_{n,3,2}\delta r_{n,2} + g_{n,3,3}\delta r_{n,3}
\end{aligned}\right\}
\tag{26}
$$

So as not to overcorrect, the velocity errors must be compared with a set of predetermined constants

$$
\left.\begin{aligned}
|\delta V_{n,1}| - h_{n,1} &= sg(\delta V_{n,1})\delta V'_{n,1} \\
|\delta V_{n,2}| - h_{n,2} &= sg(\delta V_{n,2})\delta V'_{n,2} \\
|\delta V_{n,3}| - h_{n,3} &= sg(\delta V_{n,3})\delta V'_{n,3}
\end{aligned}\right\}
\tag{27}
$$

The square of the magnitude of this correction decision vector is then

$$
V'^2_{n,1} + V'^2_{n,2} + V'^2_{n,3} = V'^2_n \tag{28}
$$

Thus, while this second statistical approach to implicit navigation permits wider variations from the reference trajectory than does the first simplified method, it also requires far more computation. The complexity of the computation for an even more advanced optimum statistical approach [7] requires the in-flight use of a central digital computer of still greater capacity.

D. Rendezvous Operations

A scheme for rendezvous operations may be based upon concepts of orbital mechanics, concepts of proportional navigation interception, or a combination of the two. For purposes of illustration, the modified proportional navigation interception approaches of Ref. [8] are presented herein. It is important to bear in mind, however, that the third possibility above probably offers significant advantages over the other two in terms of practical application.

Basically, the approaches of Ref. [8] require the use of optical sightings taken prior to and during the thrusting period for the determination of the range and closure rate information (for control of the rendezvous maneuver). Operationally, the astronaut establishes a constant line-of-sight approach to the target vehicle by arresting angular motions of this line of sight relative to the stellar background. He then controls the closure rate with scheduled braking to effect safe contact.

Optical sighting permits the astronaut to detect and measure visually the angular motion of the line of sight relative to a fixed reference system, such as a star background, for corrections. In order to control the closure rate he must have range and closure-rate information. Three techniques are offered for using optical information and a timer to obtain relative range and closure rate, while the required constant line of sight is established.

The transformation of angular sightings into range and closure rate can be readily obtained by considering the geometrical relations between the vehicles depicted in Fig. 9.

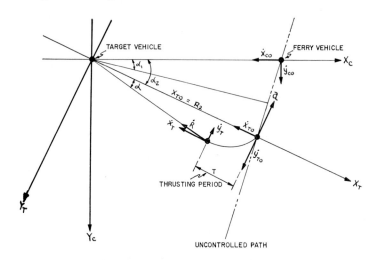

FIG. 9. Geometric relation diagram for space rendezvous techniques.

During the coasting period the range at any point may be expressed as

$$R = \frac{\ddot{y}_{co}\, t}{\sin \alpha} \tag{29}$$

Consequently, when α is small the following range relationship is obtained

$$R_1 = R_2 \frac{t_1 \alpha_2}{t_2 \alpha_1} \tag{30}$$

It can be shown through appropriate geometrical relationships and small angle approximation that during the thrusting period

$$R_2 = \frac{aT^2}{2\alpha} \tag{31}$$

and the closure rate determined from

$$\dot{R} = \frac{R_2 - R_1}{t_1 - t_2} \tag{32}$$

This technique, therefore, requires the measurement of two angular increments (α_1 and α_2) traversed by the moving line of sight while coasting, and a third angular increment (α) while a known thrusting acceleration (a) is applied to arrest the line-of-sight motion. Correspondingly, the times recorded during these angular measurements (t_1, t_2, and T) are used in Eqs. (30), (31), and (32) for determination of relative range and closure rate.

A second technique is based upon the equation of relative motion between the two vehicles. Expressed in polar coordinates

$$a_N = R\ddot{\alpha} + 2\dot{R}\dot{\alpha} \tag{33}$$

where a_N is the acceleration component normal to the line of sight to the target vehicle. The solution to Eq. (33) and the geometrical relations of Fig. 9 lead to the range determination equations

$$R_1 = \frac{aTt_1}{\alpha_1} \tag{34}$$

$$R_2 = \frac{aTt_2}{\alpha_2} \tag{35}$$

$$\dot{R} = \frac{R_2 - R_1}{t_1 - t_2} \tag{36}$$

This technique requires the measurement of the same angles and time increments during the coasting period called for by method 1 but not the third angle while thrusting.

The third technique investigated requires that the same measurements of the first technique be used as inputs to a matrix solution yielding several progressive solutions, but also requires the use of additional special purpose computing equipment.

From Fig. 9

$$\tan \alpha = \frac{\dot{y}_{T0}t - (aT^2/2)}{x_{T0} - \dot{x}_{T0}t} \tag{37}$$

or for α small

$$x_{T0}\alpha - \dot{x}_{T0}t\alpha - \dot{y}_{T0}t = -\frac{aT^2}{2} \tag{38}$$

During the coasting period

$$x_{T0}\alpha_1 - \dot{x}_{T0}t_1\alpha_1 - \dot{y}_{T0}t_1 = 0 \tag{39}$$

$$x_{T0}\alpha_2 - \dot{x}_{T0}t_2\alpha_2 - \dot{y}_{T0}t_2 = 0 \tag{40}$$

and during thrusting period

$$x_{T0}\alpha_3 - \dot{x}_{T0}t_3\alpha_3 - \dot{y}_{T0}t_3 = -\frac{aT^2}{2} \tag{41}$$

where α_1, α_2, and α_3 are angles traversed from a zero reference, and t_1, t_2, and t_3 are the respective times. Thus, three simultaneous equations with unknowns x_{T0}, $\dot{x}_{T0}$, and $\dot{y}_{T0}$ can be solved with

$$x_{T0} = R_2 \tag{42}$$

$$\dot{x}_{T0} \approx \dot{R} \tag{43}$$

Feasible instrumentation for the rendezvous operation must meet the constraints imposed by both the navigation scheme and the operational requirements of a given mission. In Section VI it is indicated that a properly designed space sextant may be employed to perform the measurements described in the preceding discussion.

V. Man as an Active Participant

Three major facets of human capability are relevant to the general problem of space navigation: man's vision, his ability to manipulate viewing and computing instruments, and his computational ability.

Two classes of constraints imposed upon the space flight navigator may be identified. These are:

(1) General limitations of human performance
(2) Limitations of human performance unique to space flight

Many data have been generated in attempts to evaluate the general limitations of human performance. The major difficulties in relating this material to the topic under consideration lie in the determination of human capabilities in the performance of specific complex operations which have not been studied, based upon performance of other nonrelevant operations. Although overall performance of the human operator may be predicted with some sort of accuracy, specific performance predictions offer great difficulty.

At this writing, useful data available on the limitations of human performance in the navigation function specifically attributable to space flight conditions are sparse. Moreover, much information considered reliable not too long ago is undergoing continual revision as additional data are accumulated.

Three data sources are commonly employed for space flight performance predictions, namely, the results of manned orbital and suborbital flights, rocket experiments with animals and biosensors, and simulation tests·

Extrapolation of data from the animal flights is relatively meaningless. Adequate simulators, when finally operational, should prove useful; but, as of now, these are not a particularly valuable source. Finally, the amount of data thus far derived from manned space flights is still too limited to be generally applicable since no significant navigation functions were performed by the astronauts. Consequently, many questions pertaining to man's performance remain unanswered.

A. The Man-Machine Interface

The basic problem under consideration within the scope of this section is the capability of the human operator to perform the space navigation function. It is unrealistic to assume, however, that the operator's function will be limited to the navigation task, or that his immediate environment will be dictated solely by the requirements of the navigation function. Within this section, the relevant navigational tasks and other factors, both functional and environmental, are summarized.

Two major aspects of space flight are near-body and midcourse flight. In general, the required types of observations and computations are dissimilar in these two cases.

A possible sequence of events which may be employed for navigation during near-body flight follows:

(1) Optical measurements and computations are performed to determine the size and shape of the orbit.
(2) If perigee is below some predetermined minimum, one of two actions must be taken:
    (a) The orbit will be corrected for a safe perigee, or
    (b) the orbit will be adjusted for a safe re-entry.
(3) If the perigee is above the predetermined minimum, no action is required and the operator will have sufficient time for other measurements and computations; in particular, the determination of his exact position in this safe orbit.

Fundamentally, only angular measurements are required for space navigation. Considering the angular separation between bodies, and the angular subtense of a body during near-body flight, specially designed instruments are required. These optical problems are compounded by restrictions on the optics due to helmet configuration and restrictions on controls due to the possible presence of gloves.

Once the necessary observations have been made with the required accuracies and at the proper times, the data must be employed in a series of computations in order to derive the necessary navigation information.

In addition to the tasks associated with the navigation function, the operator may be called upon to perform other tasks, some simultaneously. This is especially true considering possible emergency situations. Here his additional functions could include attitude control, energy system management, and communications.

B. Effects of the Space Environment

The following space environment factors are relevant to the navigation function to varying degrees: confinement, high g loads during boost, weightlessness, noise and vibration, radiation effects, temperature variation, the spacecraft atmosphere, suit limitations, spacecraft motion, and visual effects and restrictions.

1. CONFINEMENT

One of the parameters influencing the operator's reactions is the effect of the long confinements encountered during space flight. This confinement may produce degradation of performance [9]. Many of the confinement studies conducted to date are not too meaningful since they included sensory deprivation, a condition which will not exist during space flight, at least to the extremes created during these tests. Other shortcomings of some of the tests are the lack of real windows, proper incentive and motivation, and the fact that the test subjects are required to perform meaningless computations and tasks. It appears that confinement has not been a problem encountered in space flight to date (1964).

2. THE BOOST ENVIRONMENT

During the boost phase of flight, the astronaut is subjected to high g loads, noise, and vibration. The altered g force testing to date has indicated that a loading of as little as 2 g's begins to have serious effects upon the operator's performance, and by the time 4 g's are experienced, motion, judgment, performance, vision, and other functions become very poor.

Noise and vibration (as encountered during powered flight) may contribute to degradation of observational and computational performance. If the noise level is high enough during boost, a temporary hearing loss will extend into the free flight condition. This may affect later actions by the astronaut in two ways, namely, stress and auditory response.

Since the principal navigation function is initiated after injection, these effects are not of particular significance with respect to the problem under discussion herein.

3. Weightlessness

The results of the early American and Russian orbital flights have shown that weightlessness is not a particular problem. Probably the only problem, relevant to navigation, associated with the weightless environment is the physical interaction between the operator and his navigational instruments. The instrumentation must be designed so as to remain where it is positioned and its operation must not depend upon gravity.

4. The Astronaut's Well-Being

Factors such as radiation level, temperature, and humidity variation and the spacecraft's atmosphere have a profound effect upon the astronaut's well-being and consequently upon his ability to perform the navigation function. If the bounds of comfort are greatly exceeded, it is quite reasonable to expect performance to deteriorate significantly.

Although no immediate harmful effects of radiation were detected during early orbital flights, voyages of longer duration and higher altitudes may introduce this problem. High radiation levels may have an immediate effect on the operator. Under such situations it is quite probable that the astronaut would be in no condition to adequately participate in the navigation function. At such times a completely automatic system would be necessary to return the spacecraft to Earth.

The effect of variations in temperature upon task performance is difficult to predict accurately because of individual differences and particular motivation. In general, however, complex performance begins to deteriorate at 85°F in dry air. For optimum performance, humidity should not exceed 40 per cent [10]. However, an acceptable range lies between 15 and 70 per cent. This, of course, is a function of the atmospheric composition, pressure, and temperature.

At all times the carbon dioxide content of the spacecraft atmosphere must be kept within limits. If these limits are exceeded, the effect will be similar to that of extreme fatigue.

5. Suit and Helmet Limitations

Two types of restrictions are relevant: (1) reaction time limitations caused by the suit and (2) visibility limitations caused by the helmet.

Helmets interfere with visibility in several ways. The field of view is cut to approximately 80 deg to the left and right compared to a normal field of 115 deg to each side. Similar restrictions appear in the vertical plane. Head mobility is reduced. The helmet contributes to eye strain, and requires a minimum of 30 to 35 mm eye relief. Further, fogging of the visor or the interference of equipment designed to prevent fogging can affect visual acuity.

The use of optical instruments, in particular, is affected by suit and helmet limitation. Arm movement, for example, becomes somewhat restricted, and this results in problems in the control of optical and computational equipment. All of these factors must be considered in the design of instrumentation.

6. Spacecraft Motion

Tumble may be expected from a drifting spacecraft during a great portion of an extended flight. Data from several sources indicate a tumble rate of about 0.5 deg/sec maximum, and the realistic possibility of reducing this to 0.1 deg/sec.

In addition to complicating the observation and tracking tasks, tumble also reduces the time available for navigational measurements. The time available for a measurement is given by

$$\text{Time (min)} = \frac{\text{field of view half angle (deg)}}{60 \times \text{tumble rate (deg/sec)}}$$

Thus, for a 180 deg total circular field of view and a tumble rate of 0.2 deg/sec, 7.5 min are available for a measurement (assuming that the target is initially at the center of the field).

7. Visual Effects and Restrictions

Human vision has evolved in a terrestrial environment to which it is superbly adapted. To the extent that the natural selection aspects of such an environment differ from those that might have operated in a space vehicle in flight, human vision may be considered incompletely adapted to space environments.

For their impact on the visual function these environmental differences may be dichotomously divided into: (1) those inhering in the external luminous environment, and (2) those inhering in the internal environment of the organism itself, even though these internal differences, in the final analysis, are also attributable to the objective differences between the two environments. To illustrate each of these: (1) the ambient luminous environment of terrestrial daylight produced by the molecular and particulate scattering of the atmospheric envelope contrasted with the black, optically empty ambient environment of space is a difference of the first (external) kind; and (2) the effect of weightlessness and rotational changes in attitude in space upon the reflex relations between the vestibular and visual systems, contrasted with the effects of terrestrial gravity on the Earth's surface upon these same reflex relations, is a difference of the second (internal) kind.

Whatever the nature of these external and internal differences may be, they are not of such a magnitude as to preclude manned space flight. The visual problems of manned space navigation approached from a theoretical standpoint and in the light of accumulating experience are consequently concerned with ophthalmic safety and the provision of such optical and other aids to vision as may vouchsafe adequate visual capability at least, and optimal performance at most.

Some of the visual parameters which must be considered for space flight are:

(1) Man's ability to judge size, shape, and distance
(2) Judgment of relative motion
(3) Empty field myopia
(4) Excessive glare and sunblinding
(5) Nystagmic eye movements

For purposes of rendezvous and docking, items (1) and (2) assume great importance. Due to the absence of atmosphere, normal distance-size indicators (such as known buildings, trees, roads, etc.), and the excessively high contrast of solar illumination, conventional Earth-acquired methods of judgment become highly unreliable. Thus, near-body rendezvous and docking will present serious problems within the overall scope of a given space mission.

Owing to homogenous visual fields found at certain times in space flight, it is difficult for even experienced pilots to pick out targets. On space flights of long duration involving rendezvous of one vehicle with another, or requiring the visual research of space, the need of extensive aid in detection may prove necessary. This problem is known as *empty field myopia*.

There are many ways to correct for myopia in normal situations. However, its effect on visual search is not so easily compensated. In space flight the sky is essentially uniform. Search in this empty visual field is not uniform across the display area in terms of spatial and temporal distribution of attention.

Glare, one of the major visual problems in space flight, results when there is marked contrast between areas of light and darkness. The extreme brightness of objects in space, contrasted against the dark background of the sky, presents a strong contrast to the eye, causing objects to appear larger than they are and causing eye strain. This discomfort of glare is due to the overstimulation of the periphery of the retina by light which reduces the sensitivity of other, understimulated areas. This effect is emphasized if the light comes from below rather than above because the eyes are protected from overhead light by the structure of orbital ridges and eyebrows. Therefore, provided the spacecraft remains in an upright position, the major source of glare would seem to be from the cloud floor rather than the Sun. This is due to reflected light from the cloud floor. Without clouds, both the sea and snow will contribute glare. Cloud and snow-covered areas have a luminosity far greater than the full moon. Sea areas will have intense hot spots.

Exposure to intense radiation (the primary source of which is the Sun) can be extremely detrimental. In addition to producing direct radiation burns, exposure to the glare of the Sun for even short periods of time causes serious temporary visual problems, such as after-images and loss of dark-adaptation. Permanent retinal burns may be caused by exposures of a few seconds if the energy level is sufficiently high. Clearly, it is necessary to develop an adequate filtering system for an optical system to be used in space and attention must be given to the problem of forewarning the operator when an intense light source (i.e. the Sun) is to enter his field of view.

Nystagmic eye movements have been studied in detail. These movements are caused by sudden motion (e.g., tumble) and become a serious problem at rates just above $\frac{1}{2}$ deg/sec. They are caused by signals from the semicircular canals. Although a constant velocity will eliminate these eye movements after their initial response, they do persist from 10 to 20 sec after the movement ceases. Also, any head motion will bring it back again. This effect, in addition to introducing vertigo, lowers the sensitivity of the eye, and makes tracking tasks difficult.

C. The Use of Optical Instrumentation

The only accurate method for the determination of the performance and accuracy of an optical instrument or instrument technique is tests of the actual instrument. However, in the design of new systems such tests must be deferred until the instrument is fabricated. In order to obtain an evaluation of expected performance, it is necessary, therefore, to examine presently existing instrumentation concepts which are as nearly similar as possible to those which would be employed in a contemplated space instrument. This section is presented as an extrapolation of man's performance using existing optical instruments to the problems of space navigation.

1. MEASUREMENT ACCURACY

Various documents on the use of theodolites [11, 12] indicate that, while they may be read to a precision of 0.1 or 0.2 arc sec, the actual value may be in error by several seconds. Standard procedure relies on 8 to 16 double measurements to obtain enough data to enable a good average value to be computed. Typical theodolite measurements require 3 to 5 min to complete. This length of time is due to the high power of the typical theodolite (28 power), narrow field of view ($1\frac{1}{2}$ deg) and high accuracy sought.

One of the major problems in the use of a theodolite is the possibility of gross readout errors. Such errors are primarily caused by the finely divided scale. It is important in the design of an instrument to keep required instrument readout accuracy compatible with operational requirements.

It is interesting to analyze the hand-held aircraft sextant in terms of its performance and design limitations and to compare this with a space sextant. The aircraft sextant contains a two or three power telescope with a 10 to 15 deg field of view, and the alternatives of using a visual horizon or an artificial horizon. The bubble is adjustable in size and can be eliminated altogether. The readout scales are divided into 1 arc min intervals. There is an integrator (either spring wound or electrically powered) which runs for a fixed time interval (usually 2 min). The integrator is employed to average the sextant readings when the bubble is used for the horizon, and the sextant is employed on a vehicle which is subject to transverse accelerations. Use of the sextant from a stationary base eliminates the need for the integration. Now consider the accuracy of such an instrument.

Superposition of two stars, as a test routine, can be performed from a fixed base (hand-holding the sextant) to a repeatable accuracy of 1 arc min, which is just the limit of the readout. Use of the bubble for a horizon reduces the accuracy to between two and four arc minutes. Use of the bubble as a horizon under dynamic conditions, with the integrator, yields an accuracy of between four and eight arc minutes.

It is important to note, however, that a sextant for space navigation can be quite different. Since the spacecraft is mostly in free fall, a bubble horizon is meaningless. Thus, all measurements require the superimposition of two targets. No problem exists in performing such an operation while the vehicle is tumbling at a low constant rate. Higher magnification power can provide greater accuracy. It is believed, however, that analytical estimates of the performance of such an instrument cannot be too reliable. Only physical tests can provide meaningful performance specifications for a specific instrument technique.

2. TRACKING OF MOVING TARGETS

Two, alternate, modes of vehicle attitude control during the period of navigational observation can be considered. One is limit cycle stabilization. The other calls for initial orientation toward the target body (or bodies) and then free drift. The free drift is unavoidable where limit cycle operation is not feasible. The final choice between these two modes must be based (at a minimum) on simulation studies using appropriate instrumentation subsystems.

The free drift mode is caused by the inability to obtain zero angular rates by means of pulse jets. Consequently, some directionally arbitrary, constant, residual tumble rate will remain after each pointing stabilization operation. It is expected that the magnitude of these rates can be reduced to about 0.1 deg/hr.

Some information does exist which can be related to tracking under these conditions. For the space application, the targets stand still and the vehicle rotates. However, the inverse problem, i.e., tracking a moving target from a fixed base, has received considerable study. These investigations are primarily directed at improving the accuracy of manned gunnery systems, but nevertheless they have produced results which are applicable to the space navigation problem.

The conclusion has been reached by many investigators that intermittent, rather than continuous, tracking is a basic feature of human response. These intermittent corrections occur approximately every 0.5 sec with random stimuli and reduce to 0.2–0.4 sec with predictable stimuli. Some form of smoothing is usually superimposed upon discontinuous tracking [13]. This aspect of human operation is not necessarily a disadvantage to the space navigation problem, although it is not ideal for the gunnery situation. The astronaut can wait to record readings until he is aligned.

Examination of tracking performance with Directors M5A2 and M5A2E1 was reported in Ref. [14]. Twenty-four flights (12 left and 12 right) were flown by an aircraft in front of the test apparatus. The least horizontal range varied between 800 and 1050 yd. The plane flew at an altitude of 350 yd with a speed which varied between 140 and 160 mph.

TABLE II. MEANS OF MEASURES OF LATERAL AND VERTICAL
TRACKING ERRORS AND PREDICTION ERRORS FOR
DIRECTORS M5A2 AND M5A2E1

Measure	Lateral		Vertical	
	M5A2 (mils)	M5A2E1 (mils)	M5A2 (mils)	M5A2E1 (mils)
Tracking average error	1.36	1.34	0.39	0.40
Tracking variable error	0.58	0.66	0.39	0.29
Tracking constant error	1.29	1.26	0.13	0.31
Tracking rate of change	0.43	0.48	0.20	0.19
Prediction average error	1.92	1.74	0.68	0.56
Prediction variable error	1.34	1.19	0.68	0.49
Prediction constant error	1.45	1.40	0.11	0.31

Thus, the average maximum angular rate was on the order of 4 deg/sec. Table 2 of the subject report, presented here as Table II, lists the accuracy of tracking obtained under the test conditions. Instantaneous accuracies are, of course, higher.

3. READOUT ACCURACY

The proper design of readout scales plays a significant role in the reduction of errors associated with the use of optical instrumentation. As indicated previously, the primary cause of gross readout errors in the use of theodolites is the high precision (extremely fine division) of the readout scales. Operational considerations determine the accuracy required in a given measurement. These considerations should also limit the fineness of scale division.

In addition to optimizing the fineness of division, the layout of the scale is also critical. It has been shown [15] that dividing scales into decimal intervals with whole number groupings, such as 1000, 100, 10, resulted in the fewest errors. On the other hand, scale intervals in steps of 4 resulted in the most errors. Scale divisions in steps of 2 were almost as good as those in the decimal units. In general, changing the scale number by a factor of 10 does not significantly alter readout error rates. Decimal fractions such as 0.01, 0.1, 0.2, resulted in higher error rates than whole numbers like 1, 10, 2, 20.

Other investigations confirms the above conclusions. Reference [16] describes a series of tests on eleven experienced naval radar operators to determine their error rate on standard naval radar indicators. Decimal scales again provided best results. It is believed that this stems from man's natural use of decimal systems.

In general, it appears that navigation instrument readout scales should first be divided into degrees and then decimal fractions of degrees. Coarse division into degrees is based on trained man's natural familiarity with such units, and consequently his ability to check his readings for gross error by independent estimation. Fine division should be based upon decimal intervals in order to minimize error rate. Markings should be unambiguous. Finally, it is again noted that fineness of division must be based primarily upon operational considerations.

4. REQUIRED FIELD OF VIEW

Important considerations in the design of navigational equipment are magnification and field of view. Attainable accuracy is a function of magnification. The angular field of view must be so chosen as to enable the identification and acquisition of the required navigational stars. The basic problem is the antipathy between magnification power and field of view.

For identification, a wide field, and hence low power, instrument is desired; for accuracy of tracking, a high power, and hence narrow field, instrument is needed. Two approaches are generally open: the first, a compromise instrument; and the second a dual power instrument.

The periscopic sextant for aircraft navigation falls in the category of compromise instruments. It has a field of view of 15 deg and a magnification of 2 power. It is the opinion of a number of practicing navigators that the 15-deg field makes stellar identification difficult. Only if approximate aircraft position and heading, time of year, and time of day are known can one use the instrument effectively. Basic accuracy runs on the order of one to two arc minutes. It is believed that a 15-deg field would not be suitable for the location of reference stars from a spacecraft which can assume an arbitrary attitude with respect to the celestial sphere.

Submarine navigational instrumentation, on the other hand, falls into the second category. Two periscopes are employed for celestial navigation aboard some submarines at periscope depth. The first or general purpose periscope is used for identification of the star field and training of the navigational scope on the star of interest. For search and identification, a 32 deg circular field is presented at 1.5 power. For track, a 6 power view of an 8 deg field is used. When the star is tracked in the high power field of the general purpose periscope, the navigational periscope is slaved to it.

Optical design considerations encourage the reduction of the field of view as much as possible. It is important to determine the practical limit of this reduction in terms of usage requirements.

D. Man as a Computing Element

Feasibility of self-contained space navigation implies the successful solution of the equations of navigation during an actual flight. Thus, the navigation and guidance concepts of Section IV must be reduced to mechanization compatible with the constraints of a given operation. This mechanization may require automatic computers or it may employ man as a computing element in the overall navigation scheme. Clearly, man is indispensable in case of partial or total failure of the automatic equipment.

1. Man's Computational Capabilities

Many tests have been performed and much has been written concerning man's computational capabilities. Yet, practically no data applicable to the problem of space navigation exist. Almost invariably the tests have been performed on an assorted, average group of individuals who are neither trained nor motivated toward solving the sets of problems presented to them. The problems are often far removed from any semblance to the equations of aircraft or terrestrial navigation, no less those of space flight. It should be realized that the astronaut who is to navigate in space will be an above average individual with a high level of technical education and extensive training and experience in the performance of the required functions.

Some of the data contained in the literature are useful to the problem under consideration if it is properly evaluated. The evaluation may be rather remote from the objectives and findings of the original investigators. It is simply that the results of these tests imply rules or statements of limitations of computation which are outside the framework of their original problem definition and formulation. Some of these results [17–22], properly modified by the comments and suggestions of trained navigators and engineers, permit the formulation of a set of rules which are recommended as design constraints for the mechanization of manual computation.

2. Design Rules for Manual Computation

In order to design computers or computational techniques which are feasible for manual space navigation, it is necessary to derive a set of design rules which are compatible with man's abilities. The following is such a list:

(1) Of primary importance is the realization that the individual who will use any resulting computational technique is to be a trained, experienced, space navigator with a more than competent technical education. Thus, techniques must be optimum not for the average individual with no experience in their use, but rather for an individual who understands what he is doing and has familiarized himself with their usage by means of extensive simulation training.

(2) The choice of the technique employed is a function of the time available for the performance of the calculation and the stress situation of the astronaut. By way of example, the initial determination of safety for orbital flight (the geometric parameters) requires the use of a hand-cranked mechanical analog computer which accepts the basic range inputs and yields values of ϵ and d_c with no other action on the part of the operator. For noncritical situations, such as lunar midcourse flight, more time is available and one is no longer so restricted to technique.

(3) The choice of the technique employed is a function of the overall accuracy required. Analog techniques, while providing simple, real time solutions for many problems, are inadequate when high computational accuracy is required. A general rule is that analog computation can reasonably provide accuracies of one part in a thousand of full scale. When higher

accuracies are required it becomes necessary to use hand computation work sheet techniques.

(4) The choice of the technique employed is a function of the allowable error rate. (Note, of course, that the allowable error rate may be zero.) One must also specify allowable error rate in terms of significance. Thus, implicit midcourse navigation techniques which statistically employ redundant data can permit a small but finite error rate in the less significant places of the calculations. The determination of orbital safety, on the other hand, while permitting some inaccuracy at the insignificant end of the calculations, cannot tolerate any gross errors independent of error rate. This is another reason that such a problem cannot be solved by pencil and paper techniques.

(5) Design of dials (read-in and read-out) for mechanical analog computers should be based on rules already developed for this problem. The important factor is the reduction of the probability of error while retaining the necessary accuracy.

(6) A number of rules are formulated with respect to work sheet design. It must be remembered that even though they are individually listed they must be considered as an integrated whole when they are applied to a specific problem.

(a) Wherever numbers are to be inserted by hand, the number of required figures must be clearly indicated by appropriate markings in word boxes.

(b) Decimal points must be clearly indicated in each word box (i.e., by the use of double colored lines or the equivalent).

(c) Positive and negative signs should be clearly indicated for each total number (word).

(d) Transfers should be kept to a minimum. In all cases transfers must be indicated by flow lines.

(e) In general, flow lines should be used to indicate order and logic of given operations and calculations.

(f) Color should be used to distinguish types of flow lines (where necessary), outline particular numbers or operations (if greater clarity is gained by so doing), and mark decimal points and block out significant places as called for above.

(g) Longhand operations should be avoided because of possible error and a hand calculating machine should be used in their place.

(h) A proper typeface should be chosen so as to provide maximum legibility and reduce the possibility of error.

(i) In general, the instructions appearing on the worksheet should be more in the way of reminders to the operator rather than complete details on how to perform the calculation. The operator will not be seeing these work sheets for the first time when he is required to use them for space navigation.

(j) Avoid sideways addition or subtraction.

(k) Avoid upside-down subtraction when possible.

(l) If a large group of numbers, both positive and negative, are to be added, break the addition process down into parallel stages, one for positive numbers, the other for negative numbers, and then combine the sums.

(m) Keep each complete group or set of operations on one page so as to increase speed of operation, reduce possibility of transfer errors, and prevent problems of shuffling of pages (which is highly undesirable under space flight conditions).

(n) As many data as possible should be stored directly on the work sheet.

(7) Some of the rules of data storage are:

(a) Tables should form the primary means of storage of data.

(b) These tables should be sufficiently complete so as not to require any interpolation.

(c) If large quantities of data need to be stored, they should be contained in a single volume with proper cutout indexing so that time will not be lost in finding the correct section or page.

(d) Groupings should take into account man's preference for, and ease of using, decimal systems (with subgroupings of five recommended).

(e) Type form should be chosen so as to yield maximum legibility (Old Style numerals for numbers).

(f) Redundant marking of columns should be employed to reduce error rate and necessary access time.

It is also possible to formulate some general rules which are applicable to any particular problem.

(8) Recognizing other limitations and constraints as far as man is concerned, the following list of computational techniques is ordered in terms of preference based on ease of operation:

(a) Manual control of digital computer

(b) Manual mechanical analog computer

(c) Manual analog computer

(d) Slide rule computers

(e) Nomograms

(f) Computation by charts and data tables

(g) Pencil and paper hand computations

(9) Each complete technique should be self-contained, permitting the

operator to perform a required computation with the use of a minimum set of aids. In a spacecraft environment there will not be endless desk or work space available.

(10) No operation should be overly long in terms of required time, even if time is available. Fatigue developed in the performance of a set of calculations is a condition known to all concerned with technical operations.

VI. Navigational Instrumentation

A. The Spectrum of Navigation Instrumentation

The total navigation system for a given space mission is perhaps best considered in terms of the *spectrum of navigation instrumentation*. This spectrum, depicted in Fig. 10, encompasses all possible combinations of

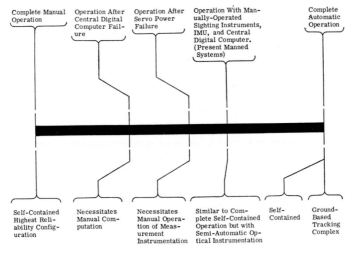

Fig. 10. The spectrum of navigation instrumentation.

spacecraft navigation instrumentation. The completely automatic navigation system is represented by the right-hand extreme. For purposes of this discussion, such an automatic system may either be wholly self-contained or include information derived from a ground-based tracking and computing complex. The navigation and guidance systems for all unmanned space programs necessarily lie at this end of the spectrum. (These spacecraft are currently characterized by having little or no flexibility of operation.)

Presently defined manned spacecraft navigation instrumentation is highly automatic and lies near the right end of the spectrum. It relies heavily upon an operative inertial measuring unit, central digital computation and automatic slaving of the optical sighting instruments. Only the sighting and tracking through the telescope and sextant are currently under manual control. In the primary mode all instrument readouts are automatic.

The completely manual navigation scheme represents the left-hand extreme of the spectrum. The need for servo power, an inertial measuring unit, or automatic computation is eliminated with such an approach. Therefore, the possible reduction in the accuracy of some of the required computation is compensated by the corresponding increase in the probability for the successful completion of the mission.

A number of other instrumentation combinations and navigational modes lie between the two ends of the spectrum. Failure of servo power, for example, need not affect the operation of the central digital computer. Similarly, failure of the central digital computer is not necessarily accompanied by failure of servo power.

In addition to use in a primary mode, manual computing instruments and manual measuring instruments may be employed to compensate for partial failures in the automatic or semiautomatic navigation equipment. A manual computing system, for example, can provide periodic verification of the operation of the automatic computer; and, in case of failure of the latter, provide the means for retaining the computational capability of the navigation system. In a similar manner, manually derived optical data may be employed with the automatic computer, where failures of servo power or failures in the inertial measuring unit have occurred.

If a total space navigation scheme is to be compatible with all of the operational situations outlined above, the following very important guideline for the synthesis of any such system must be observed.

All navigation instrumentation modes incorporated within a spacecraft designed for a specific mission should employ the same navigation concept. Only in this manner may a complete spectrum of navigation instrumentation which meets realistic operational requirements be implemented.

B. Instrumentation for Data Acquisition

The "observables" of space flight have been shown to form the basis of the various guidance concepts and their implementation. The attempt to specify a relatively simple set of navigational measurements was made with forethought of the resulting instrumentation. Accordingly, all basic phases of space navigation require only angular measurements between celestial bodies and points. (Rendezvous can also be based solely upon angular sightings, but may also be achieved through radar derived range and range rate information. For lunar landing a radar altimeter can prove useful.) These are the angular subtenses of the Moon and the planets, angles between stars and the orbiting bodies of the Solar System, and the altitudes or coaltitudes of stars relative to the Earth or Moon. The instrumentation required for these operations may assume a number of forms, each dependent upon the concept employed for the total navigation operation.

Consider a complete complement of space-borne passive data acquisition equipment, automatic, semi-automatic, and manual (excluding from the present discussion active radar systems):

(1) Space sextant
(2) Automatic horizon sensor (moderate accuracy or high accuracy)
(3) Automatic star tracker
(4) Automatic Sun sensor
(5) Gyroscopically stabilized inertial measuring unit

The above listing embodies two important considerations: (1) It frees the space navigation system from dependence upon the more exotic equipment and techniques (until their feasibility has been conclusively demonstrated in the actual operational environment) and (2) only one of the above instruments is capable of manual operation, which implies, therefore, that it can duplicate the operation of the remaining automatic equipment. The validity of this implication is discussed below.

Figure 11 presents the schematic combination of all of the above listed

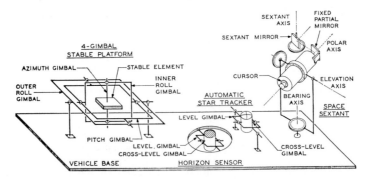

Fig. 11. Spacecraft instrumentation schematic.

components, with the exception of the Sun tracker. All of these instruments are mounted upon a common rigid base so that pointing information derived by each can be related to the others through intergimbal transformations. A four-gimbal inertial platform, capable of complete attitude freedom, is indicated. Both the horizon sensor and star tracker are in level, cross-level gimballing configurations. In the above listing, the horizon sensor was called out with alternately either high or moderate accuracy. Ideally, a high accuracy instrument is preferable. To date, however, only limited accuracies have been demonstrated in flyable hardware. Realistically, therefore, it is necessary to base an overall system upon the availability of only moderately accurate horizon sensors.

(It is important to note that a properly mechanized horizon sensor will yield two types of information. The first is, of course, a determination of the local vertical. In the process of obtaining this quantity it is also possible to derive stadimetric ranging data.)

Figure 11 also shows a pictorial representation of a space sextant equipped with both visual and automatic readouts. The gimbal suspension shown effectively isolates the sextant from the spacecraft's orientation and angular motion. Therefore, the instrument may be employed for data acquisition even under conditions of low constant arbitrary tumble rates or large excursion limit cycle operation. The sextant can be mounted behind either a flat window or an astrodome, the latter providing wider field capabilities.

The space sextant shown in the overall schematic is based upon a design put forth by the Kollsman Instrument Corporation during the early part of 1961. Figure 12 shows this instrument in greater detail.[2]

This space sextant is essentially a precision theodolite, equipped with a sextant attachment and provided with an additional degree of freedom about the theodolite line of sight. In addition, a cursor is provided for the locating of lunar occultations. The following navigational operations may be performed with this instrument:

(1) Triangulation—through the measurement of sextant angles between selected stars and bodies of the Solar System.
(2) Stadimetry—through the measurement of the angular diameters of appropriate target bodies.
(3) Occultation correlation—through the measurement of time and location of occultations.
(4) Platform alignment—if an inertial or celestial-inertial platform is to be aligned prior to re-entry or lunar landing, then automatic readouts of the sextant can provide the required information.

For purposes of illustration, it is of interest to examine the operation of the space sextant for stadimetry. In this mode it is employed to measure

[2] Although some aspects of later versions of this instrument are somewhat different from the illustrated unit, it is nonetheless representative of the type of sextant currently under consideration.

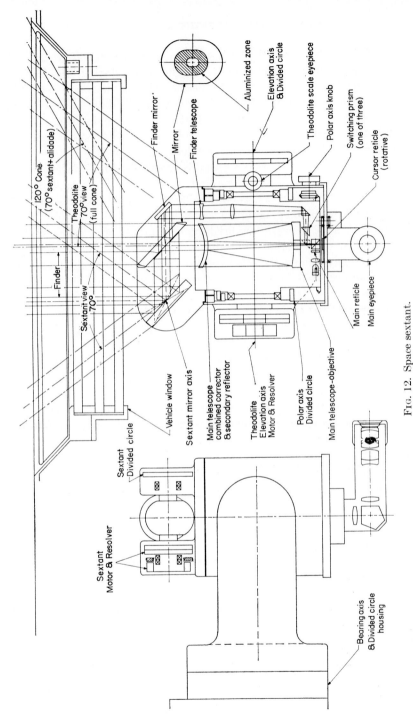

120° Cone
(70° sextant + alidade)

Theodolite
70° view
(full cone)

Sextant view
70°

Finder

Finder mirror

Mirror

Finder telescope

Aluminized zone

Elevation axis
& Divided circle

Theodolite scale eyepiece

Polar axis knob

Switching prism
(one of three)

Cursor reticle
(rotative)

Main reticle

Main eyepiece

Vehicle window

Sextant mirror axis

Main telescope
combined corrector
& secondary reflector

Theodolite
Elevation axis
Motor & Resolver

Polar axis
Divided circle

Main telescope-objective

Sextant
Divided circle

Sextant
Motor & Resolver

Bearing axis
& Divided circle
housing

FIG. 12. Space sextant.

148

the angle subtended by the full Earth's or Moon's disk. This angle may then be converted to a range measurement. In this application, the radius of curvature of the visible crescent in the case of the Moon, or of the visible airglow as in the case of the Earth, is the quantity directly measured. This measurement is read on the sextant scale.

Inherently, the stadimetric measurement is more accurate when the astronaut is closer to the body being sighted. Since this proximity is accompanied by a wide angular subtense of the body, the theodolite field of view will be exceeded. By virtue of its sextant angle feature, the configuration can be easily accommodated to the situation. In addition, the polar angle feature makes possible a visual match of the target rim and crosshair by inspection of phenomena seen only at the center of the field of view.

This operation is indicated in Fig. 13. The Moon is initially located by

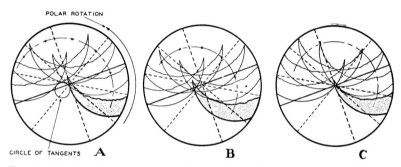

POLAR ROTATION

CIRCLE OF TANGENTS   A   B   C

FIG. 13. Stadimetric measurement. *Step 1:* Theodolite pointed approximately to track center of Moon. *Step 2:* Sextant set to bring edge to center of field with theodolite view dimmed out. *Step 3:* Polar axis set in rotation. Situation as shown above. *Step 4:* Reduction of the circle of tangents to a point by adjustment of the sextant angle. Note that this adjustment alone is sufficient for stadia ranging. To perform the adjustment precisely, a reference point is needed. This is provided in Step 5. *Step 5:* Moving the point of tangency to the reticle center by theodolite adjustment. Repeat Step 4 as needed. Moon's radius as an angular subtense is read on the sextant angle.

pointing the theodolite at the estimated center. The theodolite view is then dimmed as the sextant view is brought into the field, so as to place the bright edge tangent to the crosshair. The polar axis is then set in rotation. The Moon image will describe a circular path. The operator must now adjust the sextant angle until the visible portion of the Moon edge appears to slide past and remain tangent to one reference line in the field of view as shown in Fig. 13A. To ease determination of this condition, the theodolite axes can be adjusted so that the dotted reticle line serves this purpose. The Moon edge will now remain tangent to the dotted line at all times.

If the sextant axis has not been properly adjusted, however, the Moon edge will either remain tangent to a circular "hole" in the field, or will overlap completely (Fig. 13B). After the condition of Fig. 13C has been attained the angle of the sextant axis may now be read as the stadimetric output.

The reading is made through the main eyepiece by switching to the sextant scale presentation. The angle read is that which the radius of the body subtends.

If star sightings are to be performed automatically, a celestial tracker specifically designed for the space environment is required. Such a unit is also shown schematically in Fig. 11.

With the celestial tracker, inertial platform, and horizon sensor (if sufficiently accurate) sightings may be referenced to the inertial platform and the appropriate angular separations determined by analytical means. If a reference attitude is to be retained by the inertial platform for purposes of Earth re-entry, lunar landing, application of corrective thrust during midcourse or orbital transfer, then initial alignment is necessary.

C. Operational Use of the Navigational Instrumentation

The mission phase classification developed in Section II is particularly useful for the definition of system requirements and operations and so is repeated for reference. The phases are:

(1) Boost to injection (Earth or lunar take-off)
(2) Early midcourse flight
(3) Late midcourse flight
(4) Near-body (orbital) flight
(5) Variation of orbital trajectory
(6) Rendezvous with target in orbit
(7) Landing (including atmospheric re-entry for the Earth)

Operational considerations specify the functions of the gyroscopically stabilized inertial reference platform in the navigational and control aspects of all seven phases. These applications are examined in detail below.

1. Boost to Injection

This is the classical problem of missile boost guidance. The inertial platform is ground aligned and its accelerometer outputs are used to derive cutoff information. Greater detail on this operation may be found in publications devoted to the topic of ballistic missile guidance.

2 and 3. Early and Late Midcourse Flight

Although the platform is not needed for performance of the required navigation functions, it may be employed as an attitude reference and thrust monitor during periods of corrective thrust applications. Consequently, it should be aligned to some predetermined inertial reference in order that the vehicle's thrust axis may be oriented in the vector direction determined by the navigation computations. The accelerometers on the platform would be employed to monitor the applied thrust.

4. Near-Body Flight

During near-body or orbital flight, a local vertical reference can be of value for purposes of navigation to future points by celestial sightings as well as preparation for the initiation of the operations of orbital transfer and landing. Note, though, that if stellar sightings are used for alignment of the platform (instead of a horizon seeker, for instance) it is not meaningful to use such a derived vertical for purposes of celestial navigation.

5. Variation of Orbital Trajectory

Thrust application will have to be referenced to a chosen coordinate frame which can be retained by a stabilized platform.

6. Rendezvous with Target in Orbit

Most rendezvous guidance schemes advanced to date are based on the reduction of target line-of-sight angular rates to zero and then target closure by a simultaneous reduction of range and range rate. The angular rates of the line of sight must be measured against a fixed, but arbitrary, reference. This reference can be the inertial platform. However, the platform need not be aligned to any particular orientation.

7. Landing (with Atmospheric Re-Entry)

Attitude control is of great importance and must, at some point, be based upon the determination of the local vertical. The gyroscopically stabilized platform provides a means for "remembering" the instantaneous local vertical.

Although various mission phases have been discussed above, the detailed examination of a single operation, that of alignment of the platform to a local Earth vertical during free fall, provides sufficient insight into all such related problems that it alone can be discussed in this section without compromising the technical development of this chapter. The following section contains a theoretical description of the problem, realistic evaluation of computational requirements, and a suggested mechanization for a complete system. It is important to remember that the formulation presented is based upon a general consideration of the problem and so is open to modification based upon the special requirements of, or constraints imposed upon, a particular application.

D. Alignment of a Gyroscopically Stabilized Platform during Free Fall

Alignment of a gyroscopically stabilized platform to a local Earth vertical-orbital plane orientation may be accomplished by the use of stellar data, gyrocompassing, and horizon sightings, or combinations of these three. The approach discussed herein, based upon Ref. [23], considers only the use of celestial data and horizon sightings. The question of gyrocompassing must be treated separately.

1. Phases of Operation

Alignment by means of celestial data may be considered in terms both of equipment configuration and phases of operation. The possible equipment configurations are:

(1) Inertial platform, automatic star tracker
(2) Inertial platform, moderate accuracy horizon seeker, automatic star tracker
(3) Inertial platform, high accuracy horizon seeker, automatic star tracker
(4) Inertial platform, moderate accuracy horizon seeker, sextant
(5) Inertial platform, sextant, automatic star tracker
(6) Inertial platform, sextant
(7) Inertial platform, sextant, high accuracy horizon seeker
(8) Inertial platform, sextant, moderate accuracy horizon seeker
(9) Inertial platform, sextant, automatic star tracker, moderate accuracy horizon seeker

Figure 11 has already indicated the schematic combination of all of the above listed components.

The three phases of the alignment operation are:

(a) Coarse alignment
(b) Search and acquisition
(c) Fine alignment

The possible combinations of items (1) through (9) with (a) through (c) are shown in Fig. 14. The choice among the various approaches will be determined to a great extent by the configuration of the spacecraft (possible obscuration of portions of the heavens) and the proper functioning of each component and subsystem. The total composite of instrumentation as shown in Fig. 11 supplies a redundancy which becomes evident from an

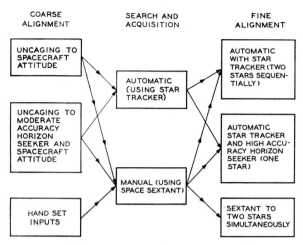

FIG. 14. System moding diagram.

examination of Fig. 14. Thus, depending upon his actual situation, the astronaut will have to resort to one or another of the possible approaches.

2. INITIAL DATA AND ORBITAL PARAMETERS

The geometry of Earth-orbital flight is shown pictorially as Fig. 15. The spacecraft is in a closed elliptical orbit about the Earth. The track of

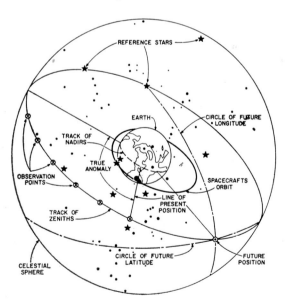

FIG. 15. Geometry of Earth-orbital flight.

the spacecraft's nadir on the Earth's sphere (or, similarly, the intersection of the Earth's sphere and the orbital plane) traces out an inertially fixed great circle which is the projection of the orbit. The track of the spacecraft's zeniths traces out a great circle on the celestial sphere, upon which the reference navigation stars also are located.

Based upon computational considerations it is desirable to store reference star data in terms of the components (in a fixed inertial space) of normalized pointing vectors to the stars. Thus, each of the required reference stars is characterized by the vector (in matrix notation)

$$\mathbf{s}(i) = \begin{pmatrix} s(i)_1 \\ s(i)_2 \\ s(i)_3 \end{pmatrix} \tag{44}$$

For practical purposes the reference space for the storage of stellar data is a North-First Point or Aries oriented frame. Transformation of the $\mathbf{s}(i)$ into the coordinate frame defined by the orbital plane and the point of perigee is accomplished by the means of the matrix $\mathbf{A}$,

$$\mathbf{s}' = \mathbf{A}\mathbf{s} \tag{45}$$

where $\mathbf{s}$ is the column matrix of the pointing vector in reference space, $\mathbf{s}'$ the column matrix in orbital plane space, and $\mathbf{A}$ is the square matrix

$$\mathbf{A} = \begin{pmatrix} A_{11} & A_{12} & A_{13} \\ A_{21} & A_{22} & A_{23} \\ A_{31} & A_{32} & A_{33} \end{pmatrix} \tag{46}$$

These nine components can be transmitted to the vehicle from the ground tracking station complex or computed in the vehicle. In either case, $\mathbf{A}$ is formed from the product of three rotational transformations, ϕ, θ, and $\psi(0)$, in that order. The first, ϕ, corresponds to a rotation about the polar axis which brings the First Point of Aries axis into the plane containing the normal to the orbital plane and the polar axis. The second, θ, represents the angle of inclination between the polar axis and the normal. The third, $\psi(0)$, corresponds to a rotation about the normal to the orbital plane which brings the transformed First Point of Aries axis into the point of perigee. Thus, the matrix $\mathbf{A}$ is given by

$$\mathbf{A} = \begin{pmatrix} 1 & 0 & 0 \\ 0 & \cos\psi(0) & -\sin\psi(0) \\ 0 & \sin\psi(0) & \cos\psi(0) \end{pmatrix} \begin{pmatrix} \cos\theta & -\sin\theta & 0 \\ \sin\theta & \cos\theta & 0 \\ 0 & 0 & 1 \end{pmatrix} \begin{pmatrix} 1 & 0 & 0 \\ 0 & \cos\phi & -\sin\phi \\ 0 & \sin\phi & \cos\phi \end{pmatrix} \tag{47}$$

If, then, the position of the vehicle in the orbital plane is defined by the parameters $\psi(t)$, the orbital angle (true anomaly), and $r(t)$, the length of the radius vector to the vehicle, the transformation of the stellar pointing vector components into the local Earth level-orbital plane space is accomplished by means of the matrix $\psi(t)$

$$\psi(t) = \begin{pmatrix} 1 & 0 & 0 \\ 0 & \cos\psi(t) & -\sin\psi(t) \\ 0 & \sin\psi(t) & \cos\psi(t) \end{pmatrix} \tag{48}$$

Depending upon whether the platform is continuously torqued to local Earth vertical or it is held to a perigee point orientation and the local vertical is obtained analytically, the transformation given by (48) will or will not be applied to the components of $\mathbf{s}'$. In either case $\psi(t)$ must be computed as a function of time.

The computation of $\psi(t)$ can be based upon some set of the constants of the orbit. If the on-board computations are to be performed by a computer functioning in a DDA mode, then the appropriate constants of the orbit can be $r(0)$ (the initial value of the radius vector, $\mathbf{E}$, the total energy of the system, and l, the angular momentum of the system. Then, solution of the following pair of equations will yield the desired function.

$$r(t) = r(0) + \int_0^t \left[\frac{2}{m}\left(E + \frac{k}{r(t)} - \frac{l^2}{2mr^2(t)} \right) \right]^{1/2} dt \tag{49}$$

$$\psi(t) = \psi(0) + l \int_0^t \frac{dt}{mr^2(t)} \tag{50}$$

Instead of the above approach, a concept developed from the technology of special purpose analog computation can be employed. The appropriate

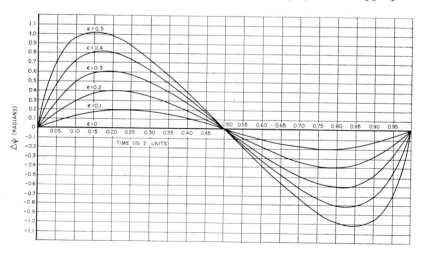

FIG. 16. The true anomaly difference in parametric form.

differential equations can be solved in parametric form for $\psi(t)$ as a function of t, time, ϵ, the eccentricity of the orbit, and τ, the period of the orbit. This functional relationship is shown graphically as Fig. 16. Actually shown is the difference between the true anomaly for a circular orbit and an orbit of eccentricity, ϵ. Note that by using τ normalized time, the family of curves is dependent only upon ϵ and correspondingly independent of orbital size.

3. INITIAL ALIGNMENT OF THE PLATFORM

Initial, or coarse, alignment can be achieved either by aligning the spacecraft with the orbital plane and uncaging to this attitude, or by slaving the platform to a gimballed horizon seeker after the spacecraft has been oriented in azimuth, or by the simple expedient of using handset attitude inputs.

For the second and most complicated alternate, the problem becomes that of developing those error signals for the torquing of the gyros which will result in the platform assuming the desired orientation. The horizon seeker vertical is denoted by the vector $\mathbf{v}$

$$\mathbf{v} = \begin{pmatrix} 0 \\ 0 \\ v \end{pmatrix} \tag{51}$$

If the platform is to be continuously precessed into a local vertical orientation, Eq. (51) will be transformed from horizon seeker space, through vehicle space, into platform space to yield the vector

$$\mathbf{v}'' = \begin{pmatrix} v_1'' \\ v_2'' \\ v_3'' \end{pmatrix} = \begin{pmatrix} e_1'' \\ e_2'' \\ v_3'' \end{pmatrix} \tag{52}$$

If the platform is to be kept fixed in inertial space and the solution for the vertical is to be analytical, then $\mathbf{v}''$ must also be rotated through $-\psi(t)$ to yield the vector components

$$\mathbf{v}' = \begin{pmatrix} v_1' \\ v_2' \\ v_3' \end{pmatrix} = \begin{pmatrix} e_1' \\ e_2' \\ v_3' \end{pmatrix} \tag{53}$$

Initial alignment is accomplished by driving either e_1'' and e_2'' or e_1' and e_2' to zero. Consider the former alternative

ϕ_2 Horizon seeker cross-level axis
ϕ_1 Horizon seeker level axis
ρ_4 Platform outer roll axis
ρ_3 Platform pitch axis
ρ_2 Platform inner roll axis
ρ_1 Platform azimuth axis

If we use the above notation, based upon Fig. 11, the transformation of $\mathbf{v}$ into $\mathbf{v}''$ is given by

$$\mathbf{v}'' = (\boldsymbol{\rho}_2)(\boldsymbol{\rho}_3)(\boldsymbol{\rho}_4)(-\boldsymbol{\phi}_1)(-\boldsymbol{\phi}_2)\mathbf{v} \tag{54}$$

Each transformation corresponding to an Eulerian rotation is represented in matrix notation below

$$(\boldsymbol{\rho}_2) = \begin{pmatrix} \cos \rho_2 & 0 & \sin \rho_2 \\ 0 & 1 & 0 \\ -\sin \rho_2 & 0 & \cos \rho_2 \end{pmatrix} \tag{55}$$

$$(\boldsymbol{\rho}_3) = \begin{pmatrix} 1 & 0 & 0 \\ 0 & \cos \rho_3 & -\sin \rho_3 \\ 0 & \sin \rho_3 & \cos \rho_3 \end{pmatrix} \tag{56}$$

$$(\boldsymbol{\rho}_4) = \begin{pmatrix} \cos \rho_4 & 0 & \sin \rho_4 \\ 0 & 1 & 0 \\ -\sin \rho_4 & 0 & \cos \rho_4 \end{pmatrix} \tag{57}$$

$$(\boldsymbol{\phi}_1) = \begin{pmatrix} 1 & 0 & 0 \\ 0 & \cos \phi_1 & -\sin \phi_1 \\ 0 & \sin \phi_1 & \cos \phi_1 \end{pmatrix} \tag{58}$$

$$(\boldsymbol{\phi}_2) = \begin{pmatrix} \cos \phi_2 & 0 & \sin \phi_2 \\ 0 & 1 & 0 \\ -\sin \phi_2 & 0 & \cos \phi_2 \end{pmatrix} \tag{59}$$

In practice, Eq. (54) need not place an excessive burden upon the computational facilities in the vehicle if resolver transformations are used. (This point is explained in greater detail in the section on mechanization.) The error signals e_1'' and e_2'' furnish the torquing signals for the gyroscopes.

4. Automatic Star Acquisition

Automatic star acquisition is accomplished by directing the celestial tracker by means of pointing commands, initiating search and then target lock-on. The pointing commands are generated in platform space, transformed into vehicle body space, and then used to provide servo control signals for the gimbal drives of the tracker. At this point, one can assume that the analytical approach to the generation of the local vertical is employed. The platform is aligned in the orbital plane to the point of perigee. The instantaneous local vertical is obtained by adding $\psi(t)$ to the platform's pitch angle.

Equation (45) gives the components of the pointing vector to a given star in the reference platform space. Successive application of the transformations $\boldsymbol{\rho}_1$, $\boldsymbol{\rho}_2$, $\boldsymbol{\rho}_3$, and $\boldsymbol{\rho}_4$ bring these components into vehicle (body) space. In matrix notation this operation is represented by

$$\begin{pmatrix} \cos \rho_4 & 0 & \sin \rho_4 \\ 0 & 1 & 0 \\ -\sin \rho_4 & 0 & \cos \rho_4 \end{pmatrix} \begin{pmatrix} 1 & 0 & 0 \\ 0 & \cos \rho_3 & -\sin \rho_3 \\ 0 & \sin \rho_3 & \cos \rho_3 \end{pmatrix} \begin{pmatrix} \cos \rho_2 & 0 & \sin \rho_2 \\ 0 & 1 & 0 \\ -\sin \rho_2 & 0 & \cos \rho_2 \end{pmatrix}$$

$$\begin{pmatrix} \cos \rho_1 & -\sin \rho_1 & 0 \\ \sin \rho_1 & \cos \rho_1 & 0 \\ 0 & 0 & 1 \end{pmatrix} \begin{pmatrix} s_1' \\ s_2' \\ s_3' \end{pmatrix} \quad (60)$$

The remaining step is the solution for μ_1 and μ_2, the level and cross-level gimbal angles of the tracker. The tracker is then directed by these commands and search initiated.

If the tracker were pointing at the reference star, then transformation of this pointing vector into vehicle space by means of the rotations μ_2 and μ_1 would yield the same vector components given by Eq. (60). Thus, in matrix form Eq. (60) is equal to

$$\begin{pmatrix} 1 & 0 & 0 \\ 0 & \cos \mu_1 & -\sin \mu_1 \\ 0 & \sin \mu_1 & \cos \mu_1 \end{pmatrix} \begin{pmatrix} \cos \mu_2 & 0 & \sin \mu_2 \\ 0 & 1 & 0 \\ -\sin \mu_2 & 0 & \cos \mu_2 \end{pmatrix} \begin{pmatrix} 0 \\ 0 \\ 1 \end{pmatrix} \quad (61)$$

The explicit solution for μ_1 and μ_2 is

$$\mu_1 = \arctan -\frac{A}{B} \quad (62)$$

$$\mu_2 = \arctan \frac{C}{D} \quad (63)$$

where

$A = s_1' \sin \rho_1 \cos \rho_3 + s_2' \cos \rho_1 \cos \rho_3 + s_1' \cos \rho_1 \sin \rho_2 \sin \rho_3$
$\quad - s_2' \sin \rho_1 \sin \rho_2 \sin \rho_3 - s_3' \cos \rho_2 \sin \rho_3$

$B = (-\sin \rho_4)[s_1' \cos \rho_1 \cos \rho_2 - s_2' \sin \rho_1 \cos \rho_2 + s_3' \sin \rho_2]$
$\quad + \cos \rho_4 [s_1' \sin \rho_1 \sin \rho_3 + s_2' \cos \rho_1 \sin \rho_3 - s_1' \cos \rho_1 \sin \rho_2 \cos \rho_3$
$\quad + s_2' \sin \rho_1 \sin \rho_2 \cos \rho_3 + s_3' \cos \rho_2 \cos \rho_3]$

$C = \cos \rho_4 [s_1' \cos \rho_1 \cos \rho_2 - s_2' \sin \rho_1 \cos \rho_2 + s_3' \sin \rho_2]$
$\quad + \sin \rho_4 [s_1' \sin \rho_1 \sin \rho_3 + s_2' \cos \rho_1 \sin \rho_3$
$\quad - s_1' \cos \rho_1 \sin \rho_2 \cos \rho_3 + s_2' \sin \rho_1 \sin \rho_2 \cos \rho_3 + s_3' \cos \rho_2 \cos \rho_3]$

$D = (-\sin \mu_1)[s_2' \cos \rho_1 \cos \rho_3 + s_1' \sin \rho_2 \cos \rho_3 - s_2' \sin \rho_1 \sin \rho_2 \sin \rho_3$
$\quad - s_3' \cos \rho_2 \sin \rho_3 + s_1' \cos \rho_1 \sin \rho_2 \sin \rho_3] + \cos \mu_1[-\sin \rho_4 \cdot$
$\quad (s_1' \cos \rho_1 \cos \rho_2 - s_2' \sin \rho_1 \cos \rho_2 + s_3' \sin \rho_2) + \cos \rho_4 \cdot$
$\quad (s_1' \sin \rho_1 \sin \rho_3 + s_2' \cos \rho_1 \sin \rho_3 - s_1' \cos \rho_1 \sin \rho_2 \cos \rho_3$
$\quad + s_2' \sin \rho_1 \sin \rho_2 \cos \rho_3 + s_3' \cos \rho_2 \cos \rho_3)]$

Performance of the above operations by means of digital computation appears to be unsatisfactory in terms of computer loading. Resolver servo instrumentation, on the other hand, can provide the desired results at a much lower cost in terms of both mechanization and complexity.

Once the reference star has been acquired, the tracker error signals are employed for tracker gimbal drive control in place of the initial pointing commands.

5. MANUAL ACQUISITION

Two modes of operation are possible. The first requires the initial locking of the tracker's gimbals in a fixed orientation and then directing the vehicle (by means of fixed crosshairs) towards the chosen reference star. Once acquisition has been effected, the gimbals are unlocked and the tracker error signals employed for servo control.

The second approach requires the use of the space sextant. The pointing vector to the reference star, established along the sextant's polar axis, is transformed into vehicle space and then used to command-orient the gimbals of the tracker. Using the following notation (based upon Fig. 11)

β_1 Space sextant bearing axis
β_2 Space sextant elevation axis
β_3 Space sextant polar axis
β_4 Space sextant, sextant axis

the transformation of the normalized pointing vector $\mathbf{p}$ into vehicle space components is accomplished by

$$\begin{pmatrix} p_1^* \\ p_2^* \\ p_3^* \end{pmatrix} = (-\beta_1)(-\beta_2) \begin{pmatrix} 1 \\ 0 \\ 0 \end{pmatrix} \tag{64}$$

Each of the transformations, corresponding to an Eulerian rotation about an axis of the space sextant is represented in the following matrix form

$$(\beta_1) = \begin{pmatrix} \cos \beta_1 & -\sin \beta_1 & 0 \\ \sin \beta_1 & \cos \beta_1 & 0 \\ 0 & 0 & 1 \end{pmatrix} \tag{65}$$

$$(\beta_2) = \begin{pmatrix} \cos \beta_2 & 0 & \sin \beta_2 \\ 0 & 1 & 0 \\ -\sin \beta_2 & 0 & \cos \beta_2 \end{pmatrix} \tag{66}$$

$$(\beta_3) = \begin{pmatrix} 1 & 0 & 0 \\ 0 & \cos \beta_3 & -\sin \beta_3 \\ 0 & \sin \beta_3 & \cos \beta_3 \end{pmatrix} \tag{67}$$

$$(\beta_4) = \begin{pmatrix} \cos \beta_4 & -\sin \beta_4 & 0 \\ \sin \beta_4 & \cos \beta_4 & 0 \\ 0 & 0 & 1 \end{pmatrix} \tag{68}$$

By equating the vehicle space components of p^* to Eq. (61), one may again solve for μ_1 and μ_2, the gimbal command angles for the automatic star tracker.

6. Platform Fine Alignment

Platform fine alignment may be accomplished in any one of three ways depending, for the most part, upon equipment availability:

(1) Automatic, using star tracker (two stars sequentially)
(2) Automatic, using star tracker and high accuracy horizon seeker (one star)
(3) Manual, using the space sextant sighted on two stars simultaneously

Detailed consideration of method (1) will prove useful in understanding all three methods and so is presented below.

For method (1), fine alignment is accomplished by comparing, for two stars, the observed star line of sight transformed into platform space with the computed star line of sight in orbital plane-perigee, point oriented, inertial space, and precessing the platform until these two vectors are coincident.

Operationally, first one star is employed and alignment, except about the line of sight, is accomplished. Then a second star, preferably about 90 deg from the first, is employed and alignment, except about this second line of sight, is accomplished. Theoretically, by Euler's Theorem,[3] the platform is thus completely aligned, since a pair of bi-axial references completely define the attitude of a rigid body. In practice, however, there is second order cross-axis coupling, for other than small angular displacements, between the gyros. The second correction can negate a part of the first; consequently, it is necessary to return to the first star and repeat the operation. Preliminary study indicates that three corrections, based upon the sequence 1-2-1 should be adequate for alignment to moderate accuracy of a high quality inertial platform. Certainly, the sequence 1-2-1-2 is definitely adequate.

In operation, after the tracker has locked on the given star, the components of the line-of-sight (LOS) vector expressed in vehicle frame coordinates are a function of the tracker gimbal angles μ_1 and μ_2. Transformation of these components into platform space is accomplished through the sequence of rotations ρ_4, ρ_3, ρ_2, and ρ_1. Thus, the components of the LOS in platform space (designated LOS_1', LOS_2', LOS_3') are given by the matrix equation

[3] Euler's Theorem: "The most general displacement of a rigid body with a fixed point is equivalent to a rotation about a line through that point. (Synge, J. L., and Griffith, B. A., "Principles of Mechanics," 2nd ed., p. 279. McGraw-Hill, New York.)

$$\begin{pmatrix} \text{LOS}_1' \\ \text{LOS}_2' \\ \text{LOS}_3' \end{pmatrix} = \begin{pmatrix} \cos\rho_1 & \sin\rho_1 & 0 \\ -\sin\rho_1 & \cos\rho_1 & 0 \\ 0 & 0 & 1 \end{pmatrix} \begin{pmatrix} \cos\rho_2 & 0 & -\sin\rho_2 \\ 0 & 1 & 0 \\ \sin\rho_2 & 0 & \cos\rho_2 \end{pmatrix} \begin{pmatrix} 1 & 0 & 0 \\ 0 & \cos\rho_3 & \sin\rho_3 \\ 0 & -\sin\rho_3 & \cos\rho_3 \end{pmatrix}$$

$$\times \begin{pmatrix} \cos\rho_4 & 0 & -\sin\rho_4 \\ 0 & 1 & 0 \\ \sin\rho_4 & 1 & \cos\rho_4 \end{pmatrix} \begin{pmatrix} 1 & 0 & 0 \\ 0 & \cos\mu_1 & -\sin\mu_1 \\ 0 & \sin\mu_1 & \cos\mu_1 \end{pmatrix} \begin{pmatrix} \cos\mu_2 & 0 & \sin\mu_2 \\ 0 & 1 & 0 \\ -\sin\mu_2 & 0 & \cos\mu_2 \end{pmatrix} \begin{pmatrix} 0 \\ 0 \\ 1 \end{pmatrix}$$

$$(69)$$

(For brevity, these matrices have not been expanded.)

The differences between the components of the vectors given by Eqs. (45) and (69), essentially the misalignment of the platform in the two axes perpendicular to the given line of sight, can be driven to zero by driving the cross-product of the vectors to zero. This approach is very similar to cross-product control of a missile during boost. The components of the vector

$$\mathbf{LOS}' \times \mathbf{s}' \qquad (70)$$

can then be used as error signals to torque the appropriate gyros so as to precess the platform into the desired orientation. These error signals (e_1, e_2, e_3) are

$$\begin{aligned} e_1 &= \text{LOS}_2' \, s_3' - \text{LOS}_3' \, s_2' \\ e_2 &= \text{LOS}_3' \, s_1' - \text{LOS}_1' \, s_3' \\ e_3 &= \text{LOS}_1' \, s_2' - \text{LOS}_2' \, s_1' \end{aligned} \qquad (71)$$

No problem with respect to the noncommutativity of finite angular rotations arises during the correction operation. All three corrections are applied simultaneously, thus resulting in only differential rotations about any one axis at any given time. Even nonlinearities in, or differences between, the three servo loops (which compromise the validity of the above statement) are not really significant because nulling servo loops are employed. Errors in the error signals are therefore unimportant because the error signal goes to zero as the servo loop is driven to null.

The study of the problem of fine alignment can be completed by a brief examination of method (3), the manual use of the space sextant. Unlike method (1), two stars are now employed simultaneously. The sums of the components of the two line-of-sight vectors are now to be compared with their equivalent computed stellar pointing vectors. The reference vector components in platform space are given by the matrix relation

$$\mathbf{A}(\mathbf{s}(1) + \mathbf{s}(2)) \qquad (72)$$

where $\mathbf{s}(1)$ and $\mathbf{s}(2)$ are the pointing vectors, in reference space, to the two chosen stars. The transformed normalized observed line-of-sight vectors in platform space are given by the matrix relation

$$(\boldsymbol{\rho}_1)(\boldsymbol{\rho}_2)(\boldsymbol{\rho}_3)(\boldsymbol{\rho}_4)(-\boldsymbol{\beta}_1)(-\boldsymbol{\beta}_2)(-\boldsymbol{\beta}_3)\left[\begin{pmatrix}1\\0\\0\end{pmatrix}+(-\boldsymbol{\beta}_4)\begin{pmatrix}1\\0\\0\end{pmatrix}\right] \qquad (73)$$

Again, the cross-product of Eqs. (72) and (73) serves as platform alignment error signals.

Method (2), the use of an automatic star tracker and high accuracy horizon seeker, is described by the combination of appropriate aspects of methods (1) and (3).

7. System Mechanization

Throughout this section references have been made to possible mechanizations of the required mathematical relationships. In some instances digital computation has been specified and in others analog techniques have been discussed. Of course, it is possible to consider a complete digital mechanization using code-wheel pick-offs on all rotation axes and sufficient computer capacity and speed to perform all necessary real time operations. On the basis of past experience it is believed, however, that a combination of both techniques, leaning possibly towards the all-analog end, would result in a "best" system. The ensuing discussion is restricted to one such approach (with alternate subsystems where appropriate) which appears to be practical and capable of meeting typical accuracy and operational requirements.

The use of resolver chains and resolver servos for most of the required coordinate transformations (all those which must be performed in real time) and of a digital computer for reference storage, arithmetic operations (which can be performed slowly), and the solution of differential equations forms the basis for the recommended instrumentation. It is believed that in this manner minimum (realistic) loading of the central digital computer will be achieved without degrading the accuracy or reliability of the overall system. Clearly, operations of the form indicated by Eqs. (62) or (63) are not practical for a digital computer and yet are simple to effect by means of resolver configurations. An interesting point to note, with respect to the accuracy of resolver chains, is that even with a large number of components (six or seven resolvers) final accuracy is usually not worse than one and a half times that of an individual resolver. This result is derived from the antistatistical error characteristics of a resolver and the ability to "spot-in" and compensate, one against another, each of the resolvers of the total chain. Thus, a chain with resolvers accurate to 20 arc sec can yield an overall accuracy within 30 arc sec.

Refer to Fig. 17 for the functional diagram of the platform alignment system in the initial automatic alignment mode. Assume that the platform

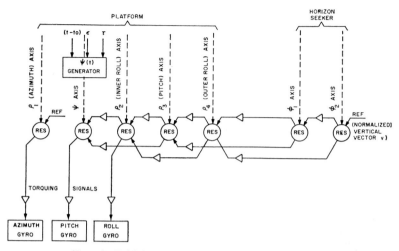

FIG. 17. Initial automatic platform alignment mode.

is to be oriented to the inertially fixed, orbital plane-perigee point coordinate space. Then Eq. (54) is not applicable and the following mechanization equation must be employed

$$\mathbf{v}' = (-\boldsymbol{\psi})(\boldsymbol{\rho}_2)(\boldsymbol{\rho}_3)(\boldsymbol{\rho}_4)(-\boldsymbol{\phi}_1)(-\boldsymbol{\phi}_2)\mathbf{v} \tag{74}$$

The reference voltage input into the ϕ_2 axis resolver represents the normalized vertical vector $\mathbf{v}$. The appropriate output signals from the ρ_2 and ψ axes resolvers are the error signals which are used to torque the gyros in order to precess the platform into the proper orientation. Azimuth alignment is accomplished by the direct orientation of the spacecraft so that the azimuth angle ρ_1 is zero. The angle $\psi(t)$ must be computed continuously. However, since analog techniques would prove quite successful here, it is realistic to state that no digital computation is required at any time during this process. Only if $\psi(t)$ were to be obtained directly from the classical differential equations would a digital computer be required. The resolvers and their isolation amplifiers serve in the other modes of the alignment operation as well, and so represent considerably less equipment in the total system than might at first be supposed.

Refer to Fig. 18 for a functional diagram of the automatic star acquisition mode. Star data storage and the $\mathbf{A}$ transformation are best mechanized by digital techniques. The A_{ij} are either transmitted from the ground or computed on-board on the basis of the elements of the orbit. The stellar pointing commands, s'_i must first be converted into analog form before serving as inputs to the resolver chain. Solution of Eqs. (62) and (63), the tracker command angles, in real time would require a digital computer of

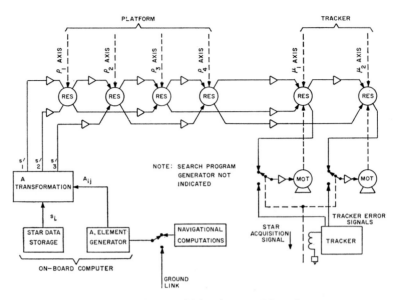

Fɪɢ. 18. Star acquisition (automatic) mode.

great capacity. Yet, the resolver chain terminated in a pair of resolver servos (Fig. 18) provides a simple, fast means of accomplishing the same objective. Once acquisition of the reference star takes place, the star acquisition relay is energized and the tracker error signals are then used for stabilization of the tracker. At this time fine alignment is initiated.

If the space sextant is employed for manual acquisition, the problem of mechanization is greatly simplified. As indicated by Eq. (64), the β_2 (elevation axis) resolver is energized by a reference voltage corresponding to the line-of-sight vector. The outputs are transformed by the β_1 resolver into vehicle space. These signals then serve as inputs to the tracker resolver system, as shown in Fig. 18.

Three methods exist for producing fine alignment of the platform. However, it is believed that a discussion of the mechanization of method (1) by itself, combined with the previous material, should produce satisfactory coverage of the entire topic. Figure 19 provides a functional diagram of the system based upon the use of tracker data for fine alignment. The series of transformations described by Eq. (69) are also achieved by means of the resolver chain. In Fig. 19 the cross-product multiplications are shown as analog operations. Here, digital techniques may be equally applicable. For digital mechanization the components of the LOS in platform space must be converted to digital form. Then the multiplications and subtrac-

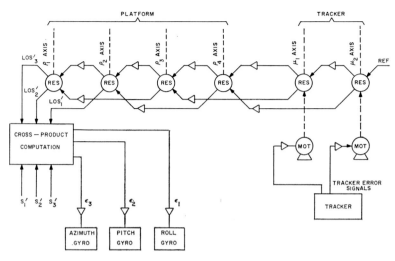

FIG. 19. Platform fine alignment by use of tracker data.

tions would be performed in the digital computer and the gyros correspondingly pulse-torqued.

E. The Celestially Locked, Astro-inertial Measurement System Concept

An interesting combination of inertial sensors, celestial trackers, and space sextant is afforded by the employment of an integrated "two-headed" astro-inertial attitude reference platform in place of conventional instrumentation. Such a unit can be aligned without the aid of computation of any sort since it exhibits a one-to-one gimbal relationship with the space sextant.

Fundamentally, the astro-inertial system described herein, and shown as Fig. 20, is realized in its optimum configuration because of the functional identity between the celestial tracker and the two-degree-of-freedom gyroscope (or two single-degree-of-freedom units). Both are two axis stabilizers and, therefore, can be mounted in a functionally symmetrical arrangement, i.e., with the tracker optical axes and the gyro spin axes collinear.

The operational requirements for the astro-inertial subsystem may be categorized according to the particular phase of flight during which it is employed. It is also necessary to consider the transition of control function between particular modes. The resulting transition classifications are: (1) manual to automatic, and (2) automatic to manual.

Upon launch, the platform is gyroscopically stabilized until star acquisition. One feasible stellar stabilization scheme is shown in Fig. 21. Upon star acquisition, the tracker-derived error signals are employed to torque the stabilization gyros. By this method, the frequency response of the platform may be maintained sufficiently high for proper stabilization, while the frequency response of the trackers can be limited to approximately 2 to 3 cps. A low frequency response for the trackers results in favorable signal-to-noise characteristic detection capabilities.

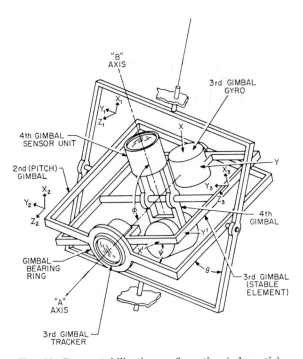

FIG. 20. Sensor stabilization configuration (schematic).

Secondary tracking and general optical equipment may be either initially aligned or slaved to the astro-inertial system by means of resolver readouts mounted on the gimbal axes. Where necessary, appropriate coordinate transformations can be achieved by a resolver chain.

Comparison of the space sextant with the platform of Fig. 20 makes obvious the one-to-one correspondence existing between the axes of the two units. Thus, the alignment of the platform to a simultaneous pair of stellar sightings is particularly simple. The interconnection between the two units is depicted schematically in Fig. 22.

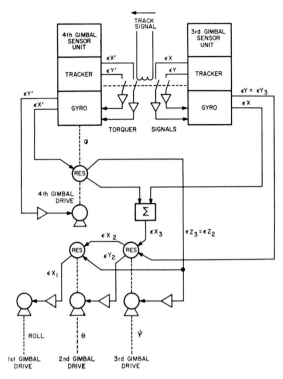

FIG. 21. In-flight stabilization scheme.

VII. Summary

The design of instrumentation for space navigation has been shown to be contingent upon the meaningful definition of the mission, proper choice of navigation or guidance concept, and optimum integration of the human operator. The possible space missions were considered and categorized in terms of the characteristics of the observables, rather than by any other method of classification. It was also indicated that the mission dictates the choice between implicit or explicit concepts of navigation. The capabilities and limitations of the astronaut were considered in the specification of design constraints for all equipment used for viewing and all equipment requiring manual manipulation. The resulting equipment configurations were related in terms of the *spectrum of navigation instrumentation*. This spectrum covered the range from completely manual to fully automatic operation. Specific equipment configurations and their operation were described.

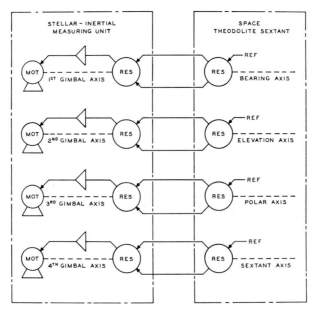

Fig. 22. Alignment of astro-inertial measuring unit to space sextant.

Acknowledgment

The authors wish to acknowledge their indebtedness to the many people at the Kollman Instrument Corporation engaged in the programs employed as background for this chapter. In particular, the equipment design contributions of Messrs. J. Meader, L. E. Sharpe, and T. Stearns, the data on human vision prepared by Dr. I. Finkelstein, and the systems contributions of Messrs. P. Reichmeider, M. Schechter, and M. Steinhacker are gratefully acknowledged.

References

1. Muckler, R. A., Hookway, S. and Burke, J. (1962). Manned control of large space boosters. *7th Symp. on Ballistic Missile and Space Technol* (Los Angeles, Calif.).
2. Results of the Second United States Manned Orbital Space Flight, May 24, 1962. *Natl. Aeron. Space Admin. Rept. SP-6* (Washington, D.C.).
3. Satyendra, K. N., and Bradford, R. E. (July, 1961). Self-contained navigational system for determination of orbital elements of a satellite. *ARS J.* Vol. 31 No. 7.
4. Goldstein, H. (1959). "Classical Mechanics," Chap. 3. Addison-Wesley, Reading, Massachusetts.
5. A recoverable interplanetary space probe. (1959). *Mass. Inst. Technol. Instrumentation Lab. Rept. R-235* (Cambridge, Mass.).
6. Frielander, A. L. and Harry, O. P. (1961). A study of statistical data adjustment and logic techniques as applied to the interplanetary midcourse guidance problem. *Natl. Aeron. Space Admin. Rept. TR-R-113* (Washington, D.C.).

7. Battin, R. H. (1961). A statistical optimizing navigation procedure for space flight. *Mass. Inst. Technol. Instrumentation Lab. Rept. R-341* (Cambridge, Mass.).

8. Analytical and preliminary simulation study of a pilot's ability to control the terminal phase of a rendezvous with simple optical devices and a timer. *Natl. Aeron. Space Admin. Tech. Note D-965* (Washington, D.C.).

9. Woodward, D. P. *et al.* (1961). General Human Factors Considerations. *Military Test Space Station Final Rept.* **3,** 173–186. The Martin Co. (Baltimore, Maryland) (*ASTIA Rept. AD 273005-L.*)

10. Human factors and remote handling in advanced systems. (1961). *Wright-Patterson Air Force Base, Aeron. Systems Div. Rept. TR 61-430* (Dayton, Ohio).

11. Haygood, R. C., and Overton, R. K. (1961). Operator accuracy in angle measurement and transfer. *Human Factors* **3,** No. 1.

12. Gossett, F. R. Manual of geodetic triangulation. *U.S. Dept. of Commerce, U.S. Coast and Geodetic Survey Special Publ. No. 247* (Washington, D.C.).

13. Adams, J. A. (1961). Human tracking behavior. *Psychol. Bull.* **58,** No. 1 (*ASTIA Rept. AD 254757.*)

14. A study of tracking on directors M5A2 and M5A2E1. (1945). *Office Sci. Res. and Development Rept. No. 5929* (Washington, D.C.).

15. Vernon, M. B. (1946). Scale and dial reading. *Med. Res. Council Unit in Appl. Psychol., Univ. of Cambridge Rept. No. APU 49* (Cambridge, England).

16. Chapanis, A. and Leyzurek, S. (1950). Accuracy of visual interpolation between scale markers as a function of the number assigned to the scale interval. *J. Exptl. Psychol.* **40,** No. 5.

17. Peterson, G. M., Evans, J. L., and Weldon, R. J. (1956). Accuracy in calculating arithmetical problems of varying complexity. *Sandia Corp. Eng. Res. Rept. SC-3896 (TR)* (Albuquerque, New Mexico).

18. Weldon, R. J., and Williams, H. L. (1956). Error reduction in computations. *Sandia Corp. Eng. Res. Rept. SC-4074 (TR)* (Albuquerque, New Mexico).

19. Relative effectiveness of presenting numerical data by the use of tables and graphs. (1946). *Wright-Patterson Air Force Base Eng. Mem. Rept. TSEAA-694-1* (Dayton, Ohio).

20. Carter, L. F. (1946). A study of the best design of tables and graphs used for presenting numerical data. *Wright-Patterson Air Force Base Eng. Mem. Rept. TSEAA-694-1C* (Dayton, Ohio).

21. Connel, S. C. (1947). The relative effectiveness of presenting numerical data by the use of scales and graphs. *Wright-Patterson Air Force Base Eng. Mem. Rept. TSEAA-694-1M* (Dayton, Ohio).

22. Type and format for sight reduction tables. (1950). *Naval Air Exptl. Station Topical Rept. XG-T-139* (Philadelphia).

23. Moskowitz, S. (1963). Stellar alignment of a gyroscopically stabilized platform during free-fall. *Am. Inst. Aeron. Astronautics Guidance and Control Conf.* (August) (Cambridge, Mass.).

Space-Related Technology: Its Commercial Use

ROBERT H. WATERMAN, JR. AND LLOYD G. MARTS

Industrial Economics Division
Denver Research Institute
University of Denver
Denver, Colorado

Many anticipate that the United States space program will provide—in addition to its primary purpose of space exploration—processes, products, and materials which will be applicable for a wide range of commercial uses.

In late 1961 the University of Denver Research Institute undertook a study to provide factual information concerning the commercial application of space technology. This study was financed by a grant from the National Aeronautics and Space Administration. The primary objectives of the study were to identify and document examples of by-products of the nation's space and missile programs and to determine the circumstances surrounding the origin of these by-products. The following questions were also investigated: (1) What future commercial benefits of space research and development (R & D) can be foreseen? (2) What is the nature of the process by which the results of space R & D are applied in the commercial economy? (3) What stimulates and what inhibits this process of application?

For the purposes of this study, the missile programs were included with the space program because they are similar in nature and thus should have similar by-products; and because it did not appear feasible to distinguish between the two types of programs.

Over 1300 persons representing industrial firms, government agencies, research organizations, universities, trade magazines, and trade organizations were contacted either personally or by questionnaire to conduct the research and were questioned about their own experience regarding the commercial benefits of space research. This chapter summarizes the findings[1].[1]

In the following material the term "missile/space" is used to encompass both the missile and space programs. Since there is no precise separation of the aircraft industry from the missile and space industries, we were rather arbitrary in deciding which programs qualified for the study and which did not. The X-15, for example, has many characteristics of an aircraft but is rocket-powered; it was included. Many missiles, such as the Navaho, are aerodynamic rather than ballistic, yet they are missiles; these were

    [1] Credit is also due John G. Welles, John S. Gilmore, and Robert Venuti, all of Denver Research Institute, who participated in the research project and co-authored the final report with the authors of this chapter.

included also. Something obviously an aircraft, such as the B-70, was excluded.

To qualify for the study, an example must have received some impetus —not necessarily in the form of direct funding—from the missile/space programs. Thus, a tangible commercial benefit evolving from a company-sponsored, yet space-motivated, R & D program would be included, just as would any commercial benefit deriving from a National Aeronautics and Space Administration or Department-of-Defense-sponsored space or missile project.

This chapter is entitled the "Space-Related Technology: Its Commercial Use." The word "commercial" is used throughout to identify the private sector of the economy embracing both industrial and consumer services. In some cases, the results of missile/space R & D had carry-over, not to any commercial sector of the economy, but to the government sector for some non-missile/space use. A few such examples were included; emphasis, however, was on commercial application.

The chapter is divided into four sections and a conclusion. Section I is a general analysis of the kinds and forms of commercial missile/space contributions that were found. The term "by-product" has been used extensively in discussing the commercial contribution of the missile/space programs, and can and has led to much confusion. It is discussed in this section.

Section II contains many of the individual examples which were identified during the study, and can be skipped if the reader wants only a summary of the study's results. Section III is a discussion of the problems inherent in identifying commercial contributions and in attempting to measure their economic impact. Section IV concludes the chapter with a discussion of the many factors which inhibit or stimulate the process of transfer of missile/space R & D results to the commercial economy.

I. Technological Transfer

Originally, we had expected to find a rather limited number of clear-cut by-products of missile/space spending; that is, products, processes, or materials which had resulted directly from missile/space work and found commercial application in substantially their original form. The term "by-product" used in this way was not new. It had been employed frequently in articles and speeches in referring to any secondary commercial contribution of government-motivated R & D.

We found, however, after using the term ourselves, that in spite of its convenience, it does not convey the scope of the missile/space contribution to the commercial economy. The term is too narrow. In the first place, "by-product" implies that the contribution comes in the form of direct, readily identifiable results of a particular effort, when in fact most contributions do not. Second, it implies a certain unexpectedness, unimportance, and lack of strong motivation for exploitation. To the contrary, some instances of contributions from the missile/space effort are expected, may prove to be of more significance than their progenitor, and their possible commercial market is as strong a motivation for development as their more apparent missile/space market.

The total contribution of the missile/space R & D to the commercial economy is broader, more complex, more indirect, and more difficult to identify than is generally imagined. The examples of contributions that may be truly called *by-products* are but a small fraction of the whole. Because of its scope and complexity, the total contribution is more significant than is generally realized, although this significance can not be measured— not in dollars, anyway.

For purposes of analysis, the examples of contributions identified during the study were grouped into six major categories: (1) stimulation of basic and applied research; (2) development or improvement of processes and techniques; (3) improvement of existing products; (4) increased availability of materials, testing equipment, and laboratory equipment; (5) development of new products; and (6) cost reduction. These categories are not mutually exclusive; one example of contribution commonly embodies several of the types of contribution mentioned.

A. Stimulation of Basic and Applied Research

Basic and applied research are stimulated in either of two ways: (1) directly through government support, or (2) indirectly through company funding motivated by missile/space needs and therefore by a potential missile/space market. Once the research is undertaken or completed, new scientific knowledge and technology become available for a broader use than that which motivated the original funding.

This type of contribution is the most significant of the six; it is also the most difficult to identify, being several steps removed from ultimate commercial application. Although it is extremely difficult to relate knowledge generated on a missile or space project to an ultimate commercial product, the product is apt to be the result of a cumulative store of knowledge, coming from many sources, of which the missile/space source is one.

The existence and apparent importance of this type of contribution, stimulation of research, leads us to the conclusion that *transfer of technology*, not by-product transfer, is the most significant missile/space contribution to the commercial economy. Therefore, in what follows we shall refer to the commercial contribution from missile/space R & D as either *contribution* or *technological transfer*, these terms being more accurate and useful than *by-product*.

Good examples of the type of contribution we have called stimulation of basic and applied research, discussed in the next section, include microelectronics, semiconductors, thermionic and thermoelectric energy conversion, magnetohydrodynamics, the plasma jet, and inertial guidance.

B. New or Improved Processes and Techniques

A second major type of contribution is in the area of new or improved processes and techniques, such as new ways of fabricating a material, forming a part, or scheduling a job. This type of contribution is different in degree rather than kind from stimulation of basic and applied research. The ultimate contribution in either case is new technology. Most of the items included in this category represent fairly direct transfers. Good examples include high energy forming, filament winding, solid state bonding, chemical milling, medical technology, and PERT.

C. Product Improvement

A third category of contribution is product improvement. It can take two forms: (1) a product developed originally for commercial use is improved as a result of being adapted to the stringent requirements of the missile/space program, and is sold in its improved form to its original commercial market, or (2) a commercial product benefits from new manufacturing, process control, or quality control techniques originally developed for the missile/space program.

Product improvement can also be considered primarily a form of technological transfer. The two most important effects of the missile/space program within the product improvement category seem to be improved reliability and improved quality. Reliability and quality are of particular importance to missile/space systems, wherein the inherent unreliability of any component is compounded by the unusual complexity of the total system. Failures of such systems can cause great damage as well as the loss of millions of dollars. Commercial systems, particularly in the field of process control, can also be complex and therefore dependent on the reliability and quality of component parts. Such parallel requirements for reliability and quality in the missile/space field and in the commercial field motivate technological transfer.

Good examples of the type of contribution we have called product improvement are contained in the Sections IIA and IIB (Instrumentation and Electronic Components). A particularly illustrative section for demonstrating the importance of quality and reliability is the one covering cables, connectors, and printed circuits.

D. Materials and Equipment Availability

In some fields the demands of missile/space programs have not produced significant improvements in basic technology; rather, the primary result seems to be the increased availability of materials, testing equipment, and laboratory equipment. Availability is brought about by extensive missile/space use. The effect is to make such fields more useful commercially. The field of cryogenics is an excellent example. Other examples will be found in the Section II,A (Instrumentation).

E. New Products

A number of products have been developed for missile/space use which later found commercial application in substantially their original form. This type of transfer probably represents most closely the usual concept of the term *by-product*.

Movements of products from the missile/space programs to commercial applications are direct and relatively clear types of transfer. Their significance individually appears to be rather small and, in total, this category seems less significant than any of the four discussed above. Part of the explanation for this is that a product, by definition, has a limited range of application, whereas the range of application of technology can be very broad.

F. Cost Reduction

The final type of contribution is cost reduction, which can take place in two ways. First, volume production for the missile/space effort can so reduce the unit cost of a product that commercial marketing is feasible. Good examples of this are the solar cell and the infrared detector, both discussed in the following section. Second, a number of companies report that they are able to staff and maintain well-equipped R & D laboratories partly because of missile/space contract research. The availability and contributions of these R & D capabilities to the commercial endeavors of these companies can be partially credited to missile/space programs and can reduce the unit cost of commercial products.

Table I presents our evaluation of the type and degree of contribution which the missile/space programs have made to various technologies. It summarizes much of the information given in the next section.

II. Examples of Technological Transfer

A. Instrumentation

1. RESISTANCE STRAIN GAGE

The modern resistance strain gage has been in use in various forms since the 1930's. The missile/space program has forced improvement, however, in many types of gages, transducers based on strain gages, bonding compounds, and readout instruments. Thus, the effect of the missile/space program has been product improvement and a consequent expansion in the sphere of application of the strain gage.

As an example, Baldwin-Lima-Hamilton's strain gages are used in missile/space work to determine stresses at exceedingly high temperatures on nose cones; to determine stresses in rocket casings, before, during, and after firing; and to determine safe stress levels in many other missile/space applications. Its gages are also used in transducers to measure force, pressure, and torque. Its force transducers are used to measure missile weight during fueling, to measure missile thrust during static firing, and to locate missile centers of gravity and main thrust vectors. Its pressure transducers are used to measure rocket chamber pressures, fuel line pressures, and other pressures vital to missile operation.

In order to keep abreast of missile/space demands, new and more precise strain gages and transducers are continually under development. As these new products become available, some are introduced into the company's commercial line. For example, the company now offers more accurate, stable, and repeatable force transducers with capacities from 10 lb to 6,000,000 lb which find widespread use in commercial weighing.

Many strain gage products manufactured by this company and used in missile/space work are improvements of products originally developed solely for commercial use. In most cases, the original commercial application benefits by this process of product improvement.

A weldable strain gage developed at Microdot, Inc., is an example of missile/space-induced improvement in strain gage bonding techniques. Using glue or epoxy as a bonding medium proved inadequate for high Mach number aircraft and missiles. As a solution, Microdot developed a strain gage which could be welded to the surface to be tested. This gage measures force, pressure, and acceleration at temperatures ranging to 750°F

TABLE I

TABULATION BY TYPE AND DEGREE OF IDENTIFIED MISSILE/SPACE CONTRIBUTION (1)

AREA OF TECHNOLOGY	DOMINANT TYPE(S) OF IDENTIFIED CONTRIBUTION						APPARENT DEGREE OF CONTRIBUTION			ANY PRESENT NON-SPACE COMMERCIAL CONTRIBUTION IDENTIFIED?		
	Stimulation of Basic and Applied Research	Development of New Processes and Techniques	Improvement of Existing Products	Increased Materials and Equipment Availability	Development of New Products	Cost Reduction	Strong	Moderate	Slight	Yes	A Small Amount	Commercial Contribution All Potential
Instrumentation												
Resistance Strain Gages	X		X	X	X			X		X		
Infrared Instrumentation	X			X		X		X		X		
Pressure Measuring Equipment			X	X	X			X				X
Temperature Measuring Equipment			X	X	X			X				X
Instrumentation Amplifiers									X			
Electronic Components and Miscellaneous Systems												
Semiconductors	X		X		X		X	X		X		
Microsystems Electronics	X								X			
Thermoelectric Refrigeration			X			X						
Connectors, Cables, and Printed Circuits			X			X		X		X	X	
Display Systems					X	X		X		X	X	
Control Systems												
Inertial Guidance	X						X			X		
Electronic Computer Systems	X		X		X			X		X		

Power Sources
  Solar Cells
  Thermionic and Thermoelectric Energy Conversion
  Fuel Cells
  Magnetohydrodynamics

Propulsion
  Cryogenics
  Fluid Transfer Systems

Fabrication
  Filament Winding
  Chemical Milling
  High Energy Forming
  Solid State Bonding

Materials
  Refractory Metals
  Maraging Steels
  Physical Metallurgy
  Superalloys
  Epoxy Resins

Medical Technology

Re-Entry Simulation - The Plasma Jet

Telemetry and Communications

Vibrational Testing

Packaging and Shipping

Management and Control - PERT

Notes: (1) Based on company statements and technological background summaries
       (2) Not applicable.

(static) and 1500°F (dynamic). Nonspace applications of the gage include measurement of piling strength on the St. Lawrence Seaway, measurement of atomic submarine hull strength, measurement of internal pressure of tanks of corrosive chemicals, and measurements inside reactor cores at Los Alamos.

There are many other examples of transfer in the resistance strain gage field: (1) A miniature displacement vibrometer uses a strain gage as the sensitive element developed by the Boeing Company for wind tunnel testing in missile/space and aircraft work. This device has commercial potential in structural testing, as in the testing of turbine engines or automobiles. (2) A palladium-silver alloy has been developed by Lockheed Electronics under a contract to find materials suitable for accurate strain measurement at very high temperatures, e.g., strain gages employing palladium on reentry vehicles. This alloy was found to have the lowest temperature coefficient of electrical resistance of any known material over extreme temperature ranges. The material appears to have commercial potential in scientific instrumentation. (3) Piezoresistivity was discovered in silicon carbide by Lockheed Electronics under the same contract. This discovery has potential commercial application, possibly as the heart of a microphone or underwater transducer. (4) An easy-to-operate indicating millivolt potentiometer was designed for use with strain gages. The device, developed by NASA's Ames Research Center, has an internal power supply, an internal temperature compensation, high sensitivity, and a self-balancing servo amplifying system, which are features not normally available on other potentiometers. (5) Strain gage input conditioning equipment and strain plotters have been developed at B & F Instruments, Inc.; both have been used commercially.

2. INFRARED INSTRUMENTATION

Some new products developed as a joint result of missile/space contribution to infrared technology and reduced cost of infrared devices made possible by mass production for missile/space markets are: (1) a telephone-like communication system based on infrared transmission principles (the instrument was first marketed to the transportation and construction industries and was later marketed as a toy); (2) a toy infrared rifle and range (the rifle is an infrared emitter, the target a detector); (3) a control device used for door automation, machine control, area surveillance, conveyor control, materials handling, and liquid level control; (4) a traffic detector.

Infrared Industries, the developing company, states that each of these products evolved from the company's participation in missile programs.

Although the theory involved in infrared spectroscopy has been understood for some time, the widespread use of this technology has a fairly recent history. This is due both to the recognition of this branch of science as a powerful analytical tool and to the usefulness of infrared in weapons, space, and military surveillance. The powerful impetus given infrared research by World War II has been continued by the missile/space program. Thus, missile work can be credited partially with the R & D instrumental in producing sophisticated infrared equipment. Of more significance, however, is the quality and quantity of instrumentation required for missile/space systems, necessitating more refined and rapid production techniques.

3. Pressure-Measuring Equipment

Missile/space requirements have greatly accelerated the development of improved pressure transducers and associated equipment. The pressure range and frequency response characteristics of existing equipment were inadequate for missile/space use. Furthermore, much improvement was needed to insure reliable pressure measurement under severe environmental conditions, such as high and low extremes of temperature, corrosive fluids and atmospheres, high levels of shock and vibration, and severe pressure overloads. Additional requirements were that they function properly under space vacuum conditions, provide relatively high-level output for telemetry systems, be as reliable as possible, and be of minimum weight and volume.

Though some improvement could be expected by further refinement and modification of existing designs, the total improvement needed, it was apparent, would have to come from some other approach.

A breakthrough, the DADEE (Dynamic Analog Differential Equation Equalizer), was reported early in 1958 by Dr. F. F. Liu and T. W. Berwin of Rocketdyne[2]. This system utilizes operational amplifiers in a configuration determined by the differential equation describing the dynamic response of the physical transducer. By this means, ringing is substantially eliminated and pressure transients having rise times on the order of one microsecond have been recorded.

The device was developed in the course of Rocketdyne's missile and space work. That company licensed this innovation to the Data Instruments Division of Telecomputing Corporation in about 1958. Telecomputing is marketing the device under the name Tranqualizer, a contraction for transducer-equalizer. Initial marketing revealed a limited demand for the product in its original form and cost. A product redevelopment program was initiated by Telecomputing and completed in the summer of 1962, resulting in several changes for the Tranqualizer. Though the sales volume of the Tranqualizer has been small to date and confined entirely to the aerospace industry, commercial use of this device is felt to be very probable.

Missile/space research requirements have also stimulated development of better methods for measuring low pressures, as in altitude measurement or wind tunnel instrumentation. In the latter field, the emphasis is on devices adaptable to automation of data-gathering systems. For example, the Ames Research Center of NASA has developed a slack diaphragm differential pressure transducer to replace the liquid monometer pressure measurement systems widely used in wind tunnel research. This device, in contrast with the liquid monometer, is compatible with high-speed data recording and electronic data reduction and its accuracy is not dependent upon considerable attention and maintenance. Potential commercial applications of this transducer include remote reading barometers, altimeters, air speed indicators, mock meters, pneumatic gaging and positioning, and industrial control.

Ames Research Center has also developed a vibrating diaphragm pressure transducer. The principal attributes of this transducer are its wide measuring range, high sensitivity, fast time response, good temperature characteristics, small size, ruggedness, ability to withstand over-pressure without damage, and its compatibility with most electronic readout and recording equipment.

Physically, the transducer may be described as a small closed cylinder partitioned into chambers by a thin metallic diaphragm. The chambers, of equal volume and connected to a common gas inlet, subject the diaphragm to equal pressure at all times. The diaphragm is driven electrostatically with a capacitance bridge, and the amount of power necessary to drive the diaphragm at constant amplitude, being directly related to pressure, is measured to determine pressure.

The transducer shows commercial promise as a pressure gage, or as a sensing element in an altimeter to measure altitudes as high as 400,000 ft. By making suitable changes in techniques and materials, it could also be used to measure magnetic damping.

Thus there has been considerable contribution to the knowledge of the whole field of pressure measurement from the development of pressure transducers and related equipment for missile/space requirements. In a very real sense, in view of prespace and present day sales volume, the requirements of the missile/space program may be said to have created a large part of the present pressure transducer industry.

4. TEMPERATURE-MEASURING EQUIPMENT

In a complex missile or space vehicle, many components, units, and systems must be monitored and controlled to keep the temperature environment within safe operational limits. Temperatures vary from the lows of cryogenics and the upper atmosphere to the highs of reentry and the combustion chamber. The safety of occupants and the proper functioning of equipment depend unequivocally upon measurement and control of temperature.

Devices which have found the majority of missile/space application are the two common types which provide electrical output—thermocouples and resistance temperature sensors. Thermocouples find application in moderately high to extremely high temperature environments; resistance sensors in moderately high to extremely low temperature environments. Specialized optical and radiation techniques are used for temperatures above the thermocouple range. Much of the incentive for development in the high temperature field has come from the metals industry, with some impetus from missile/space programs. Measurement in the moderately high to extremely low temperature range has received the most noticeable impetus from missile/space research. A probable reason is that low temperature measurement had fewer factors contributing to its development prior to missile/space research than high temperature measurement.

The platinum element resistance thermometer has been a standard laboratory device for many years, but has been miniaturized and made rugged and reliable to meet missile/space requirements. An example is a temperature transducer with platinum sensors developed by Microdot in 1959. Its original use was in rocket motor testing at the Sacramento plant of Aerojet General Corporation. Its special characteristics include small size, high base resistance, and extreme ruggedness, all of which are achieved through a sensing element of platinum film deposited on a miniature ceramic disc. The transducer has a fast response, excellent operational stability, and extreme linearity over a wide range of temperatures.

An interesting use of the transducer outside the missile/space industry is at the Sharp Memorial Hospital, California, where the device is implanted in tumors to assist in the determination of the effect of strong temperature changes on the tumor.

The Boeing Company, as a result of missile temperature measurement research, has developed a thermal parameter indicator which will measure from a distance temperatures that exceed the melting points of known thermocouples. The device will also measure the heating rate of any source, radiation or convective, and determine the heat transfer coefficient. Commercial applications of this device, although limited, are expected where temperatures over 2000°F exist, as in furnace measurement or plasma arc temperature measurement.

Precise measurement in missile/space activity has stimulated the need for measurement standards to the point that it is economically feasible for some companies to produce standards on a commercial basis, and laboratories can now purchase certain standards rather than manufacture their own. An example is the equiphase-triple-point-of-water cell developed at Trans-Sonics, Inc. The cell generates the stable temperature established by the triple point of water, at which the solid, liquid, and vapor phases of water exist in equilibrium. It is an improvement over the more commonly used ice and water bath. The primary commercial use of this cell is in standards laboratories.

Another example, developed by Genisco, Inc., is a thermal reference junction. This is a miniaturized multichannel thermocouple reference junction which can replace the fragile vacuum bottle ice bath. Commercial users of the thermal reference junction include the petrochemical industry, the food processing industry, remote weather stations, heat treating plants, and the biomedical field.

5. INSTRUMENTATION AMPLIFIERS

Going hand in glove with the accelerated development of various types of transducers has been the improvement of amplifiers to strengthen the signals generated by these transducers. Serious problems in the missile/space program include the development of reliable telemetering systems to relay data from space vehicles to command stations and the development of remote automatic controls activated by transducers. Inherent in these problems, of course, is the development of instrumentation amplifiers. Much of the space support has been directed toward reducing the size, weight, and power requirements of such amplifiers and toward making existing amplifiers more rugged by the application of recent advances in solid state electronic technology.

One particular problem in space systems is to transmit information from many transducers on a single or few radio transmission channels, for it is entirely impractical, due to weight considerations, to use a separate radio transmission channel for each transducer. The problem is solved by using an electronic or mechanical commutator which periodically scans transducer outputs rapidly so that the transmission of all channels can be considered simultaneous. Thus, the newly developed instrumentation amplifiers must have wider band width than those existent in the past in order for them to be able to respond to the rapid functioning of the commutator[3].

It seems that the most significant contribution of the missile/space program in this area is equipment availability. The instrumentation amplifier is a ubiquitous piece of industrial laboratory and testing equipment. It finds use in almost all scientific fields and branches of engineering. Therefore, it is reasonable to expect that many of the instrumentation amplifiers developed for missile/space needs will find commercial application. Several examples of this were uncovered in the study and more are expected in the future.

6. BLACKBODY INSTRUMENTATION

As part of the materials selection and testing program at NASA's Ames Research Center, it became necessary to measure the change in emissivity of various surfaces exposed to actual space flight conditions over long periods of time. These measurements were to be made in space by comparing the equilibrium temperatures of isolated test surfaces with the temperature of a blackbody reference surface having stable characteristics under space environment.

No compact blackbody was available, but a very satisfactory blackbody reference surface was devised by stacking razor blade edges together to form valleys. Incident radiation is then reflected from wall to wall of each valley a number of times, and practically all the radiation is absorbed before any is reflected outward. Because of the large number of reflections, any change in emissivity of individual surfaces has only a small effect on the overall emissivity of the reference surface. The rows of razor blades are arranged in a hexagonal pattern to minimize any directional effect which might result from the stacking of the blades.

Blackbodies are used by many laboratories as reference devices, especially in the field of spectroscopy. Those now in use are bulky and, consequently, are difficult to maintain at any reference temperature. The device developed by Ames is small and compact and has the potential of making possible the construction of small hand-held instruments.

According to Ames, its blackbody has an emissivity of about 0.93 and varies approximately 2 per cent over a temperature range of 100°F. While these characteristics are not adequate for laboratory blackbodies, Ames points out that it may be possible to obtain higher, less temperature-dependent emissivity by proper spacing of the razor blade edges.

There are other methods of obtaining high emissivity surfaces, similar to the razor blade technique, that are more amenable to theoretical treatment. One of these is a method employed by the University of Wisconsin's Meteorology Department for use in connection with its satellite radiometer program. The program required a large hemispherical blackbody which was constructed by placing many 0.38 caliber revolver cartridges as close together as possible in the form of a large hemisphere. Using this approach, the surface is composed of a large number of small blackbodies (cylinders) whose emissivities can be calculated theoretically. The only difficulty with either approach is that a certain portion of the surface will always be composed of the edge of the razor blade or the walls of the revolver cartridge, hence emissivity cannot be made to approach unity as closely as might be desired. However, because both approaches yield relatively high emissivities that theoretically should not vary with temperature, they may be refined and used to construct small, inexpensive reference surfaces.

B. Electronic Components and Miscellaneous Systems

1. SEMICONDUCTORS

The range of semiconductor devices available is very broad; no examples were identified during the study where a particular transistor or diode had first found missile/space application and then found commercial application. It would be misleading to think of the contribution of the missile/space program to the transistor field in those terms, for the contribution is more complex than that of a product here and there. The contribution is primarily that of adding new technology to the semiconductor field.

For example, Texas Instruments could indentify no specific missile/space contract which resulted directly in semiconductor devices with commercial application. But, the company goes on to add, the entire silicon transistor field would be in a far less sophisticated state of development without the advanced application to the missile/space market. In the earlier stages of their development, silicon transistors were used almost entirely in missile/space applications since they can operate at higher temperatures than germanium devices.

In a statement in the *Commercial and Financial Chronicle* in January, 1963, Texas Instruments' president, P. E. Haggerty, points out that government markets accounted for a major portion of the electronic industry's sales in 1962. Total sales amounted to $13 billion in 1962, of which an estimated $7.5 billion derived from military, space, and other government markets. With this broad support it would be difficult to find electronic developments that were not affected by government stimulus and related to the missile/space effort.

It is also of interest that technologies developed first for commercial applications are now being applied by Texas Instruments in the missile/space effort. Techniques and instrumentation developed originally for geophysical surveys are now being used to probe space. To sum up, about half of Texas Instruments' R & D has been sponsored by and directed toward the nation's defense, missile, and space needs, and much of this is applicable to commercial electronics, present and future. To look upon specific developments as "by-products" is not particularly meaningful; it is more meaningful to look upon the total effort, noting that Texas Instruments' commercial effort has missile/space application and vice versa.

The General Electric Company identifies reliability as a major contribution of the missile/space effort to the semiconductor field. Several types of high reliability transistors were developed by G.E. to meet specifications of the Minuteman missile program, and some of the knowledge gained from this effort was in turn applied to commercial and industrial transistors to improve their reliability for control and computer applications. A very small percentage, if any, of the transistors specifically developed for the Minuteman are sold for commercial or industrial use.

The Autonetics Division of North American Aviation, which also worked on the Minuteman program, states that some of the semiconductor devices developed for the Minuteman are two orders of magnitude more reliable than their pre-Minuteman counterparts. Commercial versions, incorporating much of the knowledge gained, are being sold.

Merck, Sharp & Dohme Research Laboratories discusses the effect of the missile/space program on a specific semiconductor material, gallium arsenide. The response of gallium arsenide to a space environment, to the solar radiation spectrum, and to high temperatures has led to applications of gallium arsenide in solar cells, special transistors, and diodes for operation at particularly high temperatures in the missile/space program. This, in turn, has advanced the gallium arsenide materials and application technology to the stage where the use of this material is being investigated in a number of commercial areas, e.g., tunnel or varactor diodes. Present commercial applications of gallium arsenide are in specialty devices sold in relatively small volume. The ultimate commercial potential of the material is not really known.

The Semiconductor and Materials Division of the Radio Corporation of America has done a great amount of research and development on gallium arsenide semiconductors. The original impetus for the work came from the Air Force B-70 program and later impetus came from missile-oriented, Air Force-sponsored, follow-on studies.

Gallium arsenide semiconductor work was originally pursued because gallium arsenide can operate at higher temperatures than silicon or germanium. Generally speaking, however, gallium arsenide's operating characteristics are also better at lower temperatures. As work progressed it was found that a gallium arsenide switching diode has an extremely fast switching speed. In addition, gallium arsenide varactors were discovered to have superior frequency response characteristics, and gallium arsenide solar cells demonstrated improved radiation and temperature resistance over silicon cells.

Though gallium arsenide devices have been limited to missile/space applications thus far, their fast switching speed and excellent frequency response indicate computer use and application in microwave communication.

In discussing the effect of the missile/space program on the semi-conductor field as a whole, RCA points out the particular importance of miniaturization and reliability. At first the company was interested in miniaturization for its own sake and its value in the military and space fields. Soon it realized that with miniaturization comes reliability and ease of production. As electronic circuits are miniaturized they are made as a unit rather than as a combination of separate wires and components. Less hand operations are required; fewer connections can go wrong, and production costs are reduced due to ease of assembly. Naturally, all of this has commercial application.

The Semiconductor Products Division of Motorola, Incorporated, states that in general the missile/space program makes itself felt by inducing improvements in existing devices or, in some cases, by making a device commercially salable through support of the early production phase. The company goes on to say that unquestionably the major reason for Air Force R & D in integrated circuitry is missile/space application. The current effort going into the development of integrated circuitry will definitely have an impact on the industrial and entertainment electronics market.

In summary, the missile/space program has contributed to most phases of the semiconductor field. The need for operation under severe environmental conditions, such as high ambient temperatures, has caused an acceleration in the development of new semiconductor materials. Not only do these materials have commercial application where high temperatures are encountered, but many of the materials have better characteristics at normal temperatures. The missile/space program has led to miniaturization of electronic circuits, which has particular commercial application in the computer field. Miniaturization has, in turn, led to more reliability in electronic circuits. Many missile/space programs have been directed toward improved reliability through miniaturization as well as other means. In addition, missile/space demands have led to semiconductor devices with vastly improved frequency and power characteristics.

2. MICROELECTRONICS

Microelectronics is the science of miniaturizing electronic circuitry. Although there is some confusion in terminology due to the newness of the field, microelectric circuits may be divided into three categories: (a) semiconductor-integrated circuits; (b) thin film circuits; and (c) modular circuits.

Semiconductor-integrated circuits are formed on an active substrate such as silicon or germanium. Both active elements (transistors or diodes) and passive circuits elements (resistors or capacitors) are integrated into the substrate.

Fairchild Semiconductor is one producer of semiconductor-integrated circuits. The missile/space field has created a market in which this company can produce sophisticated scientific products and achieve a fairly rapid payout on its investment in development. When the barrier of initial development cost and the resulting initially high per unit price have been surmounted, the company can develop a large market for its products in the commercial economy. In other words, the contribution of the missile/space field has been to accelerate the development of technically sophisticated products and bring them to the nongovernment user more rapidly. An example is Fairchild's micrologic series, a family of functional circuit elements in which transistors and resistors are diffused into a single chip of silicon. Micrologic was not designed specifically for the aerospace market; rather, it was designed for the computer field in general. In the early stages, however, the cost was high and micrologic elements found most of their use in missile, space, and military applications.

Thin film circuits are formed on an inactive substrate such as glass or ceramic. The passive circuit elements are formed by vacuum deposition, chemical deposition, or electro-chemical deposition, and active elements are added separately as discrete components. In the future, active elements also may be formed by a deposition process.

The research laboratories of Lear-Siegler, Inc., working on ways to mass-produce thin film microcircuits, have developed a semiautomatic machine for making these circuits. The missile/space effort has contributed toward this development in that it provides a primary market for thin film circuitry. Without this market, thin film circuitry or its mass production probably would not have reached its present level of development.

Present nonmissile/space use of the company's microcircuits include their use in autopilots, hearing aids, and commercial data transmission. Future use is expected in computers.

3. Connectors, Cables, and Printed Circuits

There have been no dramatic changes in connectors, cables, or printed circuits as a result of the missile/space program. The contribution of the missile/space program has been primarily one of product improvement. Some specific examples will serve to illustrate this point.

In about 1954, the Air Force, apparently with missile and jet aircraft needs in mind, contracted with Amphenol-Borg Electronics Corporation to develop a new line of connectors which would eventually, in the company's opinion, supersede the old AN series. Before the line was completely developed, Air Force funding ran out and the company continued development on its own in the belief that the AN series would eventually be inadequate. During the later phases of this development the Air Force began to regain interest and further specification requirements were forthcoming. Acceleration of the missile/space effort during the development period caused some basic changes in original concepts and necessitated redesign of the entire connector. The ensuing line of connectors (military specification: MIL-C-26500) was introduced in 1961. The line combines advances in materials technology for high reliability under extreme conditions with a simplified mechanical design which produces the overall connector harness and assembly cost. The connectors are used extensively in the Minuteman as well as in other missile and aircraft programs. Commercially, the connectors are used on the Boeing 727 jet aircraft and use on other future jet aircraft is expected.

Commercial use of the connectors will probably be limited, since, by their very nature, they are expensive and more than satisfy the usual demands for reliability and performance under extreme environments. However, the knowledge resulting from the development of this series of connectors has proven valuable in developing other connector lines, such as improved versions of standard lines more suited for commercial application. The latter is probably the most important contribution.

Microdot, Inc., of South Pasadena, California, was incorporated in 1950 to produce miniature cables and connectors for hearing aids requiring more reliable, smaller components. The market was limited and in 1953 the company decided to pursue the emerging missile/space and related electronics market. By the end of 1953 it had developed a proprietary line of miniaturized coaxial connectors. Between the years 1953 and 1961 Microdot expanded its product line to include multipin coaxial and multipin power connectors, solderless crimp-type coaxial connectors, coaxial switches, and some types of transducers. In addition, having found cable purchased from other manufacturers to be inadequate, the company developed its own cable.

The missile/space effort has provided the largest single market for the company's products, and rigorous requirements of the industry have forced continuous improvement in the reliability of products. Thus, the contribution is one of product availability and product improvement.

Microdot cables and connectors are being used in the missile/space industry in space communication systems, space telemetry systems, radar tracking systems, and ground test equipment. Outside of the industry, the company's products are being used in communication systems, computers, and industrial research and test programs.

There are several other examples of missile/space-stimulated developments in the cable, connector, and printed circuit field. One is a connector which makes circuit connections between two printed circuit boards without solder joints. A means of connecting printed circuits which would stand great vibration, permit dense packaging, and insure reliability was needed for missile and space vehicles. Connectors which would perform satisfactorily were not readily available and as a result, the connector now being produced by Brown Engineering Company was designed. The connectors are currently being used in the Pershing missile and Saturn space carrier vehicle.

The basic connector concept was developed by an employee of the Marshall Space Flight Center of NASA at the time this organization was part of the Army Ballistic Missile Agency. A patent application was filed by the Army Patent Office in this employee's name, and it was determined that the patent rights for the invention, with the exception of the royalty-free license retained by the government, would remain with the inventor. Brown Engineering Company has an exclusive license agreement with the inventor to manufacture and sell the connectors. The connectors are being evaluated for use in commercial digital computers, the principal potential nonspace application.

The Elco Corporation makes a product called the Varicon connector which works by the interlocking of two forked metallic strips. The product was developed about twelve years ago, and from its inception many improvements have been made, a number of which can be traced to the company's work with the missile/space industries. One of the most important changes was the miniaturization of the product. Other changes in materials, tolerances, and methods of terminating the cable in the connector have made for a more reliable product. Although the company has always had commercial sales, it feels that the missile/space effort has been the prime impetus behind the continued improvement in reliability and continued miniaturization of its product.

While working on the Polaris missile program, Epsco, Inc., required a high quality circuit board which it was not able to purchase commercially. The company developed its own and has since been established in the production of printed circuits which are now sold commercially as well as for military and space applications.

According to the International Resistance Company, the development of two of its product lines, Polystrip® and Lamoflex®, was to provide interconnection materials appropriate for miniaturization and weight reduction required by the electronics industry for the missile/space programs. Polystrip is a thin, flat, flexible cable containing rectangular conductors protected between tough plastic sheets. The two-dimensional characteristic of the harness provides advantages in installation, electric characteristics, ultimate package size, and weight. Furthermore, the flat configuration allows materially increased current capacity. Lamoflex is thin, flexible, etched circuitry or cable. The circuits are sandwiched between sheets of thermo-plastic material which is fused into a homogeneous material.

Although the majority of uses of Polystrip and Lamoflex have been in connection with the missile/space program, new applications are developing in commercial computer, communications, and appliance equipment.

4. DISPLAY SYSTEMS

The term "display systems" is used to refer to systems or devices which present information visually. Several display systems evolved or were improved with some contribution from the missile/space effort. Examples follow.

Scientists at the Aerojet Astrionics Division have developed a lenticular (made of many lenses) screen which the NASA Ames Research Center uses with a projector for simulating the outside world as viewed by the crew of supersonic aircraft or spacecraft. Most of the light projected on the screen is focused and reflected back over a narrow angle; thus a very bright picture image is obtained over a narrow field sufficient for crew training purposes. The idea originated in connection with simulation for aircraft flight and gunnery training, but it is being refined considerably in view of potential applications in the manned space program.

An advanced version of potential commercial importance to the entertainment industry is under development. It uses new principles to provide wider viewing angles and much greater suppression of ambient light. The screen can be used for movie or television projection when the available projection intensity is much lower than normally required.

Work on the development of a computer system for cataloguing pictures from the Tiros satellite led RCA to the development of a system for NBC to display, edit, and splice television pictures.

Hughes Aircraft has recently developed a small, lightweight, head-mounted cathode ray tube display. It consists of a small cathode ray tube, a single front surface mirror, an adjustable focusing lens, and a reflecting eyepiece—all of which are housed in a special tubular enclosure. The device is worn on the observer's head and presents the display information superimposed on the ambient background.

The most significant missile/space contribution to the device has been the market potential, thereby encouraging its development at this time.

Uses of the device include: (1) in space exploration to give an astronaut computer data, instrument data, or other data without necessitating undue movement. In most situations the astronaut is restricted in his ability to move about to get information; (2) in the Army, Navy or Air Force to give commanders briefings during rapidly changing battle conditions; (3) in aviation to enhance air safety by being able to watch the airborne traffic pattern while monitoring data; (4) in industry to permit assembly workers to receive wiring instructions, eliminating blueprints or printed instructions; and (5) in medicine to aid a surgeon by allowing him to detect instantly any change in the patient's condition.

5. MISCELLANEOUS

A colorimeter was recently developed by Allied Research Associates which automatically sorts cigars for uniformity of color. The development of this specialized device was made possible largely through the knowledge of solid state physics and printed circuitry gained through participation in missile/space programs. One of these is now in use and additional units are being offered at a price of approximately $100,000. A similar device, still in development, is being adapted to monitor the color of cloth as it is being dyed.

Other developments in the electronic field include: (1) an improved brushless alternator with commercial potential; (2) several improved resistors and capacitors which have commercial application or potential; and (3) an electronic scanning star tracker which has potential commercial application in celestial tracking or ship navigation.

C. Control Systems

1. GYROSCOPES AND INERTIAL GUIDANCE

Gyroscopes have been used in various forms and for various navigational and stabilization purposes since the beginning of the 19th century. The demands of the space era have led to the rapid evolution of the science of inertial guidance and to several orders of magnitude improvements in gyros and other components comprising an inertial guidance system. In the United States in 1945, Dr. C. S. Draper and colleagues at the MIT Instrumentation Laboratories began investigating the basic problems involved in the design of accurate and stable inertial guidance systems. By 1953, the first inertially guided coast-to-coast airplane flight was made under the direction of Draper and his group. Other groups working concurrently on the problem included the Autonetics Division of North American Aviation and the Army Ballistic Missile Agency [4].

The development of "Navan" gyros was started about 1945 by North American's Autonetics Division for use in the guidance system of the Navaho missile. Since their original development, Navan gyros have found many nonmissile uses. They are used in the inertial guidance systems of the Navy's atomic submarines. They are used in a precision gyrocompass developed by Autonetics (accurate to less than 30 sec of arc) which finds true north by sensing the Earth's rotation. The company has also developed a miniaturized version of the same equipment, accurate to within 60 sec of arc. Both systems are presently used for military purposes (orientation of missile launchers, artillery, and mobile radar) but use is anticipated in general surveying.

Many problems had to be overcome before missile inertial guidance could become a reality. One of the most important related to a phenomenon associated with gyros is called *precessing*. A spinning gyro reacts to any torque acting on its axis by turning about an axis perpendicular to both its own axis and the axis of applied torque. This phenomenon is one of the most useful characteristics of the gyro (the gyrocompass makes use of it), but is also the source of many inaccuracies. Frictional drag imposed by the system used to conduct power to the rotor, frictional drag of the gimbal support bearings, and forces imposed by the attitude sensing system all are unwanted sources of torque which will cause gyro error.

At the end of World War II, inaccuracy caused by undesirable precession was enough to result in an error in ICBM missile guidance of 1000 mi or more at the target. In contrast, for many applications, error enough to cause a miss of one mile at the target may be taken today as the acceptable limit of uncertainty [5]. Thus gyro error had to be reduced by several orders of magnitude. To illustrate acceptable error, undesirable torque had to be reduced to 0.073 dyne-cm. Furthermore, the deviation of the center of the mass from the gyro axis could not average more than 1/10-millionth of an inch. These constraints make recent progress in gyro development seem impossible. However, acceptable drift rate, that which would cause an uncertainty of about one mile at the target, is roughly 1/60 deg/hr, and random drift rates of 1/200 deg/hr or lower are now being maintained with some of the better gyros.

A great advantage of inertial guidance for military systems is its independence from ground radio. Being entirely self-contained, it is not subject to radio interference, nuclear radiation, or weather disturbance.

Research performed by General Precision in the field of guidance, which they say has been prompted largely by the missile/space effort, has led to significant improvements in both their gyros and inertial guidance systems. In addition to research leading to improvements in mechanical gyros, the firm is doing considerable research on bearingless gyros such as: the spinning atom, phase relationships in a laser, and spinning objects suspended in a vacuum by the field from a superconducting magnet. Most commercial aircraft use gyros, and the company feels these gyros will be improved by research done for inertial guidance. Supersonic aircraft, both military and proposed commercial, which will probably use inertial guidance systems, should also benefit.

The Sperry Gyroscope Company feels that inertial guidance owes most of its accelerated development to missile/space programs. The company also believes that inertial guidance has several advantages for commercial aircraft: (1) it requires no ground radio net, which could result in a substantial system cost saving; (2) it is more automatic than aircraft guidance systems in use today; (3) being automatic, reaction time is faster than in systems in use today; and (4) inertial guidance is almost a necessity on polar or supersonic flight.

In the past decade a science of inertial guidance has emerged. Working systems have been engineered, due primarily to the impetus of the missile/space effort. In turn, there have been several nonmissile/space effects. First, inertial guidance systems have been made technically and economically feasible for many types of military aircraft and marine vehicles. Application of inertial guidance on future commercial supersonic aircraft is anticipated. Second, the components of inertial guidance systems have been vastly improved and are available for other uses. The most significant example, of course, is the gyro which is being used in improved gyrocompasses among other things. Finally, the need for highly stable engineering materials has accelerated basic research in metallurgy to investigate and reduce creep phenomenon and thermalgradient effects in which changes in the crystal structure of a metal can cause troublesome shifts in the delicate balance of gyro rotors and support members.

2. Electronic Digital Computer Systems

The contribution of the missile/space effort to the field of digital computers and automation is primarily in the areas of miniaturization, reliability, data input-output techniques, and in some aspects of computational speed. In addition, various missile/space-oriented government agencies and companies have provided a market for sophisticated special purpose computer systems. In order to put the contribution of the missile/space effort to the computer field in perspective, an historical sketch of the field in general follows.

In 1939, Howard Aiken of Harvard University began work on the design of an "automatic sequence controlled calculator," or Mark I, which was built by IBM, completed in 1944, and donated to Harvard. The machine was significant in that it had a memory that could store numbers. In 1938, Bell Telephone Laboratories pioneered work on computing machines using electromagnetic relays for arithmetic operations. Computing requirements of World War II, particularly computation of ballistic trajectories, prompted the development of two large-scale computers by Bell Laboratories under U.S. Army sponsorship. Both were completed in 1946 and one was located at the Army's Aberdeen Proving Ground; the other was located at the Laboratories of the National Advisory Committee for Aeronautics at Langley Field, Virginia [6].

Also under U.S. Army sponsorship, the first *electronic* digital computer, the ENIAC (Electronic Numerical Integrator and Calculator), was completed at the University of Pennsylvania in 1946. ENIAC was installed in 1947 at the Ballistic Research Laboratories, Aberdeen, Maryland [6].

Howard Aiken produced a similar machine in 1948 and called it Mark II. It was installed at the Dahlgren Proving Ground of the U. S. Navy [6].

Proposals of J. von Neumann led to the construction of EDVAC, a machine which depended on acoustic delay in mercury for storage. Meanwhile, in England, M. V. Wilkes built a machine called EDSAC, patterned after EDVAC, which was the first completed to contain both instructions and numbers in storage—the first stored program computer. RAYDAC, also based on EDVAC, was installed at the U.S. Navy's Missile Test Center after completion in 1953. Two men, Mauchly and Eckert, who had participated in the design of EDVAC, formed their own company in the meantime, later to be purchased by Sperry Rand, and produced the UNIVAC I in 1951. While waiting for completion of the UNIVAC and EDVAC, the Air Force contracted with the National Bureau of Standards to produce an "interim" computer which was similar to EDVAC but simpler in design. It was completed in 1950 and became the prototype for FLAC, which was installed at the Air Force Missile Test Center at Cocoa, Florida.

Aiken completed the Mark III in 1950 at Harvard. This was the first magnetic drum computer; it was built for the Navy and installed at Dahlgren Proving Ground. OARAC, similar in design to Mark III, was built by General Electric and installed at Wright-Patterson Air Force Base in 1953. Magnetic drum storage is now used exclusively only by smaller computers, although it is sometimes used as auxiliary storage by larger machines.

Both acoustic delay memories and magnetic drum memories suffer from long access time. Williams at the University of Manchester developed a fast random access memory system using a cathode ray tube, which was adopted for the Institute of Advanced Study's computer at Princeton. The University of Illinois built both the ORDVAC, installed at Aberdeen Proving Ground, and the ILLIAC, which stayed at the University of Illinois, on the Williams tube principle. MANIAC I at Los Alamos,

ORACLE at Oak Ridge, Tennessee, and AVIDAC at Argonne Laboratories, University of Chicago, were also based on the Williams tube memories. IBM's 701 and 702 use both the Williams tube and magnetic drum memories.

The Whirlwind I, completed at MIT in 1950, used cathode ray tube storage, but dissatisfaction with this type of memory led the group at MIT to look for a substitute and magnetic core storage was invented. Magnetic core storage was immediately successful and electrostatic and acoustic delay computers were virtually abandoned.

At the apex of computer systems in 1959 was the SAGE, which was designed for Continental Air Defense. Working with tremendous reliability, this system processes streams of raw radar data, makes computations, and presents visual displays of air space occupation to Air Force personnel [7]. In the early 1960's IBM's STRETCH and Sperry Rand's LARC were the more sophisticated machines. Both systems were developed for the Atomic Energy Commission, and both were made available commercially [7].

Large systems entering the commercial market today are for the most part transistorized, and use core storage. Great speed, the ability to optimize speed by looking ahead a few steps in the program, sophisticated error detection and correction, the ability to flexibly modify instructions, simultaneous read-write-compute capabilities, and other assets of STRETCH and LARC are built into these computers [7].

Systems have recently been developed for data communication—transmission of data from outlying spots directly to the central computer. For example, RCA has announced a technique developed originally for the Minuteman ICBM which can transmit 5000 characters per second over a conventional two-wire direct cable circuit [7].

As mentioned earlier, one of the main contributions of the missile/space field to computer systems is in the area of miniaturization. Naturally, the transition from vacuum tube circuits to semiconductor circuits was an important step toward miniaturization. Not only are transistors smaller than tubes, but they lend themselves to printed circuit and modularized assembly. There is little doubt that some of the impetus for transistorization came from the missile/space effort. Coming into the picture now, with strong impetus from missile/space effort, is the microelectronic circuit. Microelectronic circuits eliminate much of the bulk, most of the wiring, and a large percentage of the solder connections of conventional circuits. (See also Section II,B,2 on Microelectronics.)

Initial use of microelectronic circuitry will be largely on spacecraft and missile computers. Once the initial cost and development barrier is surmounted, however, it is quite probable that microelectronic circuitry will penetrate the commercial computer field, as was the case with semiconductors.

Miniaturization has several advantages for computer designers besides space saving per se. First, with miniaturization comes reliability. Because microelectronic circuits eliminate much of the wiring and most soldered connections of conventional circuits, they eliminate much of the unreliability inherent in standard electronic circuitry. Second, microelectronic circuitry may eventually cut production costs of electronic circuits, since it lends itself to mass production techniques. Finally, miniaturization can mean faster computers. The effective travel time of an electromagnetic impulse in a wire is 1.7 nsec/ft. Decision elements operating in the 2.5-nsec range have been proved feasible. Thus, the length of time it takes an electrical impulse to travel one foot is on the same order of magnitude as the time a switching element takes to operate, and it is logical to conclude that miniaturization will lend speed to computers [8].

A good example of miniaturization is Remington Rand Univac Division's thin magnetic film and Microtronic circuit computers. Univac has been involved in United States missile/space programs for many years as a supplier of electronic data systems. Initially, Univac's commercial computers were used for this activity, but later government support, combined with company investment, materially advanced the state-of-the-art in computer technology. As a result, data systems with guaranteed reliability over a broad range of speed and capability became available.

One of the most promising research and development programs which Univac has recently pursued is the investigation of properties of thin magnetic films and techniques for using arrays of elements comprised of such films. Together with associated printed and electronic circuits, thin magnetic films provide both the function of memory and logic in data processing equipment. A recent development from this program is a thin film memory which meets many of the most stringent environmental requirements for aerospace application.

Historically, a primary purpose of a missile-borne computer is the solution of inertial guidance problems through the processing of data from an inertial system. As missile systems have evolved, other functions have been assigned the guidance computer, including engine control, stellar tracking, system checkout and testing, and telemetry serializing. The Univac ADD (Aerospace Digital Development) was designed by Univac to meet the increasing demand on vehicle-borne computational equipment without a corresponding increase in weight. The small, integrated package design of the ADD is achieved through a combination of thin film memories, solid state electronic elements, and welded circuitry.

Univac introduced its Microtronic computer in early 1963. It was a unit constructed largely of microelectronic semiconductor circuits. One model of the Microtronic computer weighs less than 17 lb, and has a memory of 4600 words which can be expanded to 14,000 words if necessary. The development of Microtronic products has been largely a result of the increasing requirements for small, well-integrated, high-speed, highly reliable computers for use in space vehicles.

Magnetic film memories and Microtronic techniques in making microelectronic circuitry should find future application in Univac's commercial computers due to speed, reliability, and the fact that microelectronic circuitry lends itself to mass production techniques.

Reliability is of prime importance in all systems-oriented computers—rocket guidance, air defense, manned spacecraft, or process control [7]. Many computers can stall on the detection of an error; a systems-oriented computer cannot. Reliability on the ground may be achieved through duplication. For example, the Binac, built in 1948, was duplicated throughout. However, duplication causes a weight problem in rocket guidance or manned spacecraft and on the ground it may be expensive. Advances made in enforced reliability with a minimum of redundancy for rocket and spacecraft should, therefore, be of great benefit to process control.

Univac has had a reliability program in effect since 1955. Problem areas which have been researched include: (1) probability of failure and approaches to the problem of reducing this probability; (2) consequences of failure; (3) means of identifying the cause of failure; and (4) systematic approaches to obtaining timely and effective corrective action. This program came about as a result of the importance of reliability in space-borne, as well as commercial, computers.

Input-output techniques should also benefit from missile/space technology, especially in the area of process control. For example, a system such as Nike Zeus must detect an enemy missile, compute its trajectory, and "lock-on" with its own missile. There must be some way of converting tracking data to a form that can be used by a computer and a way to convert computer output to a form that will position the missile. This is just one example; there are many others in the missile/space field where data must be converted and reconverted automatically.

California Research Corporation, for example, has helped develop systems which rapidly scan appreciable amounts of data, convert the data to digital form, process the data in a digital computer, and make automatic adjustments. These systems are used in petroleum or petrochemical refining. California Research buys most of its data acquisition and processing equipment from others, but staff members are of the opinion that the missile/space effort has contributed substantially to the development and availability of this type of equipment.

Many of the activities of the Autonetics Division, North American Aviation, have been directed toward particular requirements for inertial navigation systems and instruments, computers, data handling systems, armament systems, or flight control systems. The fund of engineering knowledge and manufacturing techniques developed in these areas has made possible similar advances in the industrial products field.

One of the product lines of the Industrial Electronics Division, Autonetics, is the RECOMP computer series, which are desk size, all-transistorized, general purpose digital computers. The RECOMP series is a transfer of missile/space technology in this sense: computer techniques developed for the Department of Defense missile computers were incorporated into the RECOMP series. Therefore, the contribution is primarily a carry-over of "know-how." RECOMP computers are used mainly for scientific computation and industrial control in the petroleum, electronics, optical, and research industries.

Autonetics' Computers and Data Systems Division is equipped through participation in the GAM 77 Hound Dog, A3J Vigilante, and Minuteman programs to provide engineering and production assistance on automatic checkout equipment that checks the missile electronics system before firing. Through building this equipment Autonetics has acquired the know-how that will facilitate design of automatic checkout equipment for any electronic system. NIFTI (Neon Indicating Functional Test Equipment) and the NAVAPI (North American Voltage and Phase Indicator) are examples of present technological transfer. The NIFTI will perform high-speed continuity checks of wiring and telephone centrals, computers, and other complex electronic equipment. The NAVAPI will test and check out networks, gyros, transformers, and other components with high accuracy.

Autonetics has had no commercial sales of the NIFTI or NAVAPI as yet, but believes that it will in the future. More important, Autonetics is of the opinion that knowledge developed with respect to automatic checkout devices will have widespread future commercial application.

A particularly interesting example of missile/space transfer, which has a long prespace history, is Minneapolis-Honeywell's self-adaptive autopilot. In the spring of 1941 a group of their engineers visited Wright Field, Ohio, with one of the company's first military products, an electronically controlled aircraft mount which would keep a camera steadily pointed downward during flight. There they learned that the Norden bomb site was not as accurate as desired because pilots, being human, were unable to maintain the exact airspeed, altitude, and course heading needed for optimum bombing accuracy. What was needed was a flight control system operating on principles similar to the camera mount.

Minneapolis-Honeywell's engineers approached the problem by duplicating a pilot's motions in an electronic device called an autopilot that contained gyros, servomotors, and a computing device. Throughout 1942 the device was improved and changed to meet the requirements of the B-17D, and in November of that year the device was successfully tested.

Meanwhile, Honeywell engineers started work on flight controls for new longer range bombers and the earliest jet fighters. The E-6 autopilot was flight-tested in February, 1946, and appeared on the B-50's and B-36's of the post-war Strategic Air Force.

Before the E-6 was in production the company was developing a later model, the E-11, for supersonic aircraft. Later the MB-3 autopilot was developed for the first operational plane to fly supersonic speeds in level flight, and other models were developed for the F-101B Air Defense Command All Weather Interceptor, and NATO's Lockheed F-104 Star Fighter.

By 1959 Honeywell concluded that the traditional design approach to flight control systems was reaching the point of diminishing returns where modifications in the basic concept produced only minor improvement. The problem was that conventional autopilots had to be electronically "taught" in advance the conditions they might encounter, and they required a constant flow of information from air data sensors on the speed and density in the atmosphere. These limitations eliminated the conventional autopilot from consideration for future craft designed for manned maneuverable space flight. The conventional autopilot just could not be programmed for the many flight variables which might be encountered in hypersonic flight through the Earth's atmosphere into space.

The answer to the problem was found in self-adaptive electronic circuitry, a computer system which adapts automatically to compensate for varying flight conditions such as speed, altitude, weight, and wind gusts. A gyro system is used as the main sensing element.

Six years of research and engineering went into the development of the adaptive flight control and Honeywell's first flight test of an adaptive system took place in March 1958 in a fully instrumented F-94C jet aircraft. Out of this early work came more tests in high performance aircraft and, finally, work on adaptive flight control systems for the X-15 and the Air Force's X-20 manned orbital glider.

A direct transfer of technology from the X-15 and X-20 research on adaptive flight control is the H-14 adaptive autopilot for light twin engine aircraft. The advantages of the H-14 include (1) simplicity, which reduces the pilot's task and cuts down the possibility of pilot error, and (2) safety, which is enhanced by the autopilot's ability to adapt to unforeseen conditions. Several aircraft manufacturers are marketing the new adaptive autopilot as a recommended system for their twin engine business aircraft.

One problem in a systems-oriented computer is how to convert analog data into digital input form which can be used by the computer. Many companies contacted during the study had built analog-to-digital conversion devices which are available commercially, have been marketed commercially, or have commercial potential. These devices usually take a position on a rotating shaft and convert it to electronic digital signals. Or, if used in reverse, the converters can take a digital signal and can convert it to a shaft position. Early models of analog-to-digital conversion equipment were in existence before the space program; the main contribution to the space program has been to improve, refine, and miniaturize such equipment.

D. Power Sources

1. SOLAR CELLS

The modern solar battery is a semiconductor device for converting radiant energy directly into electrical energy. The economic feasibility of products powered by solar batteries can be traced directly to the space program. The technical feasibility of these products, however, cannot be attributed directly to the space program, since they are based largely on technology developed prior to and independent of the space effort.

The first silicon solar cells were produced by personnel at Bell Laboratories, and some were used experimentally to power various demonstrating equipment. This development took place in the early 1950's. During the late 1950's, government agencies interested in the missile/space effort began to support research and development on these devices. In recent years thousands of solar cells have been sold to Jet Propulsion Laboratory and other organizations associated with NASA and the space effort. Hoffman Electronics Corporation has been the major supplier. As a result of these requirements for a great many high efficiency cells to be used in space satellites, a considerable quantity of less efficient cells have become available for civilian applications.

Hoffman Electronics has developed several commercial products around the solar cell. One of these is a portable radio which has been sold in quantities to the general public over the last four or five years. The price of these radios was reduced from $150 to approximately $50 as a result of decreased cost of lower efficiency solar cells which were by-products of the production of high efficiency cells for satellite vehicles. A second example is an emergency call system which provides emergency services to motorists stranded on freeways, turnpikes, and super highways. This system has been installed by Hoffman Electronics on the central portion of the Los Angeles freeway network, under a contract with the California Division of Highways.

2. THERMIONIC AND THERMOELECTRIC ENERGY CONVERSION

The ability of a hot cathode and collector in a vacuum to supply power to an electrical circuit (*thermionic conversion*) has long been known. However, the phenomenon remained an experimenter's curiosity until 1958. Since then, much effort has been expended to make practical thermionic converters. The possibility of power conversion by *thermoelectric* devices began to appear in the late 1930's and has continued to the present.

It appears that the space effort has given much support to work in the fields of thermionic and thermoelectric energy conversion. Relatively efficient and light-weight sources for small portable power conversion equipment were among the early goals of this endeavor. One example is a radio-isotope-fueled thermoelectric conversion generator designed and built by the Martin Company. This generator was first demonstrated in a terrestrial application under contract with the Atomic Energy Commission. However, research and development yielding refinements and improvements to adapt the generator to space applications has been reflected in the design of terrestrial application units and given impetus to their development. These generators are used in an automatic meteorological data transmitting radio station. The system, which constituted the first nuclear-powered remote automatic weather station, was installed on Axel Heiberg Island in the Arctic on August 17, 1961.

Current thermionic energy conversion development efforts of General Electric Company originated with the early work of Dr. Volney C. Wilson, conducted in the G. E. Research Laboratory and funded by the company. It led to subsequent company advancements in both the vacuum converter and vapor converter areas.

Currently, both government-sponsored and company-sponsored developments are continuing in the vapor thermionic converter area. Thermionic converters are considered to have possible future commercial applications as topping devices for use in central power station operations. In addition, they are foreseen as possibly applicable in systems which provide electrical power at remote, unattended locations; e.g., microwave relay stations, irrigation pumping stations, and weather transmittal and navigational beacon sites.

3. Magnetohydrodynamics

Basically, magnetohydrodynamics (MHD) is concerned with interaction between magnetic fields and high-temperature ionized gases.

The ICBM reentry problem, which became important in the 1950's, and later problems in space propulsion gave emphasis and support to studies of the high-temperature properties of gases and the interaction of conductive gases with electrical and magnetic fields. Thus, recent work performed in the field of magnetohydrodynamics, much of it related to the missile/space effort, raises the possibility of technically feasible and economically attractive plants for the large scale generation of electric energy using magnetohydrodynamic principles.

In November 1959, Avco-Everett Research Laboratory, one of the leaders in this field, and the American Electric Power Services Corporation initiated a joint study to investigate the possibility of MHD power generation and the problems involved. While it is not yet certain that a magnetohydrodynamics central station power plant can be built, there appears to be a good possibility that this development will take place, possibly within the next ten years. The principal advantage of such a plant would be the much higher thermal efficiencies. It is estimated these would approach 60 per cent, as compared with the present 40 per cent maximum.

E. Propulsion

1. Cryogenics

The theoretical background in the field of cryogenics was fairly well developed when missile/space requirements for cryogens began to be felt. In addition, some of the cryogens, especially liquid oxygen, had been in practical use for some time. However, there were some unique requirements imposed by missile and launch site specifications which led to new developments and improved techniques in cryogenics. The result was to make available more equipment, instrumentation, and techniques that are used in a variety of commercial applications. For example, missile/space requirements for large quantities of cryogenic liquids have resulted in expanded facilities throughout the industry and contributed to significant price reductions. The reduction in price for some of these cryogenic fluids has been sufficient to permit their use in additional commercial applications such as truck refrigeration systems and private research.

An example of an improved technique is the experience of the Graver Tank and Manufacturing Company. Rigid specifications and the large number of vessels involved in cryogenic liquid handling for missile base installations required improvements in cryogenic tank cleaning procedures. The company developed a chemical cleaning technique to meet the cleanliness and particle size requirement for vessels to be used for the storage of liquid oxygen, nitrogen, and other cryogenic fluids. Similar techniques are now used in cleaning cryogenic vessels for commercial use.

2. Fluid Transfer Systems

The variety of liquids used in missile/space propulsion systems have brought new problems to fluid transfer systems. The liquids may be noxious, highly reactive, intractable, or self-igniting. In addition, high and low extremes of temperature and pressure are encountered. The components used in the solutions to these problems were basically like their counterparts in earlier systems, but with special emphasis placed on reliability and weight reduction. However, commercial systems have benefited from the emphasis placed on better components, and research toward better components has been motivated largely by the missile/space effort.

Sundstrand Aviation performed an analytical research investigation on low specific speed pumps and turbines sponsored by the Office of Naval Research. The objective of this investigation was to determine which of many types of turbines and pumps would provide optimum performance in various operating conditions applicable to missile/space auxiliary power systems. The optimization technique that developed, using specific speed and specific diameter as parameters, proved to be powerful and revealing. By applying this technique of analysis to known problem areas in other fields, Sundstrand has found commercial application for a high-speed, single-stage centrifugal pump which had previously been used in missile applications. This pump design has been successfully placed in operation on commercial jet aircraft and in an oil field application.

The Hills-McCanna Company has adapted for commercial application a ball valve which was developed for use in missile launchings. This ball valve had to pass severe mechanical and hydraulic shock tests. Shortly after the successful completion of these tests, the company learned that a major builder of tank cars was interested in a bottom unloading tank car valve. This type of valve would have to pass similar tests because a mechanical failure could be costly and dangerous. With a modification of the missile valve, the company was able to offer a satisfactory tank car valve. Thus, the development went from a commercial product to a valve for a specific missile application, and back to a commercial valve offering a substantial future potential in areas where a standard commercial valve would not be acceptable. Industries using these valves commercially include petroleum, petrochemical, chemical, pulp and paper, and allied process industries.

Kieley and Mueller, Inc., has improved an existing valve to meet missile/space requirements. The improvements give higher capacities per valve, lower weight per valve, faster speeds of operation, and tighter shut-off than was previously available. The development was motivated by requirements for improved equipment to load liquid fuels and oxidizers into launch vehicles for the missile/space program. The improved valves are beginning to find commercial applications; for example, they have been sold to a commercial company which manufactures oxygen for an acetylene process.

The Harrison Manufacturing Company developed a metallic static seal for use in fittings which were giving considerable difficulty in sealing under the extremes of temperature and pressure experienced in the Atlas missile. Some commercial use of these seals has taken place in high and low temperature applications such as deep well oil drilling, liquefaction of gases, and atomic energy installations.

F. Fabrication

1. FILAMENT WINDING

Filament winding is the manufacturing process by which reinforced plastic structures are produced by winding a resin-impregnated fiber material around a mandrel of the desired shape and then heating the assembly until the resin is cured. The fiber is generally glass and the resin is usually epoxy. Two types of winding are used: helical and biaxial. Helical winding uses a varied pattern to optimize strength in either the hoop or longitudinal direction, while biaxial winding is circumferential or longitudinal with variations to accommodate specific loads.

By use of a combination of oriented filaments, maximum advantage can be taken of the strength of the fiber material. Earlier fiberglas-reinforced types of plastics were composed of woven cloth in which the kinking of the fiber during weaving introduced a sheer stress which lowered the load-carrying capacity of the composite structure. The main advantages of filament-wound structures are their high strength-to-weight ratio and their nonconducting and corrosion-resistant characteristics.

The theory of single filament winding was evolved about 1947 and testing was started [9]. However, the process was not successful until the advent of improved epoxy resins. Filament winding was used for 3000 psi pressure bottles for jet aircraft starters in 1951 [10]. Walter Kidde & Company developed a small air pressure bottle capable of holding 650 in.[3] of air under a pressure of 3000 psi to replace metal storage spheres used as actuators for aircraft hydraulic systems [9]. Aerojet-General developed pressure bottles for the Air Force's Aerobee high altitude research rocket to contain nitrous oxide at a pressure of 3000 psi, the first recorded use of a filament-wound structure in the rocket industry [11]. Subsequent rocket chambers were made by filament winding for several rockets, including Polaris and Minuteman.

After the initial work by the aerospace industry demonstrated the advantages of this type of structure, civilian applications have been forth-coming. Two companies contacted during the study, Lamtex Industries, and Black, Sivalls, and Bryson, gave good examples of contribution in filament winding from their own experience. Lamtex produces a filament-wound reinforced plastic called Hystran®. The techniques of producing Hystran have been developed and refined due largely to work Lamtex has done in making pressure vessels for rockets. Some of the major missile/space projects now using Hystran include: Minuteman, Mercury, Polaris, Pershing, Scout, Bomarc, Ranger, Nike Zeus, Vortac, Discoverer, and a variety of others.

The skills developed by Lamtex on missile/space work were turned to potential commercial activity; and such products as air tanks for truck brake systems, automotive parts, truck and railroad tank cars, and chemical tanks have been built on a limited scale. A current large volume commercial product is a filament-wound brassiere support. These are used by many manufacturers to replace conventional metal supports. Lamtex has also developed filament-wound pipe and tubing and hopes to apply Hystran for auto and boat bodies, construction materials and "hot sticks" for handling high voltage lines.

Black, Sivalls, and Bryson obtained a license in 1957 from the Young Development Company to do filament winding to make oil field tanks used in the storage of corrosive chemicals. However, in 1957–1958 when the oil industry suffered a slump, the company decided to build tanks for other users, such as the food industry, the wine industry, and the chemical industry.

Meanwhile, the company had had inquiries from the Air Force and Navy and in 1958 performed research on filament-wound rocket motor chambers. As a result, the company is now fabricating the motor chamber (the chamber containing the solid fuel) for the second stage of the Polaris missile and for the third stage of the Minuteman missile.

Though the company's commercial product line is not a direct result of missile/space programs, the company states that the program has had one significant effect on its commercial line: as a result of Polaris and Minuteman work, the company is able to draw on a much broader base of glass filament and resin technology.

2. CHEMICAL MILLING

Chemical milling is a process used to uniformly remove metal, at a controlled rate, to change the dimension or shape of metal parts. The first step in a chemical milling operation is the application of masking material (usually a neoprene) to cover the part to be milled. Next, those portions of the surface to be milled are scribed and stripped. Third, the part is immersed in a vigorously agitated basic or acidic bath where the milling process takes place. Finally, the part is removed from the bath, the chemical neutralized, and the mask removed. When more complex shapes are desired, the part may be remasked and remilled several times [12].

Chemical milling was developed in 1953 at the Guided Missile Plant of North American Aviation in Downey, California. To make a rocket casing, a cylinder made of thin aluminum sheet had to be butt-welded, but the weld failed repeatedly because of the thinness of the contact edge. Planing a thick sheet to leave a lip at the edge to be welded would be expensive, and using a thicker sheet would have created an unacceptably high weight [13].

Manuel Sanz, Chief of the North American Research Group, suggested that the cylinder be made of relatively heavy-gage aluminum and placed in a corrosive bath, covering the edges with some form of masking to protect the edge to be welded. The idea proved feasible and was the first use of chemical milling. North American applied for a patent on the process, called chemical milling, which was granted in 1956.

North American Aviation has made Turco Products, Incorporated, of Wilmington, California, its prime licensee for the process. Turco has made improvements in all phases of the process as originally developed and holds many patents on these improvements.

The advantages of chemical milling include: (1) elimination, after an intricate part has been formed, of weight unnecessary to structural strength; (2) milling after heat treatment without affecting the heat treatment condition and avoidance of warpage which could occur if parts are heat-treated after milling; (3) attainment of tolerances as close as ±0.002 in.; (4) production of inexpensive contour or tapered shapes through controlling the rate and way in which parts are immersed in the chemical; (5) uniform reduction in thickness of sheet metal, leaving strengthening ribs if necessary; (6) production of complicated shapes eliminating the necessity for welding or riveting various parts together; (7) relatively small investment in equipment; and (8) machining many alloys difficult to machine by conventional methods [14].

However, there are several disadvantages of the process. Where the work could be machined by ordinary methods, chemical milling may not compete on a cost basis. The scribing operation does not lend itself to automation. Grain size, internal stress, and contours which may trap gas bubbles all affect the rate of milling in localized areas. Internal corners are not squared but rounded, and previously locked-in internal stress may be released, causing the part to warp [14].

Aluminum, magnesium, titanium, tooled steels, stainless steels, monel, Iconel, and various superalloys have been milled chemically with success. Molybdenum, tungsten, beryllium, and tantalum have been milled, but experience is lacking [12].

Outside the missile/space field, chemical milling has been used on almost every American airframe produced in recent years. Examples include the Boeing 707 and 720, the Douglas DC-8, and the Convair 880 and 990. The application of Chem-Mill has been limited, however, outside the aerospace industry. At one time, it was used to make trunk lids for an automobile but the process was discontinued. It has been used to make parts for certain types of office equipment.

More nonaerospace business is expected by Turco in the future. The company feels that two important factors have held back its nonaerospace use to date. First, the aerospace business has pre-empted to date most of the capacity of Turco and its licensees. Second, a considerable amount of time and money must be spent in marketing the process to industries unfamiliar with it.

3. HIGH ENERGY FORMING

There are several types of high energy forming. The primary variation is in the source of energy used: explosive charge, mixture of combustible gas, or high voltage capacitor banks [15]. Explosive forming is the oldest and currently the most common type of high energy forming. In this technique the sheet of metal to be formed is placed over a female die. The cavity between the sheet of metal and the die is evacuated, and the die and piece of metal are placed in a tank of water. An explosive charge, shaped in accordance with the job to be done, is placed in the water a predetermined distance above the metal sheet. When the explosive is detonated, the metal is rammed into the die. A small amount of explosive will produce forces far in excess of the largest hydraulic press.

The process of using a controlled burst of energy in explosive forming for metalworking began in the late 19th century. At that time several patents were issued on a process using shaped explosive charges to emboss and form metals. Early uses of the process included the fabricating of ornate doorknobs and brass spittoons. The process was little used, however, until six or seven years ago when aerospace companies began exploring its possibilities as a method of shaping hard-to-form metals.

Since that time the aerospace industry has led in the development of this process. There has been a great deal of interest in its use for the fabrication of titanium and zirconium alloys, the refractory metals, and other metals having properties suitable for missiles and advanced aircraft [16].

There are several advantages of explosive forming over conventional metal forming. First, the method is particularly suitable for parts needed in small quantities because of low die and equipment costs. Only one die is necessary as compared to two for most processes. In addition, materials such as plaster, wood, concrete, soft alloys, epoxy resins, or low-carbon steel can be used for dies. Second, tolerances in the order of ± 0.002 in. are easily obtained because springback is minimized [17]. Tolerances down to ± 0.001 in. can be obtained if a longer forming cycle and better dies are used [18]. Third, explosive forming is particularly applicable to forming large shapes and unusual forms which are beyond the capabilities of standard techniques. Last, there is an excellent uniformity of the parts produced.

The disadvantages of explosive forming are that it does not lend itself to automation and that industry is reluctant to use explosives.

High-energy forming using a combustible gas mixture is similar to the explosive forming technique, but appears more adaptable when the part to be formed is thin and prone to rupture if the pressure is distributed unevenly. The metal is formed by igniting a gaseous mixture in a closed die. Igniting the gas mixture in a uniform manner causes the energy source to assume the shape of the container. This technique was developed at the Boeing Company [15].

Other high energy forming techniques which are based on the conversion of electrical energy into mechanical energy include electrohydraulic and electromagnetic processes. In the electrohydraulic method a massive electrical discharge from a bank of capacitors sends a shock wave through water to form the metal against a die. The electromagnetic technique builds up a high repulsive magnetic field in the work piece to collapse it to shape in a die [19]. Both of these techniques show promise of being more adaptable to indoor assembly-line production than explosive forming.

Electrohydraulic equipment for fabricating titanium, columbium, stainless steels, tungsten, beryllium, and their alloys, has been developed by the General Electric Company at Schenectady, New York [20].

The first electromagnetic metal forming machine for industrial use is reportedly in production at the General Atomic Division of the General Dynamics Corporation.

The Rocketdyne Division of North American Aviation started using high energy forming in 1956 in connection with tube-end forming and later started forming larger sections such as tank-ends.

North American Aviation has now set up a separate division to do high energy forming. Some work has been done for customers outside the missile/space industry. One job in particular was for the Braun Citrus Company consisting of a stainless steel feed wheel for commercial orange juice squeezers requiring very close tolerances and fabricated from a high yield point material. Another was for Watervliet Arsenal in connection with gun bore evacuators and cladding tubing.

In 1957, the Marquardt Corporation became interested in the explosive metalworking process as a possible means of reducing fabrication time and cost in the production of ramjet engines for the Bomarc missile. Marquardt reports that practically all of its current work is for use in the space and missile programs. Some experimental work is being done for nonspace uses; for example, the production of large water valves for the University of Utah. The company believes the largest number of nonspace applications of this technique will be in the forming of large tanks and similar items.

The Martin Company Denver Division's interest in explosive forming techniques was prompted by the fact that it showed promise for the production of certain large metal forms for missiles. The division's primary goal in this effort is the forming of 120-in. diameter ellipsoidal tank domes for missiles from one-piece weld-free, flat aluminum blanks. The company anticipates that the process of explosive forming can be used commercially in forming a variety of metal shapes both large and small. It should prove economical, particularly for very large metal shapes.

4. Solid State Bonding

The terms *solid state bonding* and *diffusion bonding* are synonymous. It is a process used to make metal-to-metal bonds. Clean, closely fitting surfaces are brought together under the application of pressure with or without the application of heat. In this process the parts can be visualized as "growing together," the atoms or molecules of one part diffusing into the atoms or molecules of the other. The proximity of the parts on a molecular level, and the resulting intermolecular attraction form the bond. No melting is involved.

The cleanliness of the surfaces is an extremely important factor. Soft metals can be bonded in air but only with drastic deformation to alternate the oxide film. Protection of the surfaces from atmospheric contamination by the use of ultrahigh vacuum permits bond formation with little plastic deformation, although some deformation is necessary to insure full contact. If the parts involved do not lend themselves to diffusion bonding, an intermediate metal layer may be employed to effect a better bond.

National Research Corporation, under a contract with NASA's Goddard Research Center, has been investigating the mechanism of solid state bonding primarily to prevent metals from bonding to one another in space. At room temperature in a vacuum, pieces of mild steel stuck to each other so tightly as to require forces one-fifth as great to break the joint as to break the steel itself. At 300°C, joints had one-third the strength of the steel and at 500°C were up to 96 per cent perfect. Copper stuck to itself with 60 per cent of its natural strength at room temperature.

This process suggests vastly improved means of joining metal parts such as electronic components. Unlike soldering or brazing, solid state bonding can yield a blending of two separate parts with the desirable physical and electrical characteristics of the bulk metal. This could be an important development for both the aerospace and electronic component industries. Any practical application of this process, however, will require a great deal more work in ultrahigh vacuum and bonding techniques.

The Los Angeles Division of North American Aviation became interested in diffusion bonding as a side effect of its work on honeycomb brazing for the B-70 in 1960. Work was pursued through a company-funded study program because of its possible advantages in many aspects of the aerospace programs. The Los Angeles Division currently has a contract with ASD to do research on solid state bonding and its application to hypersonic vehicles, both aircraft and winged reentry.

The process has been used by North American's Atomics International Division to clad fuel elements for nuclear reactors. Potential uses in other areas are being evaluated; for example, North American expects to see it used to weld hydrafoils on boats.

5. CLEAN ROOM

The clean room is used to provide a sterile environment in which to manufacture products that must be protected from contamination. Prior to the missile/space program, Westinghouse Electric Corporation utilized clean rooms in its atomic energy programs. In the last five years, however, the missile/space program, with its emphasis on miniaturization and reliability, has caused greater usage of the clean room in manufacturing and improvement in clean room techniques. The clean room is now being used to eliminate contamination to the submicron size and to control temperature and humidity more exactly.

Present commercial applications of clean rooms include uses in the pharmaceutical, optical, and photographic industries. Since other industries are beginning to utilize miniaturized components, it is probable that the use of clean rooms will become more commercially important in the future.

The results of a Westinghouse survey indicate that the number of clean rooms in use will increase four to five times in the next three years. Most of this immediate increase will be due to missile/space programs.

G. Materials

Early in this study a number of persons contacted indicated the field of materials should be one where we could expect to find a large, if not the largest, contribution. Although we identified some contribution, it was not as significant as anticipated. Eventually, there may well be a more important contribution, particularly in a field like refractory metals, but it is too early to define how or when this contribution will take place.

1. REFRACTORY METALS

Refractory metals were primarily a laboratory curiosity in the 1940's. Of course, there were some minor uses of refractory metals, such as tungsten in light bulbs, and small amounts of refractories were used for alloying purposes. The first major application of refractories was in the aircraft industry in 1954, when molybdenum was used in forged turbine buckets for turbojet engines and in combustion chambers for ramjet engines [21].

The need for high temperature metals for solid-fueled rocket nozzles, energy converters, ion or plasma propulsion systems, winged reentry vehicles, and recoverable boosters has stimulated much research on the most economical refractory metals—tungsten, molybdenum, columbium, and tantalum. Technology stemming from missile/space-motivated research on refractory metals should be applicable to future aircraft engines, nuclear reactors, heat exchangers, and other devices where higher efficiency can be obtained with higher operating temperature. However, though much progress has been made in the development of refractory metals, the materials are generally brittle, hard to fabricate, and difficult to weld. The most serious problem is that the refractories are quite subject to oxidation. If research motivated by the missile/space industry can help solve these problems, it is probable that the refractories will find a great deal of commercial use.

2. SUPERALLOYS

Superalloy is the name reserved for a group of iron-base, nickel-base, or cobalt-base structural materials developed for high strength at temperatures ranging from 1200°F to 2000°F. The evolution of these alloys has coincided with the evolution of the gas turbine engine in the aircraft industry. The alloys have been widely used for turbine buckets, nozzles, guide vanes, rotors, combustion liners, afterburners, and structural members and fasteners. The main contribution of the missile/space programs to the development of superalloys is that the programs have provided a market receptive to minor technological advances in the field. Continued improvement in superalloys undoubtedly does have some impact in the aircraft industry where superalloys find the majority of use.

3. PHYSICAL METALLURGY

Basic research in physical metallurgy has been going on for a number of years, but new and special demands on metals made by missile/space programs have stimulated and fostered ideas for investigation into specific areas. No specific commercial use of the knowledge gained by this research was identified, but the knowledge will be available to the nonspace sector and eventually should find commercial application, particularly in the development of new alloys.

4. EPOXY RESINS

Epoxy resins were discovered independently and practically simultaneously by researchers in the United States and Switzerland near the end of World War II. Epoxies can be any one of many resin molecules that are polymerized through the interaction of monomers. Although the basic resins contain most of the inherent qualities of a completed epoxy system, they are chemically stable, have an indefinite shelf life, and are useless by themselves. A liquid epoxy resin will remain a liquid until a compatible curing agent or hardener is added.

An early use of epoxy adhesives was in airframe construction where their applicability was proven for binding aluminum to aluminum, as well as for binding many other materials. Additional applications were recognized as missile/space requirements materialized. These more stringent requirements have stimulated the development of new formulations. Several new epoxy formulations have been developed as a result of missile/space needs. For example, the Fiber Resin Company has developed an epoxy that will harden at subzero temperatures. The material was developed originally for binding and patching steel sandwich material used in housing antimissile radar in the Arctic.

The resin has found commercial use in the repair and patching of new and old concrete in docks and flood control canals in the northwestern part of the United States. A telephone company is interested in this resin as a repair material that can be used during the winter.

5. MISCELLANEOUS

Missile/space research has also stimulated the improvement of a commercial aluminum alloy; research and experimentation on pyrolitic graphite—an isotropic refractory material with commercial potential; development of a line of dynamic materials testing equipment now being used commercially; the development of a honeycomb structural material being used on boats, buildings, and aircraft; and the origination of an all-metal insulation being used in nuclear piles.

H. Medical Technology

Several items were identified which constitute missile/space program contributions to advances in medical technology. Most of these, although not necessarily the most important ones, are in the field of medical electronics.

The impact of missile/space programs on medical electronics is extremely difficult to determine, but it appears to have been felt in two ways. First, and most directly, the programs have required the development of data gathering and telemetry systems for monitoring man's physiological responses in the harsh environment of space. These systems permit considerable freedom of action during the period of monitored response since no wires are used. Some of these devices are now finding nonspace application. Second, and more indirectly, emphasis in the space program on extreme reliability and continued miniaturization has made equipment more readily available in the field of medical electronics, where these features are also highly desirable.

A recent series of articles describes much of the current activity in medical electronics [22]. Although much of the work in the medical electronics field is being conducted by organizations not directly associated with the space program, some contribution from space activity is apparent.

In addition to electronics, there is another area where the space program has some impact on medical technology. The rigors of space flight require that only the most physically and mentally qualified trainees be accepted as astronauts. This has led to improvements in medical examination techniques. It is anticipated by those working in this field that these developments will have considerable carry-over and will be of long-term importance.

The Lovelace Foundation for Medical Education and Research in Albuquerque, New Mexico, has examined applicants for pilot and astronaut selection for a number of aerospace programs. The most extensive and widely known of these efforts was the screening of astronaut candidates for Project Mercury, the NASA manned space flight program. Similar projects have included pilot selection for other NASA programs and flight surgeon services for test pilots at the Flight Research Center at Edwards AFB, California. The Foundation also has been commissioned by NASA to develop a system for acquiring, evaluating, and disseminating all available information on bio-astronautics, the medical arts applicable to space programs. This project is now under way.

The Lovelace Foundation is immediately connected, by staff relationships and physical proximity, with the Lovelace Clinic and Bataan Memorial Methodist Hospital. This simplifies the carry-over of new medical service techniques from research to clinical application.

Unique to the Mercury astronaut program at the Foundation was the routine application of dynamic physiological tests supplementing the conventional static tests ordinarily employed in physical examinations. These tests are now being used extensively on regular clinic patients at the Lovelace Clinic and elsewhere. Another innovation was the recording of all pertinent medical data directly on punch cards by a system previously developed by Dr. A. H. Schwichtenberg of the Lovelace Foundation staff under USAF sponsorship [23].

Other techniques refined to an exceptional degree in the Mercury program included: (1) unusually exhaustive examinations, with careful coordination of many specialists, highly detailed histories, and more contact than usual between the doctor and the patient; (2) improved X-ray control procedures made important by extensive radiological examination, leading to the wide use of intensifying screens, video taping of X-ray images, and the use of higher voltage X-ray equipment resulting in shorter exposures; and (3) the use of microsampling of blood for laboratory examinations beyond the limits previously considered feasible.

Possibly the most significant future contributions of this space-related research project is the indication that this form of dynamic examination holds promise for the objective measurement of physiological aging. Follow-on research on this Physiological Aging Rating Project is under way at the Foundation, at George Washington University, and at the Civil Aeromedical Research Institute. One of the immediate needs for such measurement is determining appropriate retirement ages for pilots and others whose jobs will not permit performance deterioration resulting from aging.

More immediate feasible uses of these dynamic examination techniques are thought to include determination of the physical capabilities of people recovering from heart surgery, and evaluation of physical disability claims in Workmen's Compensation cases.

Another example of carry-over is a radioelectrocardiograph developed by Telemedics, Inc. The science of electrocardiography (analysis of the electrical activity of the heart) has been one of the keystones of cardiac diagnosis for half a century. However, because the patient had to be connected by wires to the recording apparatus, the doctor generally was limited to finding out how the heart acted while the patient was resting.

The concept of radioelectrocardiography, which is the broadcasting of the EKG waves from the patient to recording apparatus, was first published

13 years ago as a result of experimental work done by the United States Air Force. It remained in this experimental class until Telemedics, Inc., introduced its RKG *100* radioelectrocardiograph to the medical profession in May, 1961, at the annual meeting of the American College of Cardiology. This was the first radioelectrocardiograph made available commercially to physicians.

The RKG *100* is a direct descendant of the equipment originally developed by Telemedics' parent, Vector Manufacturing Company, for use in the rigorous testing program undergone by NASA's astronauts. Reactions of the hearts of the astronauts to the pressures of gravity, acceleration, and deceleration were determined with this equipment in the centrifuge at the United States Naval Air Development Station in Johnsville, Pennsylvania. Following its successful use with the astronauts, Vector established the Telemedics Division to adapt the radioelectrocardiograph for use by practicing physicians.

The RKG *100* is a miniaturized, very high frequency broadcasting system consisting of Band-Aid type electrodes, a transmitter—slightly larger than a pack of cigarettes—carried by the patient, and a receiver. The unit has a range up to 1500 ft.

The company feels that the applications for the RKG *100* are many and varied. Its prime use to date has been for the examination of more than 10,000 patients during exercise. Many of these studies have been conducted by Dr. Samuel Bellet, Chief of the Division of Cardiology of the Philadelphia General Hospital.

Beyond conducting regular exercise examinations, the RKG *100* is being used for a variety of purposes. At Lankenau Hospital in Philadelphia, patients who have suffered heart attacks are studied for a few days after their seizures to determine the effects of the slightest exertion. It is hoped that this will lead to improved recuperation programs. At Lankenau and at New York Hospital, it is used to monitor the heart during delicate X-ray procedures in the course of which a dye is injected into the heart. Since the radioelectrocardiograph broadcasts on VHF, it is above the electrical interference caused by the X-ray equipment. This is a definite advantage over conventional electrocardiographs which operate on the house wiring of the hospital and are subject to considerable electrical interference and distortion originating from appliances in the institution. At Montefiore Hospital in New York, the heart patients are monitored during rehabilitation and exercises to see that their capabilities are not exceeded. The equipment has also been used to monitor the heart during surgery, recovery, intensive care, and electroshock therapy.

A new field opened up by this equipment is that of environmental testing, or recording the cardiogram under natural conditions. The Bell

Telephone Company of Pennsylvania, which is using the RKG *100* in its Management Physical Examination Program, intends to take the EKG's of some of its executives during conferences and other office activities in which stresses might be present.

Telemedics feels that there is a genuine need for this kind of equipment in medical practice. Its research efforts are continuing toward development of related medical electronics equipment which will fill gaps now existing in other aspects of patient care [24].

A device developed at MIT provides another good example of a by-product in the medical field. This is a device used in making inertial guidance systems for the Polaris missile. It has turned out to be a valuable medical research tool for studying blood and blood viscosity. The device, an ultrasensitive instrument for measuring very small torques, was invented at the Instrumentation Laboratory of the Massachusetts Institute of Technology. The instrument continues to be used at the Laboratory in development of an advanced Polaris guidance system. In addition, for the past year it has been made available part-time to a team of MIT and Harvard Medical School researchers studying blood. In medical research, the instrument is called the "GDM Viscometer" after its developers: Gilinson, Dauwalter, and Merrill.

Major advantages of the device as a viscometer are that a test requires only a teaspoon of blood (4 cm³) and can be performed in less than a minute. Most viscometers, using different principles, require far more blood and so much time that the blood sometimes clots before a test can be completed. As a viscometer, the device is rigged to a small cup containing fluid to be tested. A rotor, turned by a torque motor, is immersed in the blood. The operator can change speeds to obtain viscosity readings at different flow rates. The rotor, when turned, exerts shear on the fluid at a known rate. The shear rate is analogous to the flow rate and causes a shear stress in the fluid which, in turn, exerts a torque on the cup containing the fluid. The GDM viscometer measures the torque on the cup with extreme precision and, knowing this torque, the researchers are able to compute the shear stress that produced it. Shear rate, known from the motor setting, and shear stress, read as torque by the viscometer, serve as the bases for computing viscosity.

The value of the GDM viscometer is that it makes possible viscosity studies at very low ranges of shear rate and shear stress. Shear rates of fluids are measured in inverse seconds while viscosity is measured in centipoise. Most viscometers can accurately measure viscosity of watery fluids like plasma under shear rates down only as low as 20 sec^{-1}. The GDM viscometer can measure viscosities under shear rates as low as 0.1 sec^{-1}— or 200 times lower than ordinary viscometers.

Ordinary viscometers have shown that viscosity of blood plasma remains constant at approximately 1.5 cp under shear rates as low as 20 sec^{-1}. Constant viscosity with reducing shear rate has led to the general assumption that plasma is Newtonian. Previous studies by Drs. Wells and Merrill, using a viscometer less sensitive than the GDM device, gave a hint about the possible non-Newtonian nature of plasma and caused them to seek a more sensitive instrument. Studies with the GDM viscometer have produced data showing that as the shear rate is lowered to the range of 0.1 sec^{-1}, plasma viscosity increases to as high as 18 cp. These are the findings that led the MIT and Harvard investigators to the conclusion that plasma is, in fact, non-Newtonian. An understanding of the non-Newtonian character of blood plasma is an important piece of fundamental knowledge. Drs. Merrill and Wells believe it might help explain some of the curious mechanics of blood circulation in capillaries, the body's smallest blood vessels.

I. Reentry Simulation—The Plasma Jet

A gas in an extremely hot, ionized state is called a plasma. It is a mixture of free electrons, positive ions, and neutral atoms; all molecular bonds are broken. A plasma may be formed in many ways. A nose cone reentering the Earth's atmosphere, for example, is surrounded by a plasma created by the high temperatures caused by air friction.

Recognizing the value of being able to create a plasma for the study of configurations and materials for the reentry of nose cones and space vehicles, the Plasmadyne Corporation built a device, called the plasma jet, which gives forth a continuous plasma stream. The original application of the plasma jet was in hyperthermal wind tunnels for reentry studies, but many commercial applications have since developed. Plasmadyne's commercial plasma jet can reach a temperature as high as 30,000°F. Above 10,000°F, even the most refractory materials are vaporized. If one introduces a powdered refractory metal or ceramic into the plasma, the plasma jet can be used to spray a protective coating onto another material, much like the spray of paint. For example, protective refractory materials are sprayed on parts of catalytic cracking systems which are subject to destructively high temperatures. Relatively inactive materials can be sprayed on piping, which is especially subject to corrosion in a chemical processing operation.

Plastics can be sprayed on electronic components to protect them from moisture and vibration by reducing the temperature of the plasma. Using a plasma jet to spray plastics eliminates the need for baking. The plasma jet can be used as a torch to make extremely clean cuts through thick metals; the high temperature of the plasma makes this possible. Because of the plasma jet's ability to deliver large amounts of heat per unit time, it has been used as a preheater in chemical processing (this is also the reason for its use in hyperthermal wind tunnels). The plasma jet has also found commercial use in the welding of difficult materials. One of the latest applications of this device is its use as a high-intensity light source. The extreme temperature of the plasma provides both brightness and total radiation levels several times those achieved by other man-made controlled sources of light. Finally, the plasma jet is being investigated as an electric rocket for deep space probes.

The development of a similar system, called the plasma gun, was begun by the Avco Corporation when the company was required to simulate the aerodynamic conditions of missile reentry for the Air Force Atlas and Titan missile programs. In order to do this, it was necessary to develop a heat source for high-temperature aerodynamic testing. The research conducted to obtain this high temperature heat source resulted in the basis for the plasma gun system. This system provides an economical means of coating many common materials, extending their useful applications and increasing their durability. In a typical application, a plasma gun spray coating will raise the melting point of the basic surface, increase its abrasion resistance, and produce corrosion and oxidation resistance. Most of the units are sold to firms in the aerospace industry, but customers also include a number of firms in industrial product industries.

J. Packaging and Shipping

Circumstances which are particularly acute in the missile/space industry have necessitated improvement in packaging and shipping containers and techniques. Components developed for the missile/space field often have a high value per pound and are easily damaged. In addition, by the time the components have reached their final application, they have been unpackaged, handled, and repackaged many times. Two problems are created: (1) the probability of damage is increased due to frequent handling, and (2) constant packing and repacking significantly increases the over-all cost. Several containers, devices, and techniques have been developed to help solve these problems. Packaging and shipping devices which have been developed originally to meet missile/space needs have transferred to commercial industry to meet similar packaging and shipping requirements. A good example is the Klimp, an L-shaped wire fastener developed by Navan Products, a subsidiary of North American Aviation. This takes the place of the nail in holding wooden shipping boxes together.

Its advantage is that it makes possible the very fast assembly and dis-assembly of packing boxes. Another example, also developed by Navan, is the Kudl-Pac. This is a container built with a hard outer plastic shell with urethane foam cushions on the inside which have interlocking faces. Its advantage is that the inner surface adapts itself to the shape of the part that is being shipped and holds it securely, protecting it from shock and damage. Both items are beginning to find wide nonspace use.

K. Management and Control—PERT

The missile program has had its effect on management and control, the most publicized example being PERT (Program Evaluation and Review Technique). PERT is an outgrowth of the sheer complexity of the Navy's fleet ballistic missile, or Polaris, program.

In 1956, a Special Projects Office (SPO) was organized by the Department of the Navy to manage the Polaris program. Numerous management tools, some old, some newly developed, were used by SPO.

However, despite the merits of these procedures, additional information was needed to assess the validity of plans and schedules, to measure research and development progress where the standard measure—money spent—did not necessarily correlate with progress, and to predict the probability of meeting objectives. In February, 1958, a team was organized by SPO to produce a system to do these things. The team consisted of personnel from SPO, consultants from Booz, Allen, and Hamilton, and specialists from the Missiles and Space Division of Lockheed. Willard Fazar, credited with guiding PERT's development, was team manager. Due to the urgent nature of the Polaris program, and the particular importance of time in the missile and space industry, it was decided to use time as a common denominator in the system. A significant change in resources or technology always means a significant change in time, and in the defense industry cost is often less important than time and may be dependent on time [25].

PERT, then, is developed around a time sequence flow diagram. Milestones representing important events in the life of the project are selected and linked graphically with arrows to portray interdependencies. Actual work activity is represented by the arrows joining the milestones, and the resulting flow diagram that is structured around these important events is called the PERT network.

Times necessary to complete each task between events are estimated in such a way as to give an appropriate measure of uncertainty. The "most likely time," "optimistic time," and "pessimistic time," are obtained from the person who is to complete a particular activity or task. A weighted average or expected time is calculated, giving the time for which there is a 50 per cent chance of completing the activity.

A basic concept of PERT is the "critical path," which is that series of interrelated tasks that will take the longest time for completion, and therefore, that chain of events which determines the minimum completion time for the total project. When the network is complex, as it was in the Polaris program, a digital computer is used to determine the critical path, to set the date for the end objective, to calculate the latest date by which each event must be finished if the project is to be completed on time, and to calculate the uncertainty involved in reaching each milestone.

As is often the case in the innovative process, a similar scheduling technique was being developed at approximately the same time as PERT, but quite independently. In the Engineering Department of E. I. duPont de Nemours and Company a group was formed in early 1956 to study the application of management science to the management of engineering [26]. A first area of attack was the planning and scheduling of construction and engineering design. Theory was developed and in 1957, the scheduling technique, called CPM (Critical Path Method), was tested on a hypothetical construction project. The construction of a new $10,000 chemical plant facility was selected for the first live test, and by March of 1958, the first part of the test was complete. The results indicated that two months could be gained at no additional cost over the time and cost as estimated by conventional means.

The developers of CPM were not aware of the PERT development efforts until early 1959, when a reporter from *Business Week* informed them of this parallel effort. As of 1959, there were many basic differences between PERT and CPM. Now, according to James E. Kelley, Jr., who is given much of the credit for the development of CPM, the techniques are similar and becoming more alike all the time.

What were the original differences in CPM and PERT? First, PERT is a system which evaluates existing schedules while CPM generates plans and schedules. Another important difference between PERT and CPM is that with PERT job durations are random variables (probabilities are assigned to the length of activities), whereas job durations are known fairly precisely with CPM. Therefore, one time estimate is made with CPM compared to PERT's three time estimates. PERT, as originally used, did not include cost as part of the model, while CPM did. Now, the Navy's SPO has developed a new system called PERT/Cost which does take cost into account. Another important difference is that PERT generates one set of schedule limits, while CPM can generate a spectrum of schedules with the minimum cost for each. A final difference is in the approach in constructing the model: PERT, as was noted, is constructed by selecting important milestones in the progress of a project and scheduling around these, while CPM is constructed around the jobs or activities that make up the total project.

It can be seen, then, that PERT is most useful for large, complex operations where the individual tasks are being done on a first-time basis or are of a research nature and, therefore, a great deal of uncertainty is involved which needs to be taken into account. CPM, on the other hand, deals with projects where the time for completion of each task can be estimated fairly accurately in advance, and the planner wants a series of schedules for the task with the cost for each. Quite naturally, PERT is more suited to the aerospace industry, defense industry, and research, while CPM is more suited to civilian construction and maintenance.

However, PERT was used to keep track of the production of the Broadway play "Morgana" [27]; at the Southwest Research Institute, a hybrid, simplified PERT/CPM was used to help schedule construction of a $20,000 shock tube [28]. H. Sheldon Phelps refers to the use of PERT in the construction of an office building, in the construction of a bakery, in a plant and facilities move, in the design, development, and production of a new product, in the changeover from pilot plant to full production, and in computer installation [29]. (The reader is cautioned, however, that CPM is often called PERT, and PERT called CPM in the literature, and some of these applications could be CPM or hybrid PERT/CPM.)

The existence of PERT, its use by government contractors, its spectacular success on the Polaris program (it was credited with saving two years over original estimates), and the publicity received therefrom, has significantly accelerated the diffusion of both PERT and CPM throughout industry. The American Management Association, Management Systems Corporation, the Service Bureau Corporation, Mauchly Associates, and several other organizations offer cram courses in PERT, CPM, or both.

Both CPM and PERT are really a way of thinking, rather than specific, well-defined techniques. Once one is familiar with PERT or CPM through reading, a class, or introduction through a government contract, he can evaluate other scheduling problems in the terms of flow network and critical path, and other techniques are easily assimilated. Here, then, in another way the missile-developed PERT can, and probably has, lent impetus to the industrial use of critical path techniques in general. The General Electric Company, for instance, introduced to PERT on the Polaris program, has now adopted either PERT or CPM in all its divisions [30].

III. Significance of Technological Transfer

This section presents additional information on the nature of the technological transfers from missile/space programs to commercial applications and provides some insight into the difficulty of formulating conclusions about the present economic significance of these transfers.

A. Identification

1. ASSIGNMENT OF CREDIT

The mere identification of transfers from missile/space programs to commercial uses is difficult for several reasons. The most obvious reason, sometimes overlooked, is that invention and innovation are continuing processes. Likewise, product development goes on continually in most firms and is subject to many influences rather than a single influence. As a consequence, it is difficult to connect newly developed technological knowledge, processes, materials, techniques, or devices with any one single cause. In most instances, there are many inputs from varied sources to new developments or products.

The question of whether missile/space or nonmissile/space use came first, or whether significant improvements were incorporated into an existing product because of missile/space experience, has been the cause of much confusion in discussions of so-called space by-products. Some items have been widely publicized as being space by-products which actually are not, largely because of this problem.

2. DUAL MOTIVATION FOR PRODUCT DEVELOPMENT

In many cases products are developed by firms, with their own funds or on a cost-sharing basis with the government, in anticipation of being able to tap both the missile/space market and the commercial market. Although developments of this nature might not take place when they did in the absence of the missile/space portion of total estimated demand, how much credit should the missile/space program have for such developments?

3. INFORMATION WITHHELD OR UNAVAILABLE

For several reasons companies cannot reveal information necessary to the identification of missile/space contributions. They include the following five major points:

  a. *Proprietary Information.* When the by-product to be marketed is in a highly competitive field, companies are obviously reluctant to release information unless the timing is right.

b. *Premature Releases.* When an item is in the sometimes lengthy process of development or market planning, many companies do not want premature publicity in the event that their product fails to live up to expectations. (Unfortunately, the reverse of this is also true. Some companies are happy to receive whatever publicity they can although it is fairly obvious that no definite commercial marketing plans exist for a potential by-product and commercial sales are, at best, very dubious.)

c. *Security.* Several firms reported that they produce to government specifications for prime contractors and, because of military security, do not know the end uses of their products. It is impossible to make any valid claims of by-products which may result from work these companies have been doing.

d. *Information Unknown.* Many suppliers do not know the end use of their product for reasons other than security. Several metals companies commented on their role in furnishing raw materials and semifabricated products, one stating that it had "not found any very definite instances where by-products have been developed, undoubtedly due to the nature of our participation simply as a basic materials supplier." Another reply, typical of several received from electronic components producers, was: "It is difficult to determine which applications are for space use and which for commercial computers." Transfers probably result from missile/space work done by firms such as these. But how can they say with certainty?

e. *Potential Patent Problem.* By-product information is sometimes withheld because a company may be in the process of filing patent applications and cannot afford to compromise its position with a premature disclosure of product descriptions.

4. Incomplete Information

In attempting to identify by-products it is necessary to rely for information on the memories of individuals concerned with particular developments within a certain organization. Often the individuals do not remember enough details for valid identification. In other cases the individuals who had been associated with the development have left the company.

B. Measurement

Measurement of the economic impact of missile/space by-products would be a desirable attainment and one in which many have expressed interest. Nevertheless there are many problems encountered, some of which are now discussed.

Measurement of general inventive *activity* is itself a formidable problem which has not been solved to anyone's satisfaction. Moving from a measure of *activity* to a measure of *economic impact* greatly increases the problem. Focusing the task to measurement of the economic impact of missile/space technological transfer reduces few problems and adds many. In this section, problems common to the measurement of general inventive activity and impact will be discussed, drawing largely on what others have written. Following this, the special problems of measuring transfer impact will be discussed.

According to Kuznets, the general problem in measuring inventive activity is that no good methods for measuring inventive input or output exist [31]. Measures of input (i.e., dollars spent on research or numbers of people engaged in research) are not adequate because large variations exist in the creativity and productivity of different inventors; furthermore, inventive activity is unpredictable. Data are more available on inventive output (e.g., patent statistics, patent office files, and occasional lists of inventions published in the literature), but these data have not been converted into efficient measures of inventive output.

Even if a measure of the output of inventive activity did exist, it would be subject to interpretation and would not be a measure of economic impact. Suppose (very hypothetically) that some figures were available for inventive output for the past year, or past several years. The figures would not tell: (1) what secondary and tertiary sales had been generated by the invention, for a *multiplier effect;* and (2) what items had been made obsolete or less useful by new inventions, for a *netting-out effect.* A notable example illustrating these two effects is the invention of the automobile. While sales of both the petroleum and steel industries were given impetus, sales of wagon wheels and buggy whips diminished. The problem, of course, is that any element of product improvement tends to change the market structures of all industries related to that product. Not only is the product changed, but the frame of reference is changed, and trying to measure the impact of product improvement is analogous to measuring a dimension with an elastic ruler.

The measurement difficulties for invention in general are not alleviated, but compounded, when the measurement of missile/space transfer impact is undertaken.

1. IDENTIFICATION

Obviously, a first step in measurement is identification of that to be measured. However, as the preceding section has explained, positive identification of examples of transfer is not always possible.

2. CLASSIFICATION

Once a by-product or technological transfer has been identified, we are far from valid measurement of economic impact. The method of measurement and the problems involved would seem to vary with the type and amount of missile/space contribution made to the item. Two types of contribution drawn from the classification explained in Section I will serve to illustrate the measurement problems: product improvement and the stimulation of basic and applied research.

Product improvement poses this problem. Suppose that a potentiometer has been improved 15 per cent in reliability and 30 per cent in accuracy as a result of research done to meet missile/space specifications. It would be unfair to attribute all the commercial sales of the improved product to the missile/space program, but a contribution has been made. What part of the total sales should be taken as a measure of missile/space contribution?

Stimulation of basic and applied research poses a different and far greater problem. The ultimate impact of this type of contribution will occur in the form of new products, processes, and materials, but this type of contribution is several steps removed from the product-process-materials level. In most cases, it is too early to know even a small fraction of the products that ultimately will be generated by this research. If some products have resulted and can be identified, one cannot even guess what portion of the ideas, knowledge, and technology embodied in the product came from missile/space-motivated research. Product generation frequently represents a synthesis of knowledge originating from many sources, and it is quite likely that the persons responsible for generating products under these circumstances have, at best, only a fuzzy idea of which knowledge truly came as a missile/space contribution. That most missile/space research is grounded on premissile/space scientific knowledge further compounds the difficulties. Thus, there is no apparent way of determining the proportionate contribution of missile/space research to an item coming from this type of transfer, and hence no way of allocating gross sales, or any other measure of economic impact, as to missile/space versus non-missile/space contribution.

3. TYPES OF MEASURE

In the examples above, sales has been mentioned as a measure of economic impact. However, other quantities might also be used: employment, profitability, or effect on gross national product. The quantity chosen would depend partially on whether attempted impact measurement had reference to the firm, the industry, or the economy.

a. *Sales.* If sales is chosen as a measure of output, several other problems arise than have been already discussed. First, some organizations consider sales data on individual product lines as proprietary and are not willing to release them. Of all the possible measures of impact, however, sales would probably be the easiest to secure; the real problem is not in obtaining the data, but in interpreting them. To use the total sales of an item of transfer as a measure of impact would present a very distorted picture if the item contains only a small degree of missile/space contribution. Unfortunately for measurement purposes, much of the total missile/space transfer takes place in the form of partial and intangible contributions to commercial products. How, then, should sales be divided to present a true picture?

The problem becomes more confusing when the components or materials manufacturer does not know what percentage of the sales of his product goes to missile/space versus non-missile/space use. For example, an alloy may have both of these uses and may incorporate some space contribution. If the percentage of non-space sales is not known, how can the gross sales of the product be allocated without doing costly market research?

Using sales as a measure becomes still more difficult when a price reduction is brought about by volume production for the missile/space program, because dollar sales may actually decrease though productivity has gone up.

b. *Employment.* Generation of employment might well be used as a measure of impact and could be very meaningful in view of today's relatively high rate of unemployment. Unfortunately, it is subject to the same difficulties as sales—and more. The increased difficulty comes in obtaining employment data; direct and indirect employment tabulated by product are figures not normally kept. If the data were obtained, however, employment would be subject to the same interpretation problem as sales. Some basis must be found for taking a reasonable percentage of the total in accord with the significance of the missile/space contribution and in accord with the percentage of non-space market. Furthermore, technology tends to increase productivity, thereby increasing or decreasing employment depending upon a number of factors, making the interpretation of the employment measure almost hopeless.

c. *Profitability or Effect on GNP.* Measurement of profitability generates some new problems of its own. Of course, sales must be measured before profitability can be determined. In addition, the sales of a product must be compared with the cost of that product (including cost of research) to determine profitability. But cost of research is a very difficult quantity to allocate to any one product due to several factors, including the uncertainty of invention, the widely varying inventive productivity of individuals, the market structure change caused by product improvement, and the variable time lag in which research yields its returns. Nor does it appear feasible to measure the effects of transfers on gross national product, due to aforementioned factors: chiefly, the multiplier and netting-out effects brought about by market structure change.

We must conclude by observing that measurement of the economic impact of missile/space transfer to the commercial economy is infeasible at this time. We do not deny that, given a readily identifiable and direct example of transfer, some sales and cost figures could be made available. We do, however, flatly state that such examples are the exceptions, not the rule. Such examples, being exceptions, prove nothing about the total amount of economic contribution which missile/space by-products are making.

C. Time Lag

There is an unknown and variable time interval between the inception of missile/space technology and its commercial application. Not only is time lag an important part of this process or any inventive process which should be recognized and analyzed, but it is also a factor which further complicates the identification of examples of transfer and the measurement of their impact.

1. BACKGROUND

Throughout the research on which this chapter is based, comments were made concerning the importance of time lag. A typical remark from industry representatives was, "We have had many important by-products in the commercial field as the result of past war and defense contracts, and I have confidence that, similarly, there will be important by-products of the U.S. space program. I would expect the time lag involved would vary with the complexity of the program." Or, "There has not been enough time since NASA got underway in 1958—or even since the inception of the Air Force missile program—for too many by-products to be forthcoming. We expect much more in the future."

The term "time lag" as used by individuals interviewed during the study was never defined explicitly but was generally used to refer to the hiatus between inception of a missile/space technological development and its commercialization. Since invention, innovation, product development, and marketing are overlapping and continuing processes, it is impossible to identify the beginning or end of time lag. However, exact definition or quantification of time lag is not as important as the recognition that time lag is an integral part of the inventive and technological transfer processes.

Some figures exist on time lag duration, however. Gilfillan studied a group of nineteen inventions selected in 1913 by a vote of *Scientific American* readership as the most useful inventions introduced in the preceding quarter century [32]. He found that the *geometric mean* interval between the year the invention was first proposed and the year of the first working model or patent was 176 years. He also found geometric mean between first machine or patent to first commercial use, 24 years; first commercial use to commercial success, 14 years; commercial success to important use, 12 years.

In our study, five to ten years was quoted most frequently as the period required to develop and market an item incorporating technology transferred from missile/space programs. There are two important differences between Gilfillan's time lag and ours, however. First, Gilfillan studied the total invention from first thought to fruition, while we studied just the secondary or by-product aspects of an invention. Second, Gilfillan studied only important inventions, while we studied all types, including many relatively unimportant innovations. In addition, many of our identified transfers are not inventions, e.g., cost reductions and product improvements.

Nevertheless, a significant time lag exists between missile/space development and commercial application. One company reported a time lag of six years from the first requirement for a materials testing machine for the missile program to a finished commercial testing machine. Another firm reported a time lag of about seven years from the first contract to develop a new series of connectors for the Air Force to the first commercial sales of the finished connector line. Still another company reported that the time lag from first automatic pilot to first work on adaptive flight control for missiles was fourteen years, and time lag from first missile work to first commercial sales of an adaptive autopilot for light aircraft was six years.

2. DURATION

That a time lag exists between a missile/space development and its commercial application is generally accepted. However, a quantitative estimate of time lag is meaningless unless viewed in light of those factors which affect its duration. Some of those factors can be summarized as follows:

(a) Nonmissile/space utilization of missile/space developments or concepts is necessarily a secondary objective. Many firms participating in missile/space developments have been too concerned with fulfilling their primary objective to give much attention to the possible commercial application of knowledge gained.

(b) The technological content of most missile/space products is many years ahead of that found in commercial or industrial products. Introducing such products before their time into an industrial or commercial system makes no more sense than using a pile-driver to pound a nail.

(c) Most missile/space products, because of their high technological content, are expensive relative to their industrial or commercial counterparts. For example, an extremely reliable resistor may be an absolute necessity in the Minuteman program but a luxury in an industrial computer. Initially such a product will receive only specialized industrial use, but as more units are used the price will come down, and as the price comes down, more units will be used. Eventually the product may find widespread nonspace use, but this process of "sliding down the cost curve" takes time.

(d) A corollary to the latter two points is that missile/space products or developments representing small improvements transfer into the commercial economy faster than those representing major innovations.

(e) The intensity of the effort put forth to commercialize a missile/space development can effect time lag. This effort can be divided into three parts: (1) adapting the product or development to commercial or industrial use from its present missile/space form; (2) producing it; and (3) actually selling the product. However, many firms which are responsible for missile/space developments are not organized to do these things, thus increasing the time lag.

Each of these factors indicates, as would be expected, that anticipated profitability is an important key to time lag.

3. CONCLUSION

We can observe that, because of the relative newness of both missile and space programs, commercial products incorporating missile/space contributions are just beginning to appear. The Air Force missile program did not really get underway until the mid-1950's; NASA was not established until 1958 (see Fig. 1). Taking time lag into consideration, it would be unreasonable to expect a large proportion of the transfers which eventually will occur to have occurred at this time.

IV. Incentives and Barriers to the Commercial Application of Missile/Space Technology

The application of missile/space technology to commercial use is influenced by many factors which can be thought of as stimuli or barriers

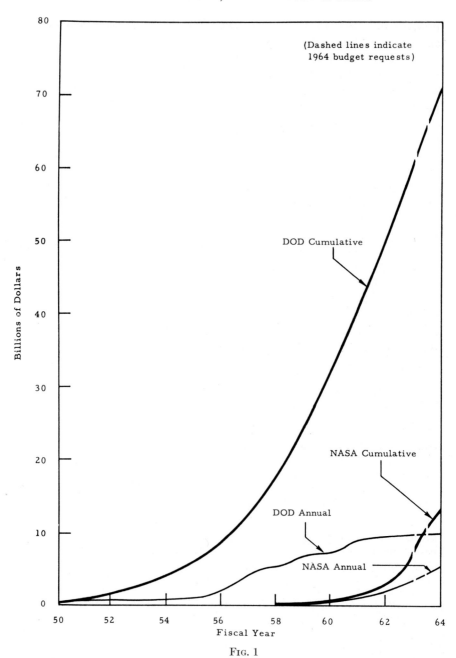

Fig. 1

to the commercial application of space-generated technology. While finding and documenting the examples of technological transfer we noted the influence which certain of these factors had on the process.

It is important to note that the same factor may act as an incentive under certain circumstances, and as a barrier under different circumstances. Furthermore, the stimuli and barriers operate at two different levels of activity, the individual and the institution (industrial firm, government laboratory, research institute, university).

We have attempted to summarize in Table II some of the incentives and barriers to help give the reader an idea of the many and varied factors which affect the commercial application of missile/space technology. Some of these factors are discussed below.

A. Government Patent Policy

No other area was the subject of as much comment and interest on the part of industry people interviewed as that of patents and government patent policy. Since the total subject of government patent policy is too broad and complex to be adequately dealt with here, the comments of industry people interviewed will be summarized. For those wishing to pursue the subject in depth, reference is made to several publications dealing specifically with the subject [33].

In a nutshell, the question is whether ownership of (title to) patents arising from government financed work should reside with the contractor, subject to a royalty free, nonexclusive, irrevocable license to the government, or should the patents be acquired by the government? At the present time there is no uniform government patent policy; the desirability of having such a policy is itself part of the dispute. The Department of Defense permits (with certain exceptions) title to reside with the con-

---

Fig. 1. Missile/space spending: Department of Defense (DOD) and National Aeronautics and Space Administration (NASA), annual and cumulative. (1) *DOD funds* available for missile development and production include not only the cost of procuring missiles for operational purposes, but also include research, developmental and capital costs involved in bringing this program to an operational status. The figures do not include military pay and costs only indirectly associated with the missile program. [*Aerospace Facts and Figures 1962*, Aerospace Industries Association of America, Inc. (Washington, D.C.: American Aviation Publications, Inc.), p. 20; "U.S. Aeronautics and Space Activities, 1962," Bureau of the Budget, Report to the Congress from the President of the U.S.; *Missiles and Rockets* 12 (January 21, 1963), p. 12.] (2) *NASA data* represent total NASA appropriations. [*Aviation Week* 78 (January 21, 1963), p. 30; Leonard S. Silk, The impact on the economy. *In* "Outer Space—Prospects for Man and Society" (L. P. Bloomfield, ed.), p. 83. Prentice-Hall, Englewood Cliffs, New Jersey, 1962.]

TABLE II. INCENTIVES AND BARRIERS AFFECTING COMMERCIAL APPLICATION
OF MISSILE/SPACE TECHNOLOGY

Individual incentives	Institutional incentives
Salary	Profits
Recognition and status	Diversification
Education	Organizational structure
Invention awards	Patents
Patents	Personnel transfers
Dissemination media	Competition
Scope of job responsibilities	Continuity of the firm
	Market preemption
	Management sophistication

Individual barriers	Institutional barriers
"Not invented here" concept	Uncertainty and risk
Reluctance to change	Reluctance to change
Time-consuming paper work	Lack of knowledge
Narrowness of interest	Government regulations
Lack of knowledge	Capital requirements
Isolation	Proprietary data
Complexity and sheer mass of data	Security regulations
	Organizational rigidities
	Incompatibilities between government and commercial R & D and production

tractor. The National Aeronautics and Space Administration, on the other hand, automatically acquires title, but may, under certain circumstances, waive rights to the contractor.

For the most part, we encountered arguments in favor of the government's *not* taking title to patents. These arguments can be summarized as follows: (1) A government owned patent has less chance of being worked than if title resides with the contractor, especially if sizable capital expenditures are required to bring the invention to market. (2) Aerospace firms which license patents to other, more commercially oriented firms, as a method of exploiting by-products are deprived of their incentive to accelerate transfer when working under a policy where government takes title. (3) Eventually, government's taking title, if uniformly applied to all government financed R & D, would result in government ownership of a tremendous number of patents. Since more than half the R & D in the United States is sponsored by the government, which in itself represents unprecedented control over technological advances, a government title policy would further strengthen the hand of government, giving it control over the use made of a significant portion of technology. This contradicts the generally accepted role of the free market (modified by the traditional patent system) as the most effective economic decision maker. (4) Government's taking title tends to act to the detriment of small firms. Without patent protection, the small firm is less apt to try to exploit a by-product than a big firm in those cases where the capital expenditure required to bring the by-product to market is sizable.

From the standpoint of commercial application, which is our main concern in this chapter, there does appear to be a fairly simple criterion with which to examine the patent question. That is: which policy will most effectively help to bridge the gap between invention and commercial application? With certain exceptions, we believe that *private* ownership of patents, as is generally embodied in the policy of the Department of Defense, best meets this criterion.

B. Security Regulations

One of the more obvious *barriers* to the commercial application of certain missile/space technology is caused by security regulations. While recognizing the necessity for security regulations, several viewpoints were expressed by a number of industry representatives which may indicate possible areas for change in the interest of better scientific communication. Generally, these viewpoints fell into the following patterns:

(1) More effort should be made to declassify technology at the earliest possible date, consistent with national security requirements, especially in the basic research field.

(2) The process through which a need to know is established should be less restrictive and less time-consuming.

(3) Companies not engaged in government work are left out of potentially valuable sources of technological information, being denied access to many government research reports.

C. Influence of Market Mix Served by Individual Firms on Technological Transfer

Whether a government contractor also serves nongovernment markets seems to have a significant effect on the contractor's ability to apply missile/space technology to commercial applications. The following comments summarize the opinions expressed by industry on this topic:

(1) Individuals in firms serving both markets (government and commercial) were especially vocal about the need for management to be commercially oriented in order to market successfully to industrial and consumer markets, commenting that a military-oriented sales depart-

ment would have great difficulty marketing commercial products. Some individuals added that the know-how gap between marketing government and industrial products was large, and the gap between government and consumer products was considerably larger.

(2) Smaller firms with both commercial and government identities would probably experience relatively more civilian applications of missile/space technology than large firms because the former could more easily effect a marriage of commercial applications with government-related technology. Large size was believed by some to be a barrier to the linkage of market and technological knowledge.

(3) A large number of potential commercial applications are probably lying dormant in some firms which have little or no commercial identity because they do not have the organization, know-how, or motivation to exploit them and because they may not have the knowledge necessary to recognize and evaluate potential commercial applications.

(4) The greater the number of commercial product lines manufactured by a government contractor, the greater tends to be the commercial application of missile/space technology. The basis for this reasoning was that the opportunities for commercial applications seem to rise somewhat proportionately with the variety of commercial products being produced, i.e., broader market coverage.

A rather obvious conclusion that one might draw from these comments is that selling to both the missile/space and commercial markets on the part of an individual firm tends to act as a *stimulus* to the transfer of missile/space technology into commercial applications.

D. Incompatibility of Government Contracting with Commercial Operations

Many and varied management problems were reported by widely differing firms which were serving both the missile/space and commercial markets or which had tried to serve both markets in the past. These problems were generally attributed to differences in the nature of, and in the conditions surrounding, the two types of work.

Areas of difference most frequently cited included: marketing, patents, production and quality control, price, accounting, finance, research and engineering, security, ownership of plant and equipment, and bidding requirements.

Typical problems in this area were expressed as follows:

(1) Missile/space products can be characterized by "reliability at any price," as opposed to "adequate reliability at a competitive price" for commercial products. This fundamental difference makes it difficult for

many technical people with government contracting backgrounds to transfer to a commercial environment and operate effectively because they tend to overdesign commercial products, thus pricing the products out of competition.

(2) The added operating costs required to meet government specifications regarding inspection, audits, reports, etc., frequently make it infeasible to produce both government and commercial products (even though quite similar) in the same facility. This has caused some companies to separate physically government and commercial manufacturing facilities.

It is obvious, therefore, that the incompatibilities between government and commercial work act as a *barrier* to the commercial use of missile/space technology.

E. Company Organization and By-Product Exploitation

An attempt has been made by several aerospace firms to bridge the gap between missile/space technological knowledge and commercial market application by developing organizational mechanisms to handle commercial exploitation of this missile/space technology. A number of factors appear to have influenced the type of organizational structures selected for this purpose, such as company size, nature of company products, type and number of customers, and desires and interests of management. There are several ways in which missile/space technology can be exploited: licensing, franchising, outright sales of inventions, manufacturing and marketing directly by originating company, or combinations of these.

Two types of formal organizational structures to exploit by-products have been adopted by several aerospace companies.

1. SEPARATE DIVISION OR SUBSIDIARY

Two aerospace companies have formed separate organizational units for the express purpose of by-product exploitation. Both of these organizations screen by-products from the other divisions of the company and carefully select those which have the best possibility for industrial or commercial application. These organizations, however, do not take by-products from an operating division if the operating division wants to make the by-product a part of its product line.

2. INTEGRATED ORGANIZATIONAL UNIT

Several companies in the aerospace industry have established separate organizational units on a lower level to handle by-products. These units tend to be more integrated into the company's main stream of business and have less autonomy than those mentioned above.

Some of these groups are fairly new and it is not yet clear what direction they will take or what the scope of their activities will be.

In general, most of the other large and medium-sized firms contacted in the course of our study which served both government and commercial markets followed the pattern of exploiting by-products within the originating division, or within another division handling similar products. In some cases, by-products are licensed to another company if they do not fit a company's product lines or for other reasons.

No special organizational mechanisms were noted in small firms contacted, and perhaps none is needed or would prove feasible.

In addition to various company organizational structures, there are other mechanisms which appear to act as gap-bridges. For example, some universities, such as the California Institute of Technology, have active research foundations which license new developments to industry for application.

F. Increasing Volume, Complexity, and Degree of Specialization of Technology

The rapid advances being made in science and technology, while presenting many opportunities, are also causing serious problems. One is how to keep up with the sheer mass of information available. Many comments were made during the course of our research reflecting the ramifications of the literature problem in industry. Generally, they fell into the following four categories:

(1) The small organization does not have the manpower required to derive full value from the information now available from various sources.

(2) Better classification of existing information and less duplication in dessemination would be helpful.

(3) The need is not for increased flow of information but for improvement in quality of the present flow.

(4) To cope with the flood of new technical knowledge, management should hire the most competent professional talent available to identify and monitor the sources of new knowledge.

Other comments ranged outside the literature problem into the implications of the increasing volume and degree of specialization of technology.

(1) There are so many poor application ideas proposed for every good idea that the good ones often get buried and are lost with the bad.

(2) Salesmen, long a source of new application ideas, are experiencing difficulty in keeping up with new technological developments within their firms and hence are not as productive as formerly in generating new uses for new products.

It is obvious that the sheer mass and relative inaccessibility of published technical literature in the missile/space field act as barriers to the commercial application of missile/space technology.

G. Pressures of Time Schedules

Closely tied to the comments reported in the previous section on the increasing volume of scientific and technical knowledge were several observations concerning problems imposed on individuals by limitations of time. These were raised in several different contexts.

(1) On the philosophical side, the dean of a business school pointed out that a major problem was developing regarding the division of a manager's time between acquisitions of knowledge and the application of knowledge to decision making.

(2) A different aspect was raised by several industry representatives to the effect that R & D people engaged on government contract work were too pressed by time schedules to give much, if any, thought to possible commercial applications of their work.

While some individuals undoubtedly are stimulated by time pressures to increase their outputs, limitations of time can also pose barriers to the transfer process.

H. Scope of Individual Interests and Knowledge

Most scientific researchers, according to a number of people contacted in industry, tend to concentrate their interests on scientific objectives. They are relatively uninterested in the eventual commercial application of their discoveries. (There are notable exceptions to this generalization, of course.) It was reported to us that many R & D space scientists never give commercial applications a second thought, since their primary objective is to produce space accomplishments as rapidly as possible.

On the other hand, a few individuals believed that the innovator himself should be the best source of information concerning commercial applications of his work. This viewpoint is contradicted, however, for several observations reported that application ideas come about only when a marriage is effected between technical knowledge and market knowledge, and the innovator frequently does not possess the latter.

In any event, most of the opinions and experiences reported by industry representatives indicated that narrowness of interest among many scientists and engineers acts as a *barrier* to the transfer of missile/space technology.

I. The "Not Invented Here" Concept

A commonly used term in industry, "not invented here," refers to the tendency on the part of the individual scientist, or working group of scientists, to be unwilling to accept freely information not developed by himself or the immediate group concerned. This tendency was regarded by a number of industry representatives as a very real barrier to diffusion of technological information, not only between different organizations but also between different divisions within a given firm. One explanation attributed this behavior to a form of pride of authorship—the wish fathering the belief that one's ideas are superior to those of another.

J. Personnel Transfers and Job Changes

A number of individuals in both government and industry expressed the belief that the flow of people within and between organizations is one of the most effective diffusers of technology.

For example, the normal promotional process of both management and skilled personnel results in intracompany transfers from space to nonspace departments in one commercial firm with large space contracts. According to an official in the firm, these people carry with them "education" received in space work, which brings into play in their nonspace assignments new ideas relating to reliability, quality control, and testing processes, as well as higher skills in mechanical functions.

It is logical to conclude that the flow of individuals from missile/space work to commercial assignments can act as a stimulus to the commercial application of missile/space technology. However, in view of the difficulties mentioned earlier which many technical people have experienced in adapting to a commercial environment from a government contracting environment, it is possible that inquiry into the operation and results of the planned flow of technical people would prove enlightening. Such an inquiry should indicate whether formal transfer programs would be worth encouraging in firms serving both missile/space and commercial markets as a means to stimulate commercial applications of missile/space technology.

V. Summary of Findings and Conclusions

The primary objectives of the study on which this chapter is based were: (1) to identify examples of present and probable future commercial contributions of the nation's missile/space programs, and (2) to determine the circumstances surrounding the origin of these contributions.

Six distinct, though not mutually exclusive, types of contribution were noted: stimulation of basic and applied research; new or improved processes and techniques; product improvement; materials and equipment availability; new products; and cost reduction.

By far the most important category of contribution is and will continue to be stimulation of research. We believe that missile/space contribution in this category has been of moderate significance to date, but will be of fairly large significance in the future. Contribution in the other categories has been of only minor economic significance. While the other five types of contribution will probably take on more significance in the future, their economic impact will always be greatly overshadowed by the impact of the first category of contribution—stimulation of research.

A time lag exists between the development of technology for missile/ space use and its transfer to commercial application. The importance of this time lag is too often overlooked. Large expenditures on missile/space programs are fairly recent and there has not yet been time for many transfers to have occurred. It is highly probable, therefore, that most of the transfer is still to occur. Supporting these two findings, over thirty broad technological areas were identified in which transfer has taken place or can be reasonably expected to occur. A total of 185 individual examples of transfer were identified, many of which have been mentioned in this chapter.

Due primarily to identification difficulties, these examples do not include all the transfers which have taken place, nor can they be considered statistically representative of what has occurred. They do, however, provide evidence of the magnitude, nature, and scope of the missile/space contribution to the commercial sector.

The six categories of contributions serve to illustrate an important point, generally overlooked. That is, the transfer of missile/space technology is not a simple, readily identifiable process, as the word "by-product" might imply, but rather is a highly complex process which takes place in a variety of ways. Missile/space R & D is but one contributor to the vast store of knowledge which is the source of technology for both government and commercial sectors of the economy. Other R & D contributors include industry, universities, and other government agencies. Because of the interaction among all these, attribution of a given technological advance to a particular source is often impossible.

The nature of the transfer process is such that quantitative measurement of the economic impact of missile/space contribution to the commercial economy is infeasible at this time.

It would be a mistake to assume that all companies can benefit directly from missile/space technological transfer. It is impossible for us to identify which firms can and which cannot profitably utilize this transfer. We can, however, offer some guidelines: (1) the firm that has a substantial portion of its sales in both government and commercial markets is in the best position to exploit transfer; (2) those companies engaged wholly or largely in government contracting, and those firms which sell wholly or largely to commercial markets, are in a difficult position to exploit transfer—except through licensing arrangements; (3) since technology is the most important type of contribution, it is reasonable to assume that this is the area of transfer where a firm trying to exploit transfer should concentrate its efforts; (4) in doing so, the firm's management should bear in mind that the more similar its commercial product line to its government product line, the easier technology will flow. This rule works both ways. Technology generated in doing commercial work also can be exploited in application to government contracts. By trying to bring its government and commercial work close together, a firm might find its efforts to exploit transfer doubly effective.

Finally, there are certain barriers and stimuli which inhibit or accelerate the process of applying missile/space technology commercially. These can be identified but their effect and interaction are not well understood. More factual knowledge on the workings of these barriers and stimuli should be of great value to those in industry and government who seek to apply fruits of missile/space R & D to the commercial economy.

References

1. Welles, J. G., et al. (1963). "The Commercial Application of Missile/Space Technology." Denver Research Institute, University of Denver, prepared for the National Aeronautics and Space Administration, September 1963.
2. Liu, F. F., and Berwin, T. W. (1958). Extending transducer transient response by electronic compensation for high-speed physical measurements. Rev. Sci. Instr., 29, 14–22; and Liu, F. F., and Berwin, T. W. (February 1958). Recent advances in dynamic pressure measurement techniques. Jet Propulsion 28, 83 ff.
3. 1962 National Symposium on Space Electronics and Telemetry (October 1962). Institute of Radio Engineers Professional Group on Space Electronics and Telemetry, PGSET Record, (Miami Beach, Florida); Inter-Range Instrumentation Group Telemetry Standards, Document No. 106-60 (December 1960). Telemetry Working

Group, Inter-Range Instrumentation Group of the Range Commanders' Conference (White Sands Missile Range, New Mexico); *1957 National Telemetering Conference Report* (May 1957). Conference sponsored by American Institute of Electrical Engineers, Institute of Aeronautical Sciences, and Instrument Society of America (El Paso, Texas).

4. Gillmor, R. E., *et al.* (1962). Gyroscope. *Encyclopaedia Britannica* 11, 47–51.

5. Draper, C. S. (March 1960). The inertial gyro—An example of basic and applied research. *Am. Scientist* 48, 15.

6. Alt, F. L. (1958). "Electronic Digital Computers," pp. 17–23. Academic Press, New York.

7. Leary, F. (April 1961). Computers today. *Electronics* 34, 64–94.

8. Harder, E. L. (May 1959). Computers and automation. *Elec. Eng.* 78, 517.

9. Mapes, D. (October 1955). High pressure air in reinforced plastic bottles. *Mod. Plastics* 33, 137.

10. *Mod. Plastics Encyclopedia, Issue for 1962* (September 1961) 39, No. IA, p. 609.

11. Epstein, G. P., and King, H. A. (March 1957). Filament winding has a fine future. *Mod. Plastics* 34, 132–2.

12. Mohler, J. B. (April 1961). Introduction to chemical milling. *Mater. Design Eng.* 53, 128–132.

13. Van Deusen, E. L. (January 1957). Chemical milling. *Sci. Am.* 196, 104–112.

14. Bertossa, R. E. (September 1960). Explosive forming and chemical milling discussed. *Metal Progr.* 78, 112–116.

15. Park, F. (June 1962). High-energy-rate metalworking. *Intern. Sci. Technol.*, pp. 12–23.

16. Greenlee, R. E., and Bickley, W. H. (Undated). Explosive metalworking . . . Its applications status. *Office of Technical Services, Report No. PB-181067*, p. 12.

17. Greenlee, R. E., and Bickley, W. H. (Undated). Explosive metalworking . . . Its applications status. *Office of Technical Services, Report No. PB-181067*, p. 9.

18. Monteil, V. H. (August 1961). How to design for explosive forming. *Metal Progr.* 80, 67.

19. Lessing, L. (September 1961). Blasting metals into shape. *Fortune* 64, pp. 127–131.

20. The research trend letter, (February 1962). *Ind. Res.* 4.

21. Levy, A. V. (1961). Use of refractory materials in air-breathing engines. *In* "Refractory Metals and Alloys," A.I.M.E. Metallurgical Society Conferences (O. M. Semchyshen and J. J. Harwood, eds.), Vol. 11, pp. 603–617. Wiley (Interscience), New York.

22. Bushor, W. E. Medical electronics. *Electronics* 34, Part I: Diagnostic measurements, January 20, 1961, pp. 49–55; Part II: Diagnostic systems and visualization, February 3, 1961, pp. 46–51; Part III: Therapeutic devices, February 24, 1961, pp. 54–60.

23. Schwichtenberg, A. H., Flickinger, D. D., and Lovelace, W. R., II (November 1959). Development and use of medical machine record cards in astronaut selection. *U.S. Armed Forces Med. J.* 10, 1324.

24. Berman, R. M. (March 1962). Radio-electrocardiography: A new technique. Paper read at a forum on medical instrumentation, The New York Society of Security Analysts, Inc.

25. Fazar, W. (December 1961). Navy's PERT system. *Federal Accountant* 11, 123–35; Fazar, W. (June 1961). Advanced management systems for advanced weapon systems. Address before the 5th Annual Military Electronics Convention of the Institute of Radio Engineers (Washington, D.C.).

26. Kelley, J. E., Jr. (April 1962). "History of CPM and Related Systems." Mauchly Associates, Inc., Fort Washington, Pennsylvania.
27. Boehm, G. A. W. (April 1962). Helping the executive to make up his mind. *Fortune* **65**, 128–132.
28. Neville, G., and Falconer, D. (October 1962). Critical path diagramming. *Intern. Sci. Technol.*, pp. 43–49.
29. Phelps, H. S. (October 1962). What your key people should know about PERT. *Management Rev.* **51**, 45.
30. Short-cut for project planning, (July 7, 1962). *Business Week*, pp. 104–106.
31. Kuznets, S. (1962). Inventive activity: Problems of definition and measurement. *In* "The Rate and Direction of Inventive Activity: Economic and Social Factors," pp. 31–35. Report of National Bureau of Economic Research, Princeton Univ. Press, Princeton, New Jersey.
32. Gilfillan, S. C. (November 1952). The prediction of technical change. *Rev. Economics Statistics* **34**, 368–85.
33. Several publications of The Patent, Trademark, and Copyright Foundation have been devoted in whole or part to this controversy: *Patent, Trademark, Copyright J. Res. Educ.* 4 (Winter 1960); 5 (Winter 1961–62); and 6 (Conference Number 1962). An excellent summary of the range of views on the question fills an entire issue of the Federal Bar Association publication: *Federal Bar J.* **21** (Winter 1961). The U.S. Senate Committee on the Judiciary has conducted extensive studies and hearings on this matter in recent years. For a summary of recent activities, see *Patents, Trademarks, and Copyrights*, 87th Congress, 2nd Session, 1962, Report No. 1481 (U.S. Government Printing Office). An excellent introduction to the broad field of patents was contributed by Fritz Machlup for the Senate Subcommittee on Patents, Trademarks, and Copyrights of the Committee on the Judiciary, *An Economic Review of the Patent System*, 85th Congress, 2nd Session, 1958, Study No. 15 (U.S. Government Printing Office). In addition, the Special Subcommittee on Patents and Scientific Inventions of the House Committee on Science and Astronautics has held lengthy testimony on the patent question. The Subcommittee's present views, together with a good discussion of what the hearings revealed, are presented in *Ownership of Inventions Developed in the Course of Federal Space Research Contracts*, House Report of the Subcommittee on Patents and Scientific Inventions of the Committee on Science and Astronautics, 87th Congress, 2nd Session, Committee Print, April 5, 1962 (U.S. Government Printing Office). A concise review of the subject has been published by the Government Patent Policy Study Committee organized by the National Council of Patent Law Associations: *Statement of Principles for the Evaluation of Federal Government Patent Policy* (National Council of Patent Law Associations, Washington, D.C., Spring 1962). A recent *Fortune* article briefly sums up the controversy and relates it to other current problems present in the patent system: Bryant, S. W. (September 1962). The patent mess. *Fortune* **66**, 111 ff.

Progress in Rocket, Missile, and Space Carrier Vehicle Testing, Launching, and Tracking Technology

Part I: Survey of Facilities in the United States

MITCHELL R. SHARPE, JR.

University of Alabama Center
Huntsville, Alabama

AND

JOHN M. LOWTHER

Chrysler Corporation
Cape Kennedy, Florida

In the past thirty years, rocket testing ranges have grown from simple and inexpensive means developed on small funds and manned by enthusiastic amateurs to multimillion dollar enterprises employing thousands of scientists and technicians. In the 1920's and 1930's, the development and testing of rockets was largely in the hands of rocket societies, very low-budget military projects, or equally low-budget research foundation programs. Typical of these operations were static firings from the trailer-mounted test equipment of the old American Rocket Society,

245

which performed many a "shoot and scoot" experiment in New Jersey; and the somewhat more sophisticated test facility of the German Verein für Raumschiffahrt at its Rakenteflugplatz in Berlin, which had to close when a leaky hydrant ran up a water bill that could not be met. By these standards the New Mexico facilities of Dr. Robert H. Goddard, financed largely by the Daniel and Florence Guggenheim Foundation, were extensive if not luxurious.

Today there are at least 75 well-equipped and instrumented rocket ranges in operation around the world. Some of these are utilized solely for the development and testing of military missiles, while others are employed only for scientific firings. The larger ones, in general, are used for both purposes. They vary in size from the vast Soviet ranges in Kazakhstan to the small sounding rocket ranges in Sweden, Argentina, and New Zealand. Depending upon capabilities, they can employ as few as 11 to as many as 27,000 persons. Annual operating budgets may run into millions of dollars for these establishments with capital assets of as much as 1.3 billion dollars.

It is the purpose of this chapter to describe, in as much detail as security measures permit, the capabilities of those missile and rocket ranges known to exist. Undoubtedly there are omissions here, but those ranges included in this chapter, which is divided into two parts, represent the best efforts of the authors in obtaining as much information as possible. Since a modern missile range is dynamic by nature and the product of continuously changing mission requirements, the physical facilities and equipment change accordingly. But the major features described herein are likely to remain accurate for some time.

Originally the authors intended to include material on the various governmental and industrial facilities in the U.S. for the static testing of solid and liquid propellant rockets. However, this coverage became impractical in view of the space alotted and the large number of such establishments in the country. Descriptions of such facilities in other countries are included in varying degrees.

The authors acknowledge the assistance received from both organizations and individuals in the preparation of this book. Among the organizations kindly contributing information were the National Aeronautics and Space Administration, Wallops Station, Virginia, Launch Operations Center, Cape Kennedy; the U.S. Air Force, Eglin A.F.B., Edwards A.F.B., Patrick A.F.B.; the U.S. Navy, Naval Ordnance Test Station, Naval Air Missile Test Center, Naval Missile Facility; the U.S. Army, White Sands Missile Range. Other organizations aiding the authors include the Sandia Corp., Albuquerque, New Mexico; Stato Maggiore della Difesa, Rome; North American Aviation, Inc., El Segundo, California; Italian Embassy,

Washington, D.C.; Deutsche Rakenten-Gesellschaft, Hannover, West Germany; Atlantic Research Corp., Alexandria, Virginia; Comisión Nacional de Investigaciones Espaciales, Buenos Aires, Argentina; Embassy of Japan, Washington, D.C.; Department of National Defence, Ottawa, Canada; British Embassy, Washington, D.C.; J. W. Fecker Div., American Optical Co., Pittsburgh, Pennsylvania; Academy of Sciences of the U.S.S.R., Moscow; Norwegian Defence Research Establishment, Kjeller, Norway; Texas Western College, El Paso, Texas; Société Anonyme de Télécommunications, Paris; Pakistan Space and Upper Atmosphere Research Committee, Karachi; National Lucht-En Ruimtevaart-laboratorium, Amsterdam; University of Tokyo, Institute of Industrial Science, Tokyo; Compagnie Française Thomson-Houston, Paris; Ambassade de France, Washington, D.C.; Physical Research Laboratory, Ahmedabad, India; Askania-Werke, Bethesda, Maryland; Department of Supply, Commonwealth of Australia, Melbourne, Australia; Ministère des Armées, Paris; and Institute of Meteorology, University of Stockholm, Stockholm.

Individuals include Mr. Vernon Monjar, the Chrysler Corp., Cape Kennedy, Florida; Sir A. C. B. Lovell, University of Manchester, Jodrell Bank, England; Mr. A. V. Cleaver, Rolls-Royce, Ltd., Derby, England; Dr. C. Ellyett, University of Canterbury, Christchurch, New Zealand; Mr. Josef Boehm, Dr. Gerhard Reisig, Mr. Hermann Ludewig, Mr. Hannes Luehrsen, George C. Marshall Space Flight Center, Huntsville, Alabama; Mr. Joseph Zygielbaum, Electro-Optical Systems, Inc., Pasadena, California; Maj. Nittotaro Mizuma, Japan Self-Defense Force; and Dr. Kurt Magnus, Technical University, Stuttgart.

The authors are especially grateful to Mr. Donald J. Ritchie, Research Laboratories Div., The Bendix Corp., Southfield, Michigan. An expert on Russian missile and space technology, Mr. Ritchie kindly gave not only his time in reviewing and enriching the text but also contributed his skill as a technical illustrator to the section on Soviet ranges.

The authors would have liked to include in this chapter material on the NATO missile range on Crete and the US Air Force range at Green River, Utah; but security regulations did not permit the release of information. Also, the Chinese missile range variously reported at Shuang Chang Tzu or in the Takla Makan desert of Sinkiang Province is not covered for the same reason.

Because of its length, this study has been divided into Parts I and II, the former appearing in this volume and the latter in Volume 7 of *Advances in Space Science and Technology*. For convenience of reference, consecutive figure, table, and section numbering is utilized as if the whole chapter appeared in one volume.

I. Introduction

Perhaps the most fitting introduction to a survey of contemporary rocket ranges is a brief description of the world's first major facility of this type: Peenemünde, the German research and development center where the first modern guided missiles were perfected.

In April 1936, final approval was given for the construction of a special rocket research and development center in the Bay of Stettin on the northern tip of Usedom Island in the Baltic Sea. Also during this month, Dr. Wernher von Braun was named technical director of the Army's portion of the joint Army-Air Force establishment. The remote, wooded area called Peenemünde was proposed by von Braun, who had noticed it during a visit to relatives living in the nearby town of Anklam. Some 20 mi² of land were purchased from the town of Wolgast for 750,000 marks, and in August 1936, construction on the new center began.

By May 1937, most of the Army operating personnel for the range had arrived from the old testing ground at Kummersdorf, some 20 mi south of Berlin. The giant static test stands, however, were not complete by this date; and this phase of the operation remained for the time being at Kummersdorf. In December 1937, the first launchings from Peenemünde, of the research rockets A-3 and A-5, were made from a concrete pad and observed from a wooden blockhouse on Greifswalder Oie. This small island, only 1100 yd long and 400 yd wide, is located some 7 mi north of Usedom.

By the end of 1937 the Air Force and Army developmental activities at Peenemünde were separated, and in November 1938 the military headquarters and the technical organization of the Army were combined into Heersversuchsstelle Peenemünde.

Throughout 1938 and 1939 technical facilities at the center expanded rapidly into two generally independent areas: Peenemünde East and Peenemünde South under the Army and Peenemünde West under the Air Force. The Army concentrated its efforts on developing the A-4 (V-2), Wasserfall, and Taifun guided missiles; while the Air Force centered its research on the V-1 flying bomb, Heinkle HE 176, and Messerschmitt ME 263 and ME 163 jet aircraft, and air-to-air missiles, jatos, and glide bombs. (During this period and for several subsequent years a rivalry grew between the two services in the development of missiles that set a pattern to be repeated almost identically in the U.S. 20 yr later, with many of the Peenemünde principals incredulously involved.)

By August 1940, all Army personnel from the Kummersdorf station were transferred to Peenemünde.

Development of the V-2 continued from 1940 to 1943, with a first and unsuccessful launching on 13 June, 1942. During this same period the Air Force at Peenemünde West worked on the development of liquid propellant jato units, the V-1, and the He 176 and Me 163 jets. Despite a perfect V-2 launch on 3 October, 1942, this developmental program and others at the center suffered from the sinusoidal funding characteristic of all large, government-sponsored research projects (regardless of national origin). In 1940, Peenemünde was placed on a low funding priority and was not restored to full top priority until July 1943.

Perhaps the most frustrating tribulations of the directors and management of the center were those that 30 yr later seem antic if not lunatic. Typical of these bureaucratic maneuvers that seriously hampered the work at Peenemünde was the jeopardy in which the V-2 program was placed because of a fantastic but unsuccessful attempt to sell the center to private industry so that it could be "managed more efficiently" (and for a profit). Equally bizarre was the delay in the program that resulted from Adolf Hitler's dream that no V-2 would ever reach England!

Despite these trials, as well as the Allied air raids (the first of which occurred in August 1943) the development of the V-2 by the Army and the V-1 by the Air Force progressed at Peenemünde. And the center eventually reached a state of activity that called for a work force of some 18,000 people and a capital outlay of 300,000,000 marks. With the release of the V-2 for production in 1943, the Army shifted its resources to the development of the radar-guided antiaircraft missiles such as Wasserfall and Taifun. While there were studies of advanced missiles, including the two-stage A-9/A-10, the antiaircraft missiles were the last major projects undertaken at the center.

Late in 1944 the Mach 5 wind tunnel was removed to Kochel, in Bavaria. Primarily the relocation was dictated by increased energy requirements that could best be met by the hydroelectric plants in that region.

Contrary to popular opinion, Peenemünde was not the production plant for the V-2. The pilot plant at the center produced only 250 missiles, which were research and development rounds. There were, according to the best accounts, 3745 production V-2s manufactured, in underground plants near Nordhausen in the Harz Mountains and elsewhere. Of these 580 missiles were used for product improvement and training missions. On 6 September, 1944, the first tactical V-2 was fired against a military target near Paris. Between that date and 27 March, 1945, there were 1115 missiles launched against England and 2050 missiles fired against targets in continental Europe.

In April 1945, Peenemünde was abandoned by its personnel as the

Russian and French forces closed in upon it. The Russian troops who captured it found the center 75 per cent destroyed and were ordered to destroy the remaining 25 per cent, which they did by dismantling usable equipment and sending it to the U.S.S.R. to be installed at the first Soviet military rocket range at Kapustin Yar. After the war, the center was partially restored by the Russians and used as a testing ground. As late as 1963, East German citizens were not permitted to travel farther north on Usedom Island than Zinnowitz, an indication that the center was still active.

Major Air Force installations at Peenemünde West consisted of the following [numbers are keyed to the accompanying Fig. A]: engineering building, 30; laboratories, 31; workshops, 32; airfield, 33; control tower, 34; hangars, 35; V-1 missile launch site, 36; seaplane port and recovery boat harbor, 37; and seaplane hangars, 38.

The Army facilities at Peenemünde East and Peenemünde South consisted of three main areas: engineering center, test area, and pilot plant production facility.

Within these areas the principal activities for the engineering center were: engineering building, 1; maintenance shop, 2; machine shop, 3; assembly shop, 4; guidance and control laboratory, 5; field test workshop, 6; materials testing laboratory, 7; liquid oxygen plant no. 1, 8; wind tunnel, 9; military headquarters, 10; officers' and scientists' messhall, 11; noncommissioned officers', enlisted men's, and technicians' messhall, 12; guest house, 13; bachelor quarters, 14; guardhouse, 15; fire station, 16; construction planning office, 17; and shipping and receiving office, 18.

The testing area consisted of: the recording and control center, 19; test stand P1 (vertical stand for complete V-2), 20; test stand P2 (small-scale, liquid propellant engines with thrusts up to 2200 lb), 21; test stand P3 (gas generators, turbines, fuel injector systems), 22; test stand P4 (jatos, later converted to a shop), 23; test stand P5 (fuel pumps, V-2 cold-flow tests, steam generator calibration), 24; test stand P6 (originally V-2 combustion chambers, later Wasserfall combustion chambers), 25; test stand P7 (control building, assembly building, three mobile static test stands, cold calibration stand, and two launching pads for V-2), 26; test stand P8 (static firing and calibration of V-2 combustion chamber), 27; test stand P9 (dual position stand for complete Wasserfall or its engine), 28; and launch site P10 for motorized V-2 firing units (controlled from P7), 29. Test stand P11 (center of three at 46) for the static testing of production V-2s from the plant at Nordhausen was practically destroyed during an air raid in July 1944 after only a few missiles had been tested on it. An additional test facility was anchored in the harbor

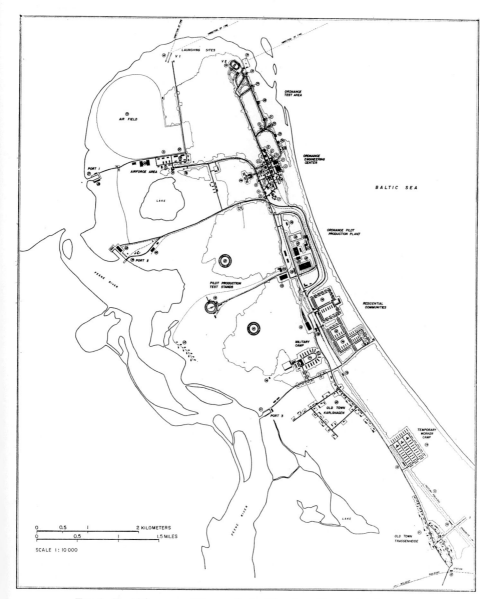

FIG. A. Facilities map of Peenemünde. (Courtesy Hannes Leuhrsen).

at Peenemünde. Code named *schwimmweste* (lifebelt), it was a floating stand for the Wasserfall.

The Army pilot production facility included: assembly building no. 1, 39; assembly building no. 2, 40; carpentry shop, 41; repair and maintenance building, 42; warehouse, 43; storage facility, 44; vacuum drying plant (for graphite rudders), 45; static test stands for V-2 engines, 46; storage bunkers for completed missiles, 47.

Utilities serving the establishment as a whole included: power and heating plant, 48; liquid oxygen plant no. 2, 49; coal supply port, 50; general supply and construction materials port, 51; factory railroad, 52; railroad stations, 53; sewage treatment plant, 54; and water treatment plant, 55.

Housing and cantonment areas were: civilian quarters, 56; military camp, 57; camp for Air Force temporary workers, 58; camp for Army temporary workers, 59; Karlshagen (civilian workers), 60; Trassenheide (civilian workers), 61; and railway station at Trassenheide, 62.

From the rangehead at P10 and P7 and a site on Greifswalder Oie for motorized V-2 units, missiles were launched in a northeasterly direction up the Baltic Sea but relatively close to the Pomeranian coast. In addition, the Air Force had a V-1 launching site near the village of Zempin, south of Zinnowitz. The overall range area and major instrumentation sites are shown in Fig. B.

Three Doppler tracking networks (the "Campania" system) tracked missiles fired from Peenemünde. Of the two smaller networks, one had a transmitter and receivers located on the mainland at Lubmin, at Peenemünde, and on the islands of Ruden and Greifswalder Oie. The other consisted of three stations and a transmitter near Stolpmünde. The large tracking network had four stations: Peenemünde, Bornholm island, and two down-range sites on the Pomeranian coast at Köslin and near Stolpmünde.

During 1941 and 1942 a special Doppler tracking system for the V-2 was employed. It consisted of a ground transmitter, code-named Naples 1, which sent a signal of 36 Mc/sec to a transponder and frequency doubler in the missile code-named Ortler. The system was the forerunner of the Dovap systems currently used on several of the world's large missile ranges.

The telemetry system used at Peenemünde was the "Messina 1," 3-channel (one of which could be commutated), AM/AM type, which operated on a frequency of 61.8 Mc/sec. One of the receiving stations was located in a lighthouse near Koserow, southeast of Zinnowitz; but the primary receiving equipment was located in the guidance, control, and telemetry building at Peenemünde. A "Messina 2" system was under development when the war ended. It was a 9-channel, FM/FM telemetry system.

Optical tracking instruments at the range consisted of both cinetheodolites and phototheodolites. Askania cinetheodolites with 12-cm lens of

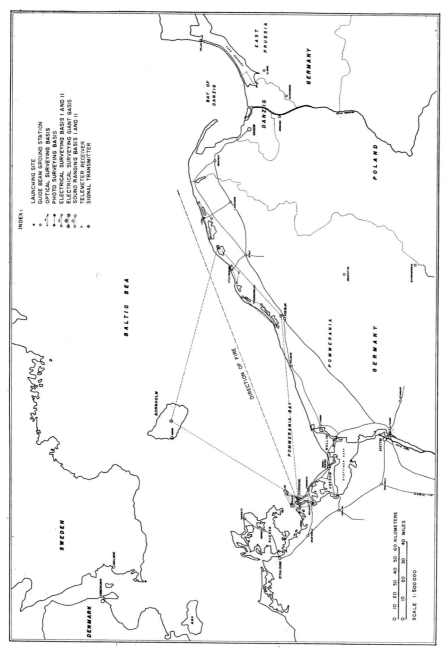

Fig. B. Tracking facilities at Peenemünde. (Courtesy Hannes Leuhrsen.)

1-meter focal length were used to track the V-2 to an altitude of 20 mi. Accuracy was of the order of 15 ft., with sites at Peenemünde and on the islands of Ruden and Greifswalder Oie. The phototheodolite network consisted of two Zeiss ballistic cameras designed to establish the trajectories for heavy artillery guns. These were located on a 4.8-mi baseline on the mainland.

To aid in locating missile impacts in the Baltic Sea, two acoustical sound-ranging networks were sited on the Pomeranian coast: one at Rugenwalde, 100 mi from Peenemünde, and one at Leba, 150 mi from the rangehead. While these installations gave fairly accurate data on surface impacts, they were of little use in fixing air bursts (which were not infrequent.)

United States Ranges

A. Atlantic Missile Range

The Atlantic Missile Range, one of the largest missile proving grounds in the world, extends from Cape Kennedy (formerly Cape Canaveral), Florida, into the Indian Ocean beyond the Cape of Good Hope. Operated and maintained by the Air Force Missile Test Center (AFMTC), a part of the Air Force Systems Command (AFSC), the 9000-mi range is used by the National Aeronautics and Space Administration and by the military services engaged in long-range missile testing and aerospace operations. AFMTC headquarters is located at Patrick Air Force Base (PAFB), 15 mi south of Cape Kennedy. Its overall mission is the accomplishment of long-range missile research and development flight tests and the collection and evaluation of missile flight data.

Cape Kennedy was chosen as the range launching point for two reasons: the existence of a base, the inactive Banana River Naval Air Station (now Patrick Air Force Base) and the almost unlimited over-water flight potential toward the southeast where conveniently located islands could be utilized as missile tracking stations.

In 1947 the Air Force was directed to develop the range, and a joint services group was formed for this purpose; meanwhile, negotiations were started to establish tracking stations in the Bahamas and West Indies Islands. By 1949 legislation had been enacted that established a long-range missile proving ground, and construction was begun at Cape Kennedy for missile launching pads and instrumentation sites. The Joint Long Range Proving Ground became the sole responsibility of the Air Force in 1950.[1] Today the range consists of three major areas: the

---

[1] The first missile launching was in July 1950, when a German V-2 with a Wac Corporal second stage was fired, using a sandbagged trench as a firing point and an Army tank as a blockhouse. During the countdown a 10-ft alligator appeared at the launching bunker, adding greatly to the existing tension.

launching and test area at Cape Kennedy, the administrative area at Patrick Air Force Base (18 mi south of the Cape proper), and a series of downrange tracking stations on islands in the South Atlantic [1].

1. Geographical Factors

The rangehead and adjacent area at Cape Kennedy is located halfway between Jacksonville and Miami, and 50 mi east of Orlando (28° 26′ N 80° 35′ W). Not considered a part of the Florida mainland proper, the Cape is an island bordered by the Atlantic Ocean on the east and the Banana River on the west (Fig. 1).

Cape Kennedy has an average temperature of 72.5°F. August is the warmest month, and high humidity prevails throughout most of the year. Precipitation is highest from May through October, and the average rainfall is 41 in. The water table varies from 2 to 6 ft, and the highest natural elevation on the Cape is 10 ft.

The natural terrain of the Cape's 15,000 acres is flat, sandy, with thick undergrowth and palmetto scrubs. Additional acquisition by NASA increased this area to approximately 103,000 acres in 1963.

2. Launching Facilities

The most extensive launch facilities in the free world are located at the rangehead or Cape Kennedy Missile Test Annex (CKMTA) as it is officially known. The Air Force, Army, Navy, and NASA conduct ballistic and spaceflight operations from launch complexes that employ a wide variety of launch techniques and equipment. Along the northeast border are launch complexes (Fig. 1) for the large carrier vehicles and missiles, such as Saturn, Atlas Centaur, Atlas, and Titan 2; the southeastern area is used primarily to launch the smaller missiles such as Minuteman, Thor, and Polaris.

Due to the extensive nature of these facilities and the expansion of NASA activities in the area, this section is subdivided into military and NASA installations. Within these arbitrary divisions, sub-divisions are made into present and projected launch facilities, since much of the nation's future space efforts will originate from the Cape Kennedy and nearby Merritt Island areas.[2]

[2] AMR is also one of a series of seven ranges in the US and Canada that are organized into a rocket synoptic meteorological network. The center of launching activity for this function of the range is at Complex 21 (Fig. 1), formerly a launching pad for the Mace guided missile, and Complex 23, easternmost of the range's launching pads. Typical rockets launched include the Nike Cajun, Arcas, and Hopi Dart, as well as specially designed smoke-releasing rockets using the M5E1 jato as a motor.

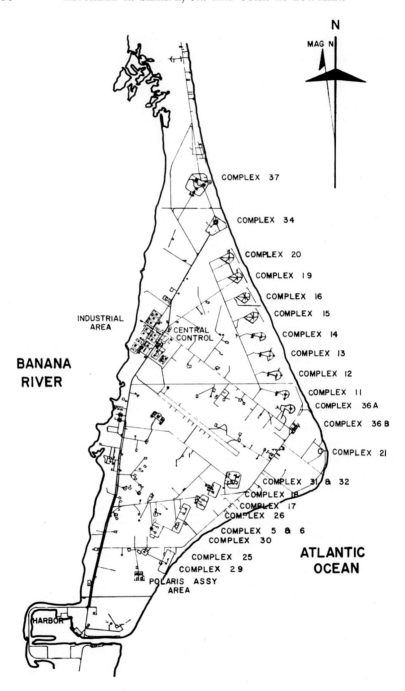

a. Military Launching Facilities, Present

The principal user of CKMTA is the Air Force. The major Air Force complexes are used to launch the Atlas, Titan, and Minuteman (Fig. 2) intercontinental ballistic missiles.

A typical Atlas launch complex at Cape Kennedy is a tear-shaped arrangement of facilities surrounded by a security fence (Fig. 2). The launch pad, located near the center of the complex, is a raised concrete platform adjacent to a fixed umbilical tower. Radiating from the launch site are a drainage trough, access ramp, water main, and tracks for the mobile service structure. Other facilities within the complex include a ready room, blockhouse, propellants and oxidizer storage areas. A camera road within the complex, with several camera pad locations, parallels the security fence. Titan 1 and 2 launch complexes (Fig. 3) at Cape Kennedy are similar in their appearance and functions to the Atlas complexes.

The Navy Polaris testing operations at CKMTA are from above ground sites, from surface ships that simulate the launch sequence and operation of a submarine launch, and from submerged submarines off-shore from Cape Kennedy. In the latter operations, the Polaris is launched by an air ejection system that propels the missile from its launching tube, up through the water, and to a point above the surface where its solid propellant motor ignites.

b. Military Launching Facilities, Projected

Two Titan 3 complexes, which will figure prominently in Air Force space exploration plans, are being constructed at Cape Kennedy. Each complex (Fig. 4) will have one 600-ft diameter circular launch pad, with launch platform; a 240-ft diameter circular launch pad, with launch platform; a 240-ft tall mobile service tower; and a 170-ft tall umbilical tower with sliding platforms. Liquid hydrogen facilities will be shared by both complexes [2].

c. NASA Launching Facilities, Present

The NASA Launch Operations Center (LOC) at Cape Kennedy is responsible for the overall planning and supervision of the integration,

---

FIG. 1. Atlantic Missile Range, Cape Kennedy, Florida. Launching complexes shown are: 37, Saturn 1/1B; 34, Saturn 1-B; 20, Titan 1; 19, Titan 2; 16, Titan 1; 15, Titan 1; 14, Atlas; 13, Atlas; 12, Atlas; 11, Atlas, Atlas Agena; 36A, Atlas Centaur; 36B, Atlas Centaur; 21, sounding rocket pad; 31, Minuteman; 32, Minuteman; 18, Blue Scout; 17, Thor Delta; 26, Jupiter; 5, Redstone; 6, Mercury Redstone; 30, Pershing; 25, Polaris; and 29, Polaris. Courtesy NASA.

FIG. 2a. Atlas launching complex at AMR showing blockhouse to rear of launcher and missile.

Fig. 2b. Minuteman silo at AMR. U.S. Air Force photographs.

FIG. 3. Titan 2 being launched at AMR. U.S. Air Force photograph.

checkout and launch of NASA space vehicles at AMR. The Center reports directly to the Office of Manned Space Flight, Washington, D.C.

Three launch complexes and associated electronic and optical tracking stations (Fig. 5) are or have been used by LOC at the Cape: Redstone Complex 26, Jupiter-Juno Complex 56, and Saturn Complex 34. Saturn Complex 37 and Centaur Complex 36B have recently been constructed. In addition, LOC has joint usage of Atlas Agena B Complex 12 (Fig. 6) and Centaur Complex 36 with the Air Force Systems Command. One of the busiest pads, 17A, at the Cape, for the Thor Delta, is shown in Fig. 7. The Center is also responsible for planning and establishing the Saturn 5 Complex for the Manned Lunar Landing Program.

FIG. 4. Titan 3 space carrier vehicle launching site at AMR. Courtesy United Technology Corp.

Saturn Launch Complex 34 (LC-34), seen in Fig. 8, is operational and has proven its capability in several operations. It incorporates all the features and most of the refinements of the typical, large launch complex. For these reasons, and because detailed information on military installations are of necessity often limited, a detailed description of LC34 is presented [2].

LC34 is the largest operational launching site in the free world, and is the first planned solely for the peaceful exploration of space. A 45-acre, multimillion dollar facility, it is one of several LOC-operated launching facilities. The size and complexity of it is evident in its component facilities.

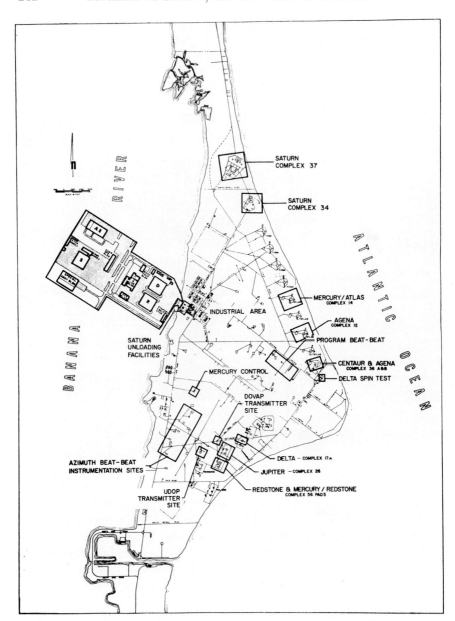

Fig. 5. NASA launching facilities at AMR. Courtesy NASA.

The launch control center, or blockhouse (adapted from the Atlas blockhouse) is a domed building, 120 ft in diameter, with approximately 10,000 ft^2 of protected floor space on two levels and an additional 2150

FIG. 6. Atlas Agena space carrier vehicle on launcher at LC12. Courtesy NASA.

ft^2 of unprotected space in an equipment room not occupied during launchings. The inner dome is made up of 5-ft thick reinforced concrete, and the building is designed to withstand a blast pressure equivalent to the explosion of 50 kilotons of TNT at a distance of 50 ft. The first floor is used for tracking and telemetry operations; while launch supervision, monitoring, and recording operations take place on the second floor (Fig. 9). The operating area can be observed from a small room behind a glass

Fig. 7. Thor Delta space carrier vehicle on launcher at LC17A. Courtesy NASA.

window; and exterior, prelaunch activities in the area can be observed from a balcony on top of the building. A periscope in the blockhouse permits close-up views of the launcher during liftoff.

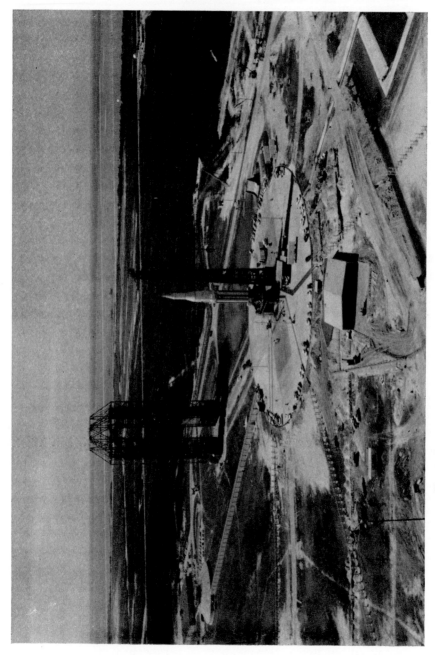

FIG. 8. Saturn space carrier vehicle pad LC34. Courtesy NASA.

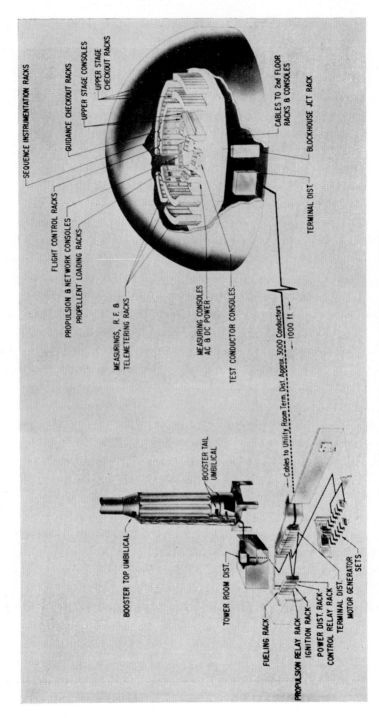

FIG. 9. Saturn blockhouse at LC34. Courtesy NASA.

The 310-ft tall launch service structure (Fig. 10), weighing 2900 tons, is used to erect and checkout Saturn 1 and 1B carriers at the launch pad. Each leg of the structure contains enclosed operating and checkout equipment, seven fixed platforms at various elevations, and five hori-

Fig. 10. Service structure for Saturn at LC34. Courtesy NASA.

zontally retractable platforms. When anchored to steel piers by hy-draulically operated steel pins, the structure and vehicle can withstand winds up to 125 mph. When checkout is complete, the structure is moved to a parking area approximately 600 ft from the launch pedestal.

The 430-ft diameter launch pad is constructed of 8-in. thick rein-forced concrete and is paved with refractory brick where thrust heat is most intense. It has a perimeter flume for drainage of surface water and possible propellant spillage.

The launch pedestal (Fig. 11) is located in the middle of the pad and is used to support and retain the vehicle during checkout and launch

FIG. 11. Launch pedestal for Saturn at LC34. Courtesy NASA.

operations. Bolted at the top of the 42-ft square and 27-ft tall pedestal are eight steel arms, four to support the vehicle, and four to support and restrain it until proper engine operation and thrust are achieved. These arms are automatically controlled during the launch sequence.

The two-way steel flame deflector diverts engine flow into predetermined directions. When not being used it is parked on rails adjacent to the pedestal.

The umbilical tower, seen in Fig. 8 to the right of the Saturn, provides electrical, hydraulic, and pneumatic lines to the vehicle. The 240-ft structure has hydraulically controlled swing arms that connect it to the vehicle and swing out of the way during launch.

Beneath much of the launch pad is the 215-ft long and 38-ft wide automatic ground control station, which acts as a distribution point for all measuring and checkout equipment, power, and high-pressure gas. Cables from this station to the launch control center are routed through a roofed cableway, also seen in Fig. 8.

The RP-1 fueling facility consists of storage and transfer equipment, protective revetments, foundation, and wall. Two 30,000-gal, cylindrical storage tanks are employed; and the transfer system contains two 1000-gpm pumps, a circulation pump, a filter-separator unit, an eductor system, valves, piping, controls, and a support pad. The system is fully automated and is controlled from the launch control center.

The liquid oxygen system has two storage tanks, a main tank with an inner and outer sphere with an outside diameter of 43 ft, and a smaller tank for replenishing oxygen which boils off the vehicle before launch. The two spheres of the main tank are separated by 4 ft of "perlite," a mineral insulating powder. The facility is barricaded by an earth revetment on the side that faces the launch pedestal.

The liquid hydrogen facility contains a vacuum-jacketed, spherical tank; pneumatic and electrical consoles; and necessary plumbing and valves.

The high-pressure gas facility (Fig. 12) provides the helium and nitrogen gases required by the carrier vehicle. Helium is boosted from 3000 to 6000 psi, and nitrogen is converted to gaseous from liquid form before it enters the carrier. Helium bubbles through the liquid oxygen tanks to keep the oxygen from forming different temperature strata. Nitrogen is used to purge fuel and liquid oxygen lines, engine and instrument compartments, and to operate pneumatic components.

The skimming basin is a 104-ft × 180-ft concrete-paved vat used to collect fluids spilled on the pad and to prevent their intrusion into drainage canals.

A water supply system is located on the pad and throughout the serv-

FIG. 12. High-pressure gas facility at LC34. Courtesy NASA.

ice structure where it is available at all work levels for fire protection. The pad flush system is ready to wash away any spilled fuel, and the quenching system on the pedestal is available for use against accidental fires in the Saturn "boat-tail" or engine compartment and to extinguish flames in the engine compartment if engines are cut off after ignition but before liftoff. In addition, four 3500-gpm nozzles are located around the pad as a general protective measure.

The operations support building, with 30,000 ft² of floor space, is used for general shop and engineering activities in direct support of launch operations. It is located to the rear of the blockhouse, as shown in Fig. 8.

Camera stations are located around the launch pedestal to permit remote-controlled photographic coverage of launchings.

The communications systems at LC34 includes a comprehensive voice communications network of approximately 200 stations. A closed-circuit television loop is also available to monitor, checkout, and observe launch operations.

NASA-LOC Launch Complex 37 (LC37) for launching the Saturn 1 and 1B carrier vehicles is even larger than LC34. With two launch pads, A and B, LC37 covers 120 acres at the north end of Cape Kennedy. Each of the pads have individual automatic ground control stations (AGCS), launch pedestals, and umbilical towers.

Each of the 50-ft × 122-ft AGCS buildings are directly below a portion of each umbilical tower. The buildings, which are unoccupied during launch operations, are multileveled, with three stories above ground. Each umbilical tower has a 32-ft square base and a height of 268 ft, which can be extended to 320 ft. Each launch pedestal measures 47 ft square and has a 12-sided cutout, 32 ft in diameter in its center for exhaust escape. Triangular platforms on top of the pedestal enlarge the work area to a 55-ft square base.

Common to both sites are the launch control center; operations support building; propellant storage and transfer facilities; and a mobile, self-propelled service structure, which moves on 1200 ft of rails between the two pads.

The launch control center has a blast resistant dome 12.5 ft thick and measures 110 ft in diameter and 37 ft tall in its interior. The principal launch functions, tracking operations, and observations are conducted from within the launch control center. Propellant storage and transport facilities include a liquid oxygen system with two tanks, a 125,000-gal spherical storage unit, and a 28,000-gal cylindrical replenishing tank. A 43,500-gal cylindrical tank is used for storage of RP-1 fuel, and a spherical tank is available for storage of liquid hydrogen. The high-pressure gas facility includes a 35,000-gal liquid storage tank and facilities for storage of helium.

The launch service structure is probably the largest movable structure in the world. Its basic height of 300 ft can be extended to 330 ft; and, with a derrick mast rising above the extended superstructure, it reaches a height of 375 ft. The derrick and 90-ft boom, which top the structure, have hook capabilities of 10, 40, and 60 tons for assembly of vehicle stages. Fixed platform levels and adjustable service platforms permit convenient assembly of the vehicle. All levels are serviced by two high-speed elevators. The structure can move at speeds up to 40 fpm, using a variable speed drive that supplies traction to 32 of the 48 three-foot diameter wheels. A 1000 kVa substation in the structure supplies electrical power. When the service structure is in its operational position at the launch site, the load is removed from the wheels by hydraulic equalizer rams; and the structure is lowered to the foundation anchor assemblies and locked in place. The process is reversed before moving the structure to its parking position.

Mercury-Atlas LC14, Atlas-Centaur and Atlas-Agena LC36, Thor-Delta LC17, Jupiter LC26, and Redstone and Mercury-Redstone LC5 and 6, the other launch facilities used by NASA, utilize launch equipment similar to that used by their military counterparts.

d. NASA Launching Facilities, Projected

NASA LC39 (Fig. 13) will launch the Saturn 5 carrier vehicle for NASA's Apollo lunar spaceship program and subsequent space explorations. Designed for a high launch rate, present plans call for a continuing launch program of 12–15 launches a year beginning about 1968.

Employing a new concept in launch operations, LC39 will depart from the traditional practice of performing assembly, service, and checkout operations on the launch pad. Instead, the vehicle will be assembled, serviced, and will have undergone many of its prelaunch checkout operations before it reaches the launch site. The feasibility of this concept is due, in part, to an advance in state-of-the-art development, which permits digital checkout of the vehicle. All three launching pads can be monitored by a single launch control center, above ground and remote from thrust effect, rather than from heavily reinforced blockhouses near the pads [2].

Three facilities, which emphasize most the unique character of LC39, are the Vertical Assembly Building (VAB), illustrated in Fig. 13, the Launcher Umbilical Tower (LUT), and the crawler. The VAB, 524 ft tall and anchored to bedrock 170 ft below the ground level, will have high and low bay areas equipped to erect and checkout the Saturn 5.

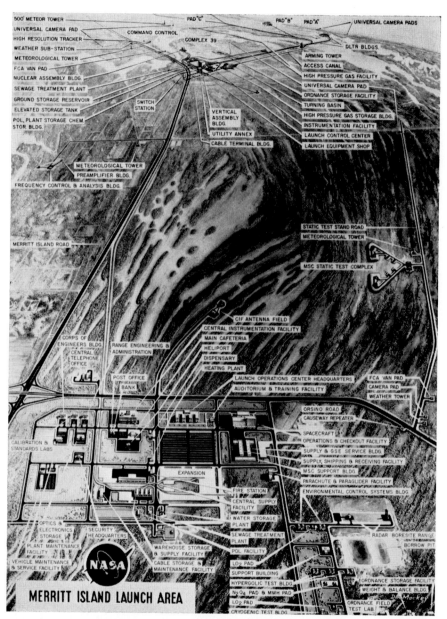

FIG. 13. Concept of Saturn LC39. Courtesy NASA.

The LUT is a launcher platform, with its own umbilical tower, upon which the vehicle will be placed while still in the VAB. The crawler, shown in Fig. 14, is a platform with tractor-type treads. It will pick up the LUT, with its vehicle, and transport them to the launch site. Another feature of LC39 is the arming tower, also mobile and capable of being moved by the crawler, which will affix solid propellant rockets, necessary

FIG. 14. Concept of launcher platform and umbilical tower at Saturn LC39. Courtesy NASA.

pad explosives, and other pyrotechnics to the vehicle at the launch pad.

Located adjacent to the VAB and connected to it by an enclosed bridge will be a 4-story launch control center. The first floor will contain offices, cafeteria, and a dispensary. The second floor will house the telemetry, ground receivers, and recording apparatus; and the third and fourth floors will have four firing rooms, one for each of the high-bay areas in the VAB. Each firing room will have a set of control and monitoring equipment designed to permit launching of a vehicle and simultaneous checkout of others. The equipment is identical in all four rooms, and each room will have its own computer facilities.

Fueling operations and final checkout of the Saturn 5 will occur prior with the vehicle on the launch pad. Nearly 3000 tons of kerosene (RP-1), liquid oxygen, and liquid hydrogen will be required for each vehicle, the latter two items supplied from the largest cold-storage fuel containers ever built, with capabilities up to 900,000 gal.

After each launch, the LUT will be returned by the crawler to the VAB. This mobility will permit immediate readiness of the pad for the next launch.

The Saturn 5 carrier vehicle, to be launched from LC39, is the most advanced of the Saturn configurations. Over 350 ft tall, with the Apollo spacecraft and ejection tower in place, the three-stage carrier will perform the lunar orbital rendezvous space-flight leading to the landing of two men on the Moon and their return to Earth.

Beyond Saturn and LC39 lie the post-Saturn class of carrier vehicle and the launch means it will need. Perhaps another new concept will be required. One point has already been established in regards to it; like the Saturn 5, it will probably be launched from NASA-acquired land in the Cape Kennedy area.

3. TRACKING AND COMMUNICATIONS

AMR is linked by a series of tracking stations (Fig. 15). Range Station 1 is Cape Kennedy. Range Station 2 is located about 100 mi south of the Cape at Jupiter Inlet, Florida. Range Station 3, at Grand Bahama Island, is the first off-shore station. Stations 4 through 9 are located respectively at Eleuthera Island, San Salvador, Mayaguana, Grand Turk, the Dominican Republic, and Puerto Rico. Station 9.1, at Antigua in the Leeward Islands, was primarily established for ballistic missile programs. Station 10, at St. Lucia, used previously for cruise missile tests, by 1963 had become inactive. Station 11 is located on the Brazilian Island of Fernando de Noronha; and the last island tracking station is at Ascension Island, over 5000 mi from the rangehead in the South Atlantic Ocean.

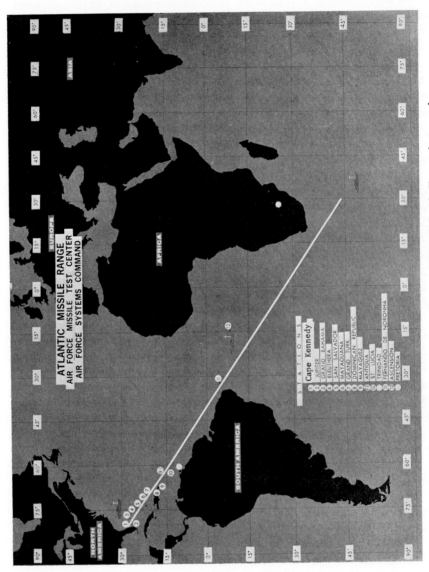

FIG. 15. Downrange tracking stations of AMR. U.S. Air Force photograph.

In addition to the numbered stations, other sites are used for range operation, including a temporary instrumented site at Pretoria, South Africa, and fully instrumented tracking ships (Fig. 16), which extend the range into the Indian Ocean impact area beyond the Cape of Good Hope.

Typical of these ships is the ARIS (Advanced Range Instrumentation Ship) *General H. H. Arnold,* which was placed in service in 1963 [3]. Its usual position between Ascension Island and the east coast of Africa makes it the southernmost tracking station of AMR. Its primary tracking mission is gathering data on re-entry vehicles and participating in decoy and penetration aid programs, although it can track Earth satellites and, to a lesser extent, interplanetary probes.[3]

With its C-, X-, and L-band radars, the ship is able to track multiple targets with multiple frequencies and is considered to be equal to or better than land-based tracking stations at the range. The 30-ft diameter C-band radar is of the pulse-compression type and tracks both the beacon and the skin of the target. It has two identical transmitters, which alternately transmit horizontally and vertically polarized beams. Each produces a 1-Mw, 30-μsec pulse. One is used for tracking while the other receives data. The C-band radar is also slaved to the X-band and L-band radars.

The X- and L-band radars share a 40-ft diameter dish antenna. This antenna has a slew rate of 30 deg/sec and a look angle of 130 deg. Both radars have dual transmitters.

Telemetry reception aboard the ship is by means of a 30-ft diameter antenna that can receive wide-band signals in the range 200 Mc/sec to 1000 Mc/sec or narrow-band at 13,000 Mc/sec. It can be slewed at a rate of 0.5 rad/sec. The antenna can also acquire a target and direct the C-band radar to it. Signals from it are recorded in predetected form on tape and retransmitted in this form.

Specialized communications equipment aboard the ship includes a radar video recorder that combines signals from all three tracking radars in digital form and stores them on tape. A separate tape recorder is used for the telemetry radar. Tracking data are sent to AMR by means of specially developed helical antennas at a rate of about 50 bits/sec.

Voice, data, and teletype in the range 15 kc to 400 Mc/sec are provided by 10-kw and 2.5-kw, HF transmitters using helical antennas.

The ship also has its own timing central, which is synchronized within 0.01 sec of AMR time and is accurate to 5 parts in 10×10^9 per day.

---

[3] A sister ship, the *General Hoyt S. Vandenberg,* features the Sintrak radar—the most powerful radar system at AMR. It can transmit and receive six types of pulses and record 3,600,000 bits/min.

FIG. 16. ARIS tracking ship *General Hoyt S. Vandenberg*. Courtesy Sperry Gyroscope Co.

The system uses WWV signals from Washington, D.C., or LF signals from the National Bureau of Standards station NBA in Washington, D.C. or Boulder, Colorado.

Meteorological data are supplied the ship by conventional instrumentation, balloon radio sondes, and Arcas sounding rockets launched from the ship. Weather data up to 125,000 ft are thus available.

The ship's data processing system consists of a Univac 1206 digital computer and a central data conversion unit. Raw data from the various radars are placed in digital form by the converter for entry into the computer. Thus, the ship can send raw data or computations to AMR over a HF radio link.

An interesting feature of the ship is its inertial guidance system, similar to that used in the nuclear submarines that serve as launching pads for the Polaris missile. This unit gives a reference for all radars and telemetry antennas aboard and furnishes an input to the computer on pitch, yaw, and roll motions of the ship during tracking operations. During data reduction the ship's motions are subtracted from the antenna pointings with the result that data are, for all practical purposes, received by a station on a stable platform.

The size of each downrange station is dependent on the type and quantity of tracking instrumentation required at a particular location. A typical station will have approximately 140 full-time technicians and maintenance personnel. Each station is commanded by an Air Force officer.

The instrumentation for gathering missile flight data at AMR is designed to precisely determine missile performance at any given moment during flight. This instrumentation represents a major portion of the Air Force Missile Test Center's billion-dollar capital plant investment, and its use has reduced to a fraction the number of launches required from prototype to operational missile in a given program.

The AMR communications system includes point-to-point, ground-to-air, ship-to-shore, and intrastation contacts. A variety of equipment is employed, including submarine cable, AM and HF-SSB radio, troposcatter, VHF and UHF radio, microwave links, hardwire, telephone, and teletype. They are used for range administration and operation, for test data transmission and reception, and for missile command and destruct [4].

AMR point-to-point communications include submarine cable, HF radio, tropospheric scatter, microwave, and hardwire. A single, 12 channel (each 4 kc wide), coaxial submarine cable with a 10-gage copper center links Cape Kennedy, Pt. Jupiter, GBI, Eleuthera, San Salvador, Mayaguana, and Grand Turk Island.

AMR is also equipped with a special, air-transportable communications system known as Atrax. All equipment for it is contained in trailer vans, which can be quickly flown from one point to another. Basically, Atrax is a HF-SSB, VHF, UHF, microwave, and wire communication system designed to provide general communications support for the range. It can be used independently or as an addition to the primary system. The HF-SSB equipment has both omni- and unidirectional antennas to provide optimum radio communications for ranges up to 3000 mi. Telephone circuits are serviced by both automatic and manual exchange switchboards.

From Grand Turk to Antigua, by way of Puerto Rico, there is a 60-channel, two-way cable that replaces the formerly used tropospheric scatter and SSB radio links. The 715-mi cable provides 60 one-way channels for carriers operating in the 24-kc to 264-kc range for down-range transmissions and 60 channels in the 312-kc to 552-kc band for uprange transmissions. All channels are 4-kc wide. Forty two-way repeater amplifiers are used to compensate for cable attenuation and are placed at 17.5-mi intervals along its length [5].

HF-SSB radio transmitters (45 kw) and receivers are located at Cape Kennedy, Antigua, Ascension, and Pretoria. In addition to the two rhombic antennas used with each transmitter, one for day and the other for LF night use, rotatable-log antennas are located at Cape Kennedy, Ascension, and Pretoria, with an antenna switching system at Cape Kennedy that permits any one transmitter to be connected to any other antenna. In addition, a low-power, HF-SSB transmitter provides a communications link from Trinidad to Cape Kennedy. A quadruple diversity tropospheric scatter system, AN/MRC-85, provides a 382-mi link between Grand Turk Island and Puerto Rico's East Island.

Three microwave links in use at AMR provide for operation and data transmission for the Mistram site at Valkaria, Florida; interisland communications for the Grand Bahama Island (GBI) area; and for tying Ramey AFB, Puerto Rico, into range communications.

Ground-to-air and ship-to-shore communications are provided by HF, SSB, VHF, and UHF radio links. Transmitter locations and radiated power are indicated in Table I.

The range is divided into three areas for control of ship-to-shore communications, including frequency assignment, maintenance and range test information distribution. Control points for the areas are Cape Kennedy, Antigua, and Ascension.

Intrastation communications [6] between buildings and facilities at all AMR stations depend on outside cable distribution systems. Long

TABLE I. AMR GROUND-TO-AIR AND SHIP-TO-SHORE TRANSMITTERS

Location	HF/SSB	VHF (watts)	UHF (watts)
Cape Kennedy	10 and 2.5 kw	50	50
GBI	2.5 kw	50	50
San Salvador	50 watts	50	50
Grand Turk	50 watts	50	50
Antigua	10 and 2.5 kw	50	50
Ascension	10 and 2.5 kw	50	
Pretoria	45 kw		

runs use nonloaded 19-gage or loaded 22-gage cable to reduce current loss. Special, wideband, 16-gage pairs are also used at AMR.

All AMR stations have automatic dial telephone systems. Stations on the subcable can manually patch their systems in, thus permitting calls from any telephone on one station to any telephone on another station.

Two major interphone systems extend throughout Cape Kennedy and into several of the downrange stations. They are the *green* phone system and the missile operations system. The green system consists of 10- or 20-line manual key panels. Battery operated, it provides a rapid and power-failure proof means of communications from instrumentation supervisors to operating personnel and other supervisors. The Mops intercom system permits net selection by use of rotary selector switchers on the end instruments. When an operator selects a net, he is heard by all other operators on the same net. Access to a given net can be limited to certain operators.

Public address systems are also used in certain areas within AMR stations for dissemination of information and instructions to personnel. A public address system in Central Control at Cape Kennedy is principally used to provide countdown information, but it can also be used in emergency situations for relaying safety precautions.

a. Radar Tracking and Telemetry [6–8]

Radars currently employed at AMR include AN/FPS-8; AN/FPS-16; AN/MPS-25; AN/FPQ-6; AN/TPQ-18; Mod II; Mod IV; AN/FPS-43 (XW-1); and AN/FPS-44.

The AN/FPS-8 is a standard L-band, early-warning, air-surveillance radar used at Cape Kennedy to assure range clearance and for acquisition of target data. The AN/FPS-16 is a C-band tracking radar used for missile instrumentation. Sites are located at Cape Kennedy, Patrick AFB, Grand Bahama Island, San Salvador, Ascension Island, and aboard the tracking ships.

Two units of the AN/MPS-25, a trailer-mounted version of the AN/FPS-16, are also located downrange. The AN/FPQ-6 is an advanced C-band tracking radar with a transportable version designated AN/TPQ-18. Both types have a built-in flexibility, and display of both skin and beacon information permits a quick switch to skin track in the event of a transponder failure. One AN/FPQ-6 is located at Patrick AFB and one at Antigua. An AN/TPQ-18 is located at Cape Kennedy, with four additional units scheduled for location at downrange stations.

The Mod II is a modified SCR-584 radar. It is primarily used to provide a tracking network system for range safety. The Mod IV is a modified, X-band, Nike-Ajax target tracking radar. It provides real-time missile position display for range safety on all launches from engine ignition to about 40,000 yd.

The experimental, L-band, AN/FPS-43 (XW-1) tracker and AN/FPS-44 (XW-1) make up what is referred to as the Trinidad radar. It has two transmitters that can be switched between two antennas. The scanning radar has a 165-ft by 330-ft parabolic torus reflector with a rotary organ-pipe scanner, and the tracking antenna has an 84-ft parabolic dish with a conical scan assembly for automatic tracking. Appropriately, the Trinidad radar is located on Trinidad Island.

In addition to the data received from radar systems, AMR employs various CW systems, including Azusa, Glotrac, Mistram, and Udop.[4]

Two Azusa systems are used at AMR, the Mk I at GBI and the Mk II at Cape Kennedy. The Mk I is a single-site, short-baseline system with two perpendicular baselines at their mutual midpoints. The Mk II system is similar to the Mk I but differs in its refinement of circuitry design and in the addition of cosine rate baselines for better direction cosine data. The antenna layout is also somewhat modified.

The Glotrack system is so named because of the original intention of using it as a global instrumentation system; however, program changes restrict its use to the AMR uprange area. It uses data from the Azusa Mk II system, pulse-radar systems, and range and range-rate equipment to make highly accurate target velocity and position measurements.

Mistram systems are presently installed at Valkaria, Florida, and at

---

[4] The range also possesses an infrared (ir) tracking system with one station at Cape Kennedy (Station 1), for studying ir characteristics at launch, and the other at Ascension Island (Station 12), for recording ir reentry phenomena. Equipment at Station 1 is housed in vans, while that of Station 12 is in a permanent structure. Each station has a 3-channel ir radiometer, 6-channel uv photometer, and multielement ir scanner. These are mounted on a Nike Ajax radar pedestal. A Mk 51 optical tracker is used until the ir sensors lock on the target. Movie cameras are also coaxially mounted with the tracking sensors.

Eleuthera Island [9]. After determining a vehicle's position and velocity, Mistram provides real-time readout of the information (Fig. 17). It uses CW phase comparison techniques to measure range from a central station and range differences across orthogonal baselines. Range is measured by counting the wavelengths the signal travels to the vehicle and back to the central station, and range difference is determined by the difference of the wavelengths traveled by the signals from the vehicle to each end of the baselines. Position is fixed by the range and range differences, and the rates at which the range and range differences are varying determine velocity. The Mistram stations are arranged in the shape of an L (Fig. 18). Mistram I in Valkaria has the central station at the vertex of the L and the four remote stations ranged along the baselines of the L at 10,000-ft and 100,000-ft distances. The two stations at the shorter distance are connected to the central station by 3-in. diameter circular waveguides, while the stations at the 100,000-ft distances use airlink transmissions. The Mistram II system at Eleuthera eliminates the two 100,000-ft stations. Both Mistram I and II have microwave antenna towers at the vertex and at the 100,000-ft distances.

Udop (UHF Doppler) is a trajectory-measuring system that uses phase comparison techniques to determine missile position and velocity; the data obtained, however, are ambiguous and must be resolved with a known position. The system contains UHF-VHF transmitters, a missile transponder, four or more receivers, and a central recorder. There are two Udop systems at AMR, one at Cape Kennedy and the other at GBI. The former is operated by the NASA while the downrange system is operated by the Air Force.

Telemetry transmitted by missiles and spacecraft at AMR include FM/FM, PAM/FM/FM, PDM/FM, PDM/FM/FM, and PCM [6]. Each telemetry transmitting system has a corresponding telemetry receiving system. Four types of telemetry antennas are used at AMR: seven-turn helix, trihelix, quadhelix, and parabolic (TLM-18). Each is directive for gain, and the latter two are additionally directive for tracking.

The antenna output is routed through preamplifiers and multicouplers to receivers (FM/FM, PAM/FM/FM, PDM/FM/FM, and PCM/FM) that amplify and detect the signals to produce composite signals that are replicas of the signals used to frequency modulate the on-board transmitters. Either of two bandwidths can be selected by a switching arrangement, with the wider bandwidth normally used for FM/FM signals.

The discriminators (FM/FM, PAM/FM/FM/, PDM/FM/FM) are band-pass filters, one for each subcarrier channel, that pass the band

Fig. 17. Operations room of Mistram I system at AMR. Courtesy General Electric Co.

frequencies about the center frequency of the subcarrier oscillators and detect the resultants to produce the signals fed to the subcarrier oscillators in the vehicle.

Commutated (PAM, PDM) data are applied to parallel switches sequentially closed for short intervals. Each gate delivers discrete levels representing one channel of commutated data when the closures are synchronized with the pulses applied to the input. In PDM systems, a PDM/PAM converter produces a PAM signal from the PDM input.

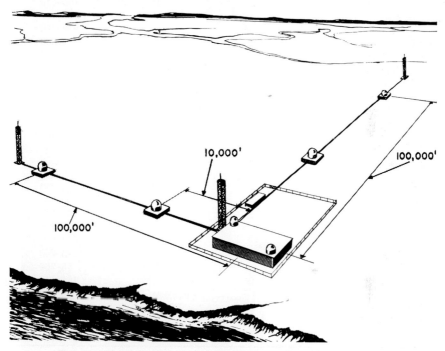

FIG. 18. Mistram system at AMR. Courtesy General Electric Co.

The oscillograph and the pen recorder are used for data display. A seven-track, record-reproduce unit is the standard magnetic tape recorder for composite signals at AMR. Plug-in record and reproduce amplifiers are provided for analog, FM, and PDM operational modes.

Table II lists the telemetry receivers in use at AMR and their operating charactreistics [6].

b. Optical Tracking and Camera Coverage [6, 10, 11]

Optical coverage at AMR includes ballistic cameras, cinetheodolites, fixed metric cameras, tracking telescopes, and documentary cameras.

TABLE II. AMR TELEMETRY RECEIVERS

Station	Quantity	Modulation	Frequency range (Mc/sec)
Tel 2 Cape Kennedy	54	FM	215–260
Tel 3 Cape Kennedy	14	FM	215–260
	3	AM-FM	50–260
	2	AM-FSK-SSB	0.5–32
	3	AM	0.54–54
	2	AM-PM-AM	105–140
			370–410
			920–960
GBI	24	FM	215–260
San Salvador	12	FM	215–260
Grand Turk	8	FM	215–260
Antigua	13	FM	215–260
	2	FM	126–137
	2	FM-PM-AM	105–140
			370–140
			920–960
	1	AM-FM	20–230
Ascension	12	FM	215–260
	1	AM-FM	55–260
	1	AM-FM	20–230
	3	AM	0.54–54
	2	FM-PM-AM	105–140
			370–410
			920–960
Pretoria	4	FM	215–260
	1	AM-FM	55–260
	2	PM-AM	105–140
			920–960
Aircraft	6	FM	215–260
Ocean Range Vessels	32	FM	215–260
		PM-AM	920–960

The ballistic camera is a precisely constructed instrument with an optically flat photographic plate of high-quality glass mounted on a machined case. Four types of ballistic cameras are used at AMR: the Wild BC-4, six of which have shutters that can be synchronized with range timing; the BC-600, with lens specifically designed to pass light at 4800 Å for use with a strobe light; and the MBC-1000 camera, which is trailer-mounted, although it requires removal from the trailer for operation. There are also six specially designed, 600-mm focal length, $f/2$ Northrop cameras (Fig. 19) used for flight path photography of missiles and space vehicles. The camera weighs 1660 lb and has a unique louvre-type shutter that operates in 2 msec. The range has a total of 29 ballistic camera sites on the Florida mainland and 30 at downrange stations.

The standard cinetheodolite (Fig. 20) in use at AMR is the Askania Kth53, which consists of a double-width frame, 35-mm camera and two sighting telescopes. The cinetheodolites cover missile launches from 0 ft

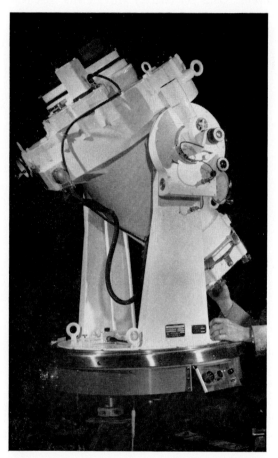

Fig. 19. A 600-mm ballistic camera used at AMR. Courtesy Northrop Corp.

to 100,000 ft and provides backup for the fixed cameras from 0 ft to 10,000 ft. When photographic conditions permit, it is also a main source of missile position data from the 2000-ft to 100,000-ft altitudes. Eight cinetheodolites are located in permanent astrodome-tower installations from north of Cape Kennedy to south of Patrick AFB. The Model E Contraves instrument is also used.

The CZR-1 and RC-5, standard metric (ribbon frame) cameras at AMR, are functionally identical except for minor details in design. Mounted on trailers, they provide azimuth and elevation angles during a missile's first 2000 ft of flight from which position and roll and attitude data can be obtained. For each launch, the cameras are located on the previously surveyed pads selected to provide the best coverage.

FIG. 20. Askania cinetheodolite used for missile tracking at AMR. Courtesy Askania-Werke.

The two types of tracking telescopes used at AMR are the Igor (Intercept Ground Optical Recorder), shown in Fig. 21, and the Roti Mk 1 and Mk 2 (Recording Optical Tracking Instrument). The Igor photographs missile flights as a function of time; as the camera records images seen through the main objective telescope, timing marks are exposed along the film's edge. The Roti makes time-correlated, high-resolution, long-range photographs of objects in space. Both instruments contain automatic focus and exposure controls, employ similar siting telescopes, and are provided the same three methods of tracking: slaved to the target acquisition bus; slaved operation with operator override (with servo error voltages given to the operator for correction); and operator

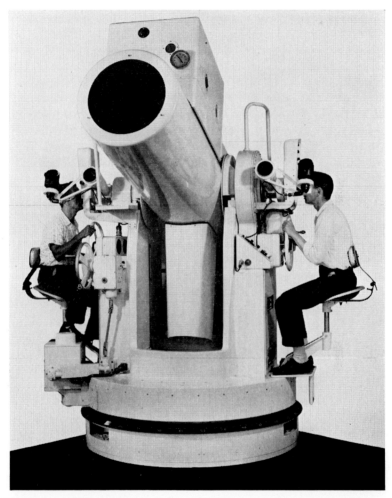

FIG. 21. Igor tracking telescope used for optical tracking at AMR. Courtesy J. W. Fecker Div., American Optical Co.

tracking by joystick (Roti) or handwheel (Igor). The Roti has a modified, Navy Mk 30, 5-in. gun mount, while the Igor employs a modified, Navy Mk 27, 5-in. gun mount. Both telescopes are protected by astrodomes with curved overhead doors that slide open for operation. Igor sites are located at False Cape (north of Cape Kennedy), Cocoa Beach, and Patrick AFB. Roti sites are located at Williams Point, Melbourne Beach, and Vero Beach.

Documentary photography is provided by intermediate focal length trackers and high-speed cameras.

The intermediate focal length trackers are mobile instruments using 35-mm or 70-mm motion picture cameras to obtain documentary photographs and attitude data at short and medium slant ranges. Three types are in use at AMR: the M-45 Ifltt, the Iflot, and the Mk-51. The M-45 Ifltt (Intermediate Focal Length Tracking Telescope) is a single-operator, power-driven tracker with two photo-optical systems of 40-in., 48-in., 60-in., or 120-in. focal length. A 16-mm camera can be used with 40-in. lenses and a 35-mm camera with all lenses. The Iflot (Intermediate Focal Length Optical Tracker) was designed to replace the M-45 Ifltt. In addition to the 16-mm and 35-mm cameras, it can use a 70-mm camera with 80-in. lens. The Mk-51 is a Navy radar gun director, modified to be used as an Ifltt, which uses either 16-mm or 35-mm cameras. Five M-45 Ifltt, seven Iflot, and five Mk-51 trackers are available at AMR.

Three different 16-mm and 35-mm cameras are used for high-speed photographs at AMR: the Mitchell, the Fastax, and the Milliken. The 16-mm and 35-mm Mitchell camera has a 4-lens turret with lenses of 25-mm to 152-mm local length. It can be mounted on a tripod, tracking telescope, or special mount. The 16-mm and 35-mm Fastax is a rotating-prism, high-speed camera with an external magazine that holds a 500-ft, daylight-loading spool employing a 115-volt, 60-cps takeup motor. The 16-mm Milliken camera is used for close-in photographs of vehicle performance at liftoff and for other uses where cameras of small size and study construction are needed.

Three types of 70-mm documentary cameras are used at AMR: Mitchell, Photosonics, and Flight Research. All three have an intermittent type of film movement with film registration pins and a vacuum back for accurate positioning of the film during exposure. They are used for documentary photography in the launch area and can be mounted on intermediate and long-focal-length telescopes.

4. SUPPORT ACTIVITIES

Technical support activities at AMR include a range timing system, a countdown sequencing system, a supervisory control system, meteorological instrumentation, real-time and post-launch data handling, photographic services, and frequency control and analysis.

To facilitate locating nosecones or payloads that impact in the water downrange, AMR uses a Mils system similar to that of PMR. Like the system on the west coast, the one at AMR employs an array of hydrophones distributed throughout the impact area. They are connected by cable to shore stations. The hydrophones respond to explosions caused by charges ejected from payloads upon impact. Time delay of underwater signals and triangulation permit relatively precise locations. A modified form of the system is also used and is called Star (Ship Tended Acoustic Relay). The major difference between Star and Mils is that the former uses a ship anchored in the area to interrogate transponders on the hydrophones for determining its own position. Then the hydrophones function as they do with the Mils system.

The AMR timing system has a central timing station at Cape Kennedy, downrange central timing sites, and subcentral units at other stations. Two dual, time-based generators, one for backup, produce time codes and repetition rates at the Cape Kennedy central timing station. The codes and rates are clocked and compared; and when an inconsistency between signals exists, an alarm is triggered and the backup generator may be manually switched in. The system is synchronized with Loran-C signals and is correlated to within ±25 μsec of Greenwich Mean Time. The basic clock consists of steerable, crystal, 100-kc frequency standards with a stability better than 5 parts in 10^{10} and an accuracy better than 1 part in 10^9.

Downrange central timing systems, each with one dual time-base generator, are located at GBI, San Salvador, and Grand Turk, where they receive a synchronizing signal sent down the subcable from Cape Kennedy. Using the subcable signal, these generators can by synchronized to within ±40 μsec correlation or to within ±5 msec correlation using a signal from WWV. When subcable or WWV signals are used, the transmission delay times are compensated for in the calibration. Correlations to within ±100 μsec can be obtained by off-cable stations using Loran-C equipment.

Subcentral terminal equipment at remote instrumentation sites produces the same waveforms that are available at central timing. The signal from Cape Kennedy to subcentral terminal equipment at remote instrumentation sites is periodically measured in a loop circuit for transmission delays that are compensated for to make available timing within 50 μsec of synchronization with Cape Kennedy.

When complete, the AMR Range Control Center will be housed in a three-story building and will permit the simultaneous countdown of two missiles and global range support of an orbital vehicle. The center will be staffed by 64 personnel with 137 consoles from which they can simultaneously control 800 range instrumentation status displays, four missile trajectory plots, three impact prediction plotters, three separate countdown and timing systems, and a 10,000-mi network of communications within the AMR complex.

The center contains command and control equipment for participation in worldwide activities supporting missile, orbital, and interplanetary probe launchings.

A trajectory display system provides operator control of the automatic display of four missile trajectory plots on any one of four programmable backgrounds. Trajectory data are received from tracking stations and fed into the system through a high-speed computer.

Console operators can also simultaneously control status displays for launch operations instrumentation and lead range tracking stations. More than 600 status indications are provided for the automatic or manually controlled status displays of radar, optical, and telemetry systems. Information from some 200 worldwide tracking stations are fed into the Range Control Center for display as lead range status.

Other displays include the position of all aircraft within the vicinity of AMR, the disposition of recovery forces, and the locations of emergency control teams. Large status displays are used to show the condition of support test operations during launch.

In addition, some 34 television monitors and a number of cameras are provided for viewing launch pad operations, meteorological displays, in-flight missile performance, and other remote operations.

Capability is also provided for operator control of countdown, in coordination with blockhouse personnel, during launch operations for as many as three missions. A further capability exists for the generation and control of three simultaneous "first motion counts" and nine "interval time counts."

To expedite the recovery of payloads, real-time impact prediction plots and radar data displays of recovery aircraft are also provided.

All console operators are provided with voice communications tied into the overall AMR network. Certain of the consoles have, in addition, facsimile equipment that permits the transmission of written data; and all consoles have provisions for recording communications for later transcription.

Personnel and equipment are divided into three operational areas within the building. The entire forward wall of each area on the first floor is taken up by trajectory and status displays. The major display here is an 8-ft × 8-ft trajectory display and a launch status panel as well as a lead range status panel with the status of the 200 instrumentation sites. Associated with these is a large timing display indicating countdown time, first motion time, and local time. Other display areas in the center will be similar but on a lesser scale.

The first floor level of each of the three operational areas contains the AFMTC coordination consoles and the Project Office Consoles, which can accommodate eight missile launch agency representatives. The second floor level, overlooking the main operations theater, has three identical Range Operations Areas, each equipped with a large window from which console operators can see the wall display below. Each also has a Superintendent of Range Operations console and a Lead Range Superintendent Console, which are the central control stations for simultaneous launching missions and control for the status display system. In the rear of the second floor level are four auxiliary control areas: aircraft, recovery, emergency, and support test control, each of which maintains its own status and display. The third floor of the Range Control Center is a gallery for visitors viewing the operations.

The new Range Control Center is constructed along the lines of the Manned Space Flight Control Center (Fig. 22) from which the Project Mercury flights were monitored and controlled at AMR. However, for future manned flights in Projects Gemini and Apollo, these functions will be directed from NASA's Integrated Mission Control Center at Houston, Texas [12].

Meteorological instrumentation at AMR provides accurate and pertinent weather data to range users. Daily observations are made throughout the range, then sent to Patrick AFB where they are studied, used to prepare forecasts, and sent to the National Weather Bureau in Washington, D.C. for general distribution. A total of 215 weather instruments and devices are in use throughout the range, with the greatest number, 54, at Cape Kennedy. The instruments and systems include Rawin sets, wiresonde sets, storm detection radars, cloud height sets, ceilometers, ceiling light projectors, wind measuring sets, wind sets, theodolites, temperature-humidity measuring sets, psychrometers, hygrothermographs, mercurial barometers, barographs, meteorological rocket systems, a data processing system, and a weather information network and display system. The most common device is the AN/GMQ-11 wind measuring set, with a total of 42 located throughout the range. The instruments and systems are located at Patrick AFB, Cape Kennedy, Valkaria, GBI, Eleuthera, San Salvador, Grand Turk, Antigua, Ascension, and on ocean range vessels and ARIS ships.

Range users are supplied a variety of data by the AMR real-time handling system, including vehicle performance and position information, target acquisition messages, and critical data quality validation. It is also used to retrieve data for post-launch processing.

In addition to data handling, AMR provides its users with post-launch data handling information in the form of quick look (usually within 6

FIG. 22. Manned Spaceflight Control Center as set up for Project Mercury at AMR. Courtesy NASA.

hr after a test), or limited data analysis, and a final post-launch report, including a complete data reduction.

Final trajectory data reduction on a given test is performed at Patrick AFB. Data from Mistram; Udop; Azusa Mk I and II; Mils; Glotrac; AN/FPS-16, AN/FPQ-6, and AN/MPS-25 radars are recorded in real-time in transmission-modulated form. The recording is made on tape, which is rewound and played back on the data converter, one system at a time. Reduction is also made on telemetry data, analog signals from Mils hydrophones, and film from the various camera stations. The end result of the final trajectory data reduction is the Flight Test Report. A best estimate of trajectory (BET), reduced from Azusa, Udop, theodolite, and BC-4 camera data, is also computed and becomes a source of information for the Flight Test Report.

Post-launch data reduction equipment at Patrick AFB includes both the BCD and binary, solid-state, stored-program digital computers. Three 1401 decimal computers serve as data gathering, formatting, editing, and printing devices, thus releasing the one IBM 7094 binary computer for maximum computations.

Also used are data converters, plotters, telemetry automatic reduction equipment (Tare), Doppler Automatic Reduction Equipment (Dare), plate previewer, comparators, microscopes, film readers, attitude readers, and theodolite readers.

The Air Force Missile Test Center (AFMTC) Cine Processing Laboratory provides range users with two types of service: one which processes metric and engineering film to produce 16-mm, 35-mm, and 70-mm negatives in black and white and color; and the other which processes a documentary film to produce 16-mm and 35-mm negatives in black and white or color. Immediate processing of the various films used to gather data on a missile test is furnished by 11 machines. They process color negative-positives, color prints, color reversal films, Anscochrome-type films, high-contrast negative films used by cinetheodolites and tracking telescopes, normal contrast negative film for report films and other documentary use, and print films.

Frequency Control and Analysis (FCA) monitors critical frequencies to look for interference and also checks to see that operational frequency assignments and schedules are maintained within their limits. FCA facilities include fixed stations as well as mobile and semimobile vans. The mobile vans are used at the launch sites during missile launchings and, on short notice, can be driven elsewhere to make special measurements or to chase down interference sources. The two fixed stations are located at Cape Kennedy and Patrick AFB; the latter, operating as a Space Electronic Detection System (SEDS), is equipped to monitor and track radiating satellites. Similar support is provided by a semimobile van at Antigua. If required, C-131 aircraft are available for airborne monitoring assignments.

Administrative support services include a security police force, fire department, medical department, and a commissary. The security police check Cape personnel identifications, regulate traffic on the 65 mi of Cape roads, protect Cape installations from theft, and maintain road-blocks at access roads to danger areas during launch operations.

The fire department has equipment on standby alert for every launch, test, and fueling operation at the Cape. It inspects for fire hazards in launch complexes and buildings, conducts special courses in firefight-ing, and performs the regular duties of a municipal fire department. The medical department is staffed with specially trained doctors, industrial nurses, medical technologists, and driver-orderlies who maintain 4 dis-pensaries and 13 ambulances on the Cape. During every launching, test, or fueling operation, medical department personnel stand by with the "impact crew." Industrial hygienists working for the medical depart-ment conduct tests for noise levels, blast effects, toxic gas conditions, and radiological content. The commissary at Cape Kennedy maintains a cafeteria, 2 snack bars, and 11 mobile snack wagons. A commissary warehouse is also maintained where food and sundries are packaged and shipped to downrange stations. Perishable foods are transported by Air Force aircraft, while the remainder goes by commercial freighters and subsequently by LSM-8's to the more remote stations and sites.

Maintenance support at Cape Kennedy includes electrical, plumbing, carpentry, masonry, painting, building maintenance, air conditioning, heating, metal and machine shop, water and sewage, vehicle maintenance, sanitation control, and custodial services.

NASA also maintains many support activities at AMR and will en-large upon them as the manned spaceflight programs of Projects Gemini and Apollo become operational [13]. The manned spacecraft facilities at Cape Kennedy were developed for and during the series of suborbital and orbital flights of Project Mercury. They included a white room for capsule assembly, an altitude chamber, a biomedical area for astronauts, a radiation tower, and storage, service, and supply areas.

Manned Spacecraft Center facilities for future projects such as Gemini and Apollo will, for the most part, be located at the Merritt Island Industrial Area (MILA) and will include an operations and checkout building; supply, shipping, and receiving building; fluid test complex; weight and balance building; parachute and paraglider build-ing; radar boresight range; supply and ground support equipment service building; ordnance storage building; static test complex; and an ordnance field test laboratory. The static test complex will be used to test the propulsion systems of future spacecraft, and the test laboratory will provide a facility for testing small pyrotechnic devices.

B. Pacific Missile Range and Vandenberg Air Force Base

The Pacific Missile Range (PMR) is a Department of Defense National Range operated by the U.S. Air Force.* It was established in June 1958 when the Secretary of the Navy, under the direction of the Secretary of Defense, combined the Naval Air Missile Test Center, Pt. Mugu, California, and the Naval Missile Facility, Pt. Arguello, California, and designated them the Pacific Missile Range. From an operational viewpoint, PMR can be considered as a complex of four separate ranges controlled and coordinated by a headquarters located at Pt. Mugu, California, as shown in Fig. 23. These installations are the *Sea Test Range*, an area some 175 mi by 300 mi off the coast of Southern California, in which air-to-air, air-to-surface, surface-to-air, underwater, and short-range surface-to-surface missiles are tested. Missiles are launched from Pt. Mugu facilities.

The *Polar Satellite Range*, from which are launched the Discoverer satellites and other military satellites with polar orbits, originates at Vandenberg Air Force Base and Pt. Arguello. The *Ballistic Missile Range* is an area extending from Vandenberg Air Force Base and Pt. Arguello launching pads to impact areas in the Marshall Islands, for ICBM's, and an area some 1500 mi offshore for IRBM's. The *Antimissile Range* is centered at Kwajalein Atoll with supporting facilities located on neighboring atolls such as Ennylabegan, Guegeegue, and Roi-Namur [13, 14].

Vandenberg Air Force Base, a separate and distinct installation of the U.S. Air Force, is located just to the north of Pt. Arguello. It was established in 1956 when the Air Force Ballistic Missile Division took over an unused Army post named Camp Cooke. In 1957 it was renamed Cooke Air Force Base and construction began on the missile base; on 1 January 1958, the Strategic Air Command activated its first ballistic missile division, and the name was changed to Vandenberg Air Force Base. PMR assumed a tracking function for the base on 17 December 1958.

Other PMR installations are located in Hawaii, the Marshall Islands, Wake Island, Midway Island, Canton Island, Howland Island, and Baker Island. Stations are also located at Point Sur, Point Pillar, San

* See note added in proof page 433.

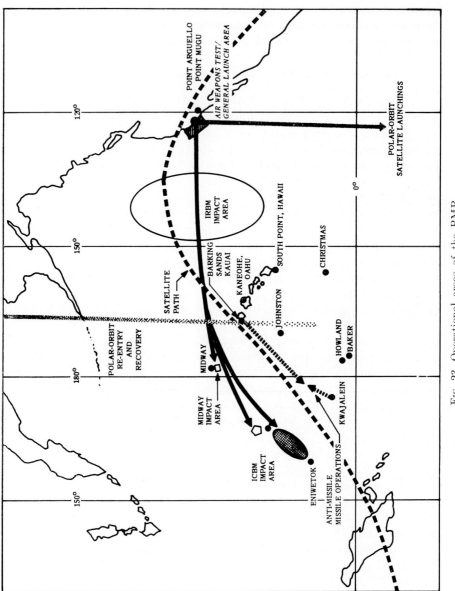

FIG. 23. Operational areas of the PMR.

Clemente Island, Anacapa Island, Santa Cruz Island, Santa Rosa Island, San Miguel Island, and San Nicolas Island. The Hawaiian stations are located at Kancohe, Kokee Park, Barking Sands, and South Point.

1. Geographical Factors

The primary rangehead at PMR is at Pt. Arguello, 34°37′N 120°35′W, an area covering 30 mi^2 located on the coast of southern California, 165 mi north of Los Angeles. It is on dry, sparsely vegetated, and windswept Burton Mesa with shale bedrock indented by gullies and pockets of sand, clay, and gravel. Vandenberg Air Force Base (34°38′N 120°32′W, elevation 360 ft) is contiguous, with its southern boundary joining Pt. Arguello. This Air Force installation with its 108 mi^2, including 32 mi of coastline, is the second largest in the U.S.—exceeded only by Eglin Air Force Base, Florida.

2. Launching Facilities

Facilities for westward launches are located both at Point Mugu (Fig. 24) and Point Arguello (Fig. 25); however, the principal launch activity is at Point Arguello and Vandenberg AFB, where intermediate range and intercontinental ballistic missiles are fired. The facility at Pt. Mugu conducts more localized launch experiments.

At Pt. Arguello, Launch Complex 1 and 2 in Fig. 25, consists of two conventional Atlas launching pads for vehicles generally associated with military satellites. Each pad has its independent oxidizer and fuel storage supply and service tower. However, one blockhouse controls both pads, having individual control rooms. Some distance to the northwest of these pads, there is a complex for Hawk and Terrier surface-to-air missiles. To the south of this facility is a launching complex (Launch Pads on Fig. 25) for the Argo D8, solid propellant space carrier vehicle used in space biological experiments. This installation also permits the launching of sounding rockets such as Nike Cajun. Farther to the south are other launching pads (not shown on Fig. 25) from which sounding rockets such as Nike Viper and HAS are launched. The HAS launcher is unique, having been designed by the Sandia Corp. It is trainable through 360° of azimuth and 90° of elevation. The tubular steel, straight-rail device is manually pointed in azimuth but has an electrically powered winch for elevation [14–16].

NASA utilizes the facilities at Pt. Arguello for launching such space carrier vehicles as Thor Agena B, Atlas Agena B, and Scout. The Scout launcher, Probe Launch Complex on Fig. 25, is of the older type, first used at NASA's own launching range at Wallops Island, and shown in Fig. 51.

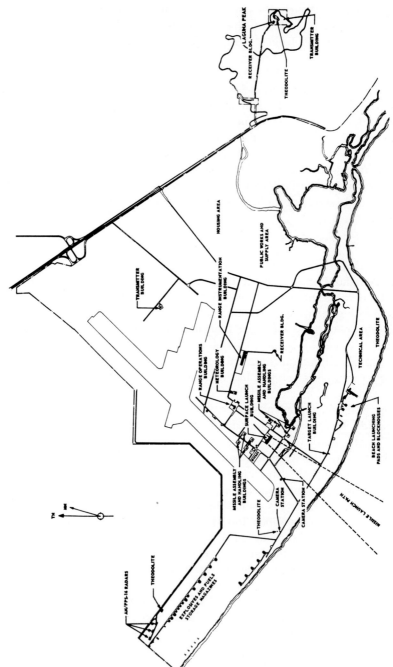

Fig. 24. Facilities at Naval Missile and Astronautics Center, Pt. Mugu, California, PMR. U.S. Navy photograph.

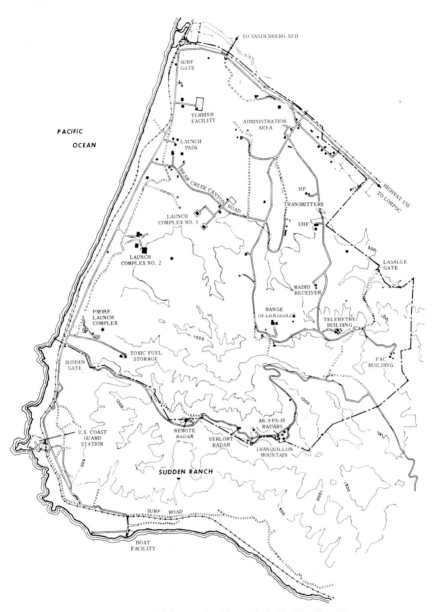

FIG. 25. Activities at the Naval Missile Facility, Pt.
Arguello, California, PMR. Courtesy U.S. Navy.

A unique launcher at the Naval Missile Facility is a modified, twin, 5-in. gun mount that can launch rockets weighing up to 20,000 lb. Using gun-pointing systems, the launcher is directed by remote control. Rockets are attached to a horizontal boom and elevated by steel cables winched through the top of the gun mount [17].

Of all the launchings made at PMR, almost 70 per cent are from facilities located at Pt. Mugu. It is from this rangehead that small, tactical missiles such as Sparrow, Hawk, Terrier, Tartar, Bullpup, Sidewinder, and Talos are launched. Target drones (such as the Q2C and KD2R) and meteorological sounding rockets (such as Cricket and Nike Cajun) are also fired from Pt. Mugu. Extensive use also is made of the Sea Test Range for air-to-air, air-to-surface, and short range surface-to-surface missile tests [18].

At Pt. Mugu a 125-ft by 350-ft steel and concrete building functions as a self-contained launch complex for surface launched missiles. The roof of the building is the launching pad, while the interior rooms contain launch control equipment, auxiliary equipment, and emergency repair facilities. The first 3300 ft of missile flight from the roof is over land, permitting the recovery of booster equipment and instrumentation devices ejected after booster burnout. The rooftop pad can simultaneously accommodate three 20-ft × 50-ft launchers and five smaller launchers (20 ft × 30 ft). Each launcher location is supplied electrical power and high-pressure gas by cable and piping outlets, some of which permit direct connection of communication and control cables from the missile launch control rooms. All interior areas are connected by corridors. Also located on top of the building is an AN/UMQ-5 wind measuring set that transmits surface wind measurements to the missile launch control rooms below. Each launcher location has built-in, remotely operated firefighting equipment. Maximum vehicle thrust from the roof is 150,000 lb.

Launcher pad No. 2 at Pt. Mugu, a special submarine simulator device, contributed greatly during the development of the Polaris missile.

Four launching pads and two concrete blockhouses (designated launcher controls Able and Baker) located near the beach (Fig. 24) are available for special projects. Pads 1 and 2 are controlled by launcher control Able, which is equipped with windows affording a direct view of the pads. Pad 1 has two catapults, while pad 2 has restrained-firing facilities. Each pad is supplied air at 2400 psi from 15-ft³ bottles, and each has its own water deluge system. Pads 3 and 4 are controlled by launcher control Baker, which has two periscope-type windows for indirect viewing of the pads and one window shielded with safety glass for direct viewing. Each pad has carbon dioxide fire extinguishers and a water deluge system electrically controlled from the blockhouse.

Located just to the west of the beach launching facilities is a target drone launching complex. Catapults in this area can launch drones weighing as much as 600 lb with velocities at the end of the catapult of 90 knots. Other installations associated with the facility include a small storage hangar, fuel and oil dumps, and an electronics shop.

Flight test control for all firings from Pt. Mugu is directed from the Range Operations Building located to the northeast of the Surface Launch Building. This three-story, concrete structure contains all the electronic means for directing, tracking, and controlling aircraft and surface craft within the Sea Test Range and insuring area clearance prior to a missile launch. To accomplish these functions, the roof of the building is a site for height-finding and surveillance radars. It also houses the SCR-584/615 radars, range safety computers, plotting boards, and other display devices. During operations within the Sea Test Range, this facility can conduct three simultaneous tests, while a similar installation on San Nicolas Island can at the same time conduct an additional test.

Vandenberg Air Force Base serves a double purpose. Its primary task is support in the research and development of military ballistic missiles and space systems. In this role it works very closely with the PMR, especially that part of it located at Pt. Arguello. The secondary role is that of a tactical launching site for ICBM's of the U.S. Strategic Air Command, a role it can assume in less than 24 hr after receiving an alert. Within the base is at least one type of launching site for operational, long-range, Air Force missiles. Because of its favorable geographical position, Vandenberg also lends itself to the launching of Earth satellites into polar orbits without the necessity of having carrier vehicles pass over populated areas. Thus the base finds extensive use as a launching site for certain military satellites placed into such orbits.

Vandenberg AFB employs three types of ICBM launchers: the "soft" emplacement, which employs a service tower (Fig. 26); the horizontal "soft," an above-ground shelter in which the ICBM's are maintained in horizontal position until ready for fueling and launching; the horizontal semihard, a below-ground Atlas E shelter hardened against possible attack (Fig. 27); and the silo emplacement, for Atlas F (Fig. 28), Titan 1 and 2, and Minuteman missiles. The Atlas and Titan 1 missiles are raised to ground level by hydraulic elevators before they are fueled and launched, while the Titan 2 and Minuteman ICBM's are stored and launched from the bottom of their silos.

The Titan 1 silo is deceptively simple above ground. Little shows except the massive, 750-ton steel and concrete cover of the silo. The area is fenced and contains electronic sentries to warn of intruders (Fig. 29). Smaller covers protect miniature silos that contain the radio guid-

FIG. 26. "Soft site" emplacement at Vandenberg AFB, California, PMR, for launching Earth satellites by Thor Agena B and D space carrier vehicles. Courtesy North American Aviation, Inc.

ance antennas. Beneath the ground, however, is a maze of tunnels totaling 3000 ft; a complete storage system for liquid propellants; and a special three-level, steel-enclosed, control room. The silo itself is 174 ft deep and has working platforms at nine levels. The special elevator that lifts the missile is stressed for 170 tons [19].

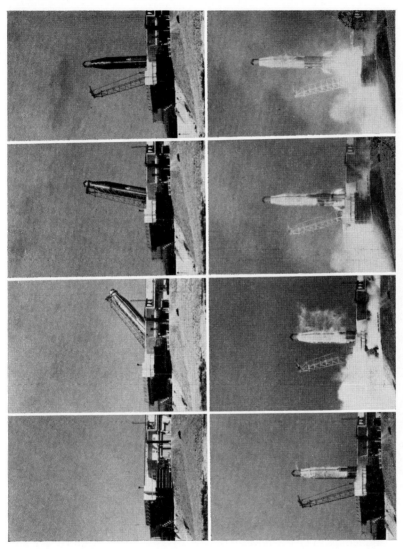

FIG. 27. Atlas E launching from a semihardened, coffin site at Vandenberg AFB. U.S. Air Force photograph.

Titan 2 silos at Vandenberg are an improvement on the Titan 1 models. They do away with the elevator that lifts the exposed missile above ground for propellant topping off and launch. The missile is fired directly from the bottom of the silo, exhaust flames and gases being vented to the atmosphere through special ducts leading to the surface (Fig. 30).

Fig. 28. Atlas F silo at Vandenberg AFB. Courtesy North American Aviation, Inc.

The Minuteman silo is a further refinement in the technique of underground launching. The 88-ft deep silo is covered by a 90-ton steel and concrete cap, which protects the site from all but a direct surface or low-altitude nuclear warhead burst. The two-acre area around the site is surrounded by a fence and special electronic devices in the soil that detect both humans and animals attempting to tunnel into it. Should intruders appear, warning lights flash in the control center.

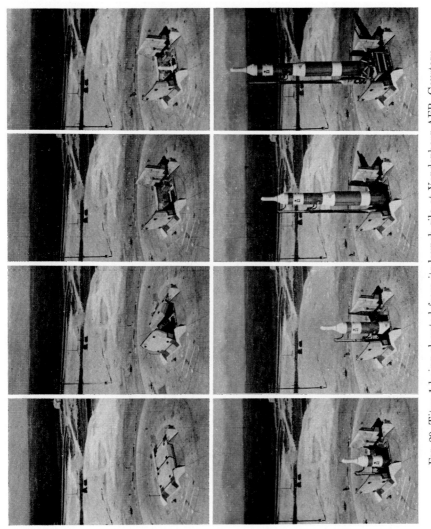

FIG. 29. Titan 1 being elevated from its launch silo at Vandenberg AFB. Courtesy The Martin Co.

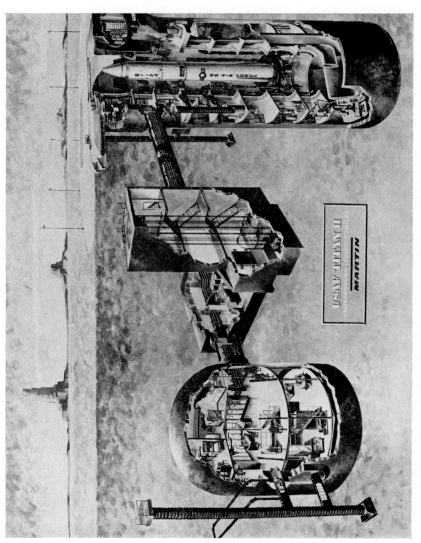

FIG. 30. Titan 2 silo at Vandenberg AFB. Structure on left is control center, where crew members live and work. The center structure permits access to the site, and the missile silo proper is on the left. Courtesy The Martin Co.

The control center for the Minuteman silo is also designed to survive nuclear bursts. It is an ellipsoidal shaped room buried 32 ft underground. Protected by a seven-ton steel door that can be opened only from the inside, the launching control personnel are protected from both blasts and saboteurs. The room is suspended on special shock absorbers to help dampen the effects of atomic near-misses. It also has a novel, emergency escape hatch. In the event they are trapped, launching personnel can ascend a 36-in. diameter tunnel, which is filled with sand up to a point 5 ft beneath the soil. By opening the escape hatch the sand drains into the control room, permitting the personnel to climb to the point beneath the soil from which they dig their way out with shovels [19, 20].

A summary of launching facilities found at PMR is given in Table III.

TABLE III. SUMMARY OF LAUNCHING FACILITIES OF PMR

Location	Number	Type
Pt. Mugu	1	Surface launcher
	1	XM-1
	1	XM-1-1
	1	XM-2
	1	Submarine simulator
	1	Zero length launcher
	2	Meteorological launcher
	3	Undesignated launcher
Pt. Arguello	2	Argo D8 and Nike Cajun
	1	Tumbleweed
	2	Terrier and Hawk
	4	Atlas
San Nicolas Island	1	IGY project pad
	1	Sergeant
Barking Sands, Kauai, Hawaii	1	Regulus 1
	1	Meteorological launcher
Kwajalein Island	4	Nike Zeus missile launcher
	1	Speed ball
	1	Meteorological launcher
Vandenberg AFB	7	Thor (two for Thor Able Star)
	9	Atlas (three silos, three coffins, three tables)
	3	Titan 1 silo
	3	Titan 2 silo
	6	Minuteman silo
	2	Titan 3[a]
Eniwetok Atoll	1	Regular-type launcher pad
	1	Meteorological launcher

[a] One tentatively scheduled for 1965 and the other for 1966.

3. TRACKING AND COMMUNICATIONS

Communications requirements of the Pacific Missile Range are co-ordinated from Pt. Mugu, where a Communications Central is the point of convergence for voice, signal data, control, and teletype lines from operational facilities all over the range. Communications systems employed include Kineplex, multiplex, facsimile, teletype, telephone, air-to-ground and ship-to-shore, and closed-circuit television. Inputs include standard frequencies, timing and computer inputs, drone control and destruct, signals status indications, countdown information, and the automatic programmer data. The facility has a radio circuit on the patchboard that includes 18 party lines, 12 with 12-party loops and 6 with 20-party loops.

Also located in Communications Central is a 400-line private automatic exchange (PAX), which permits selected personnel to have a private telephone dialing system. On another section of the patchboard, through the use of UHF transmitters, receivers, and multiplex systems, the conference loops, base administrative phone system, PAX system, and teletype circuits are extended to Santa Cruz and San Nicolas Island, which have comparable equipment at their end [21].

A receiver building containing 110 receivers (HF, VHF, and UHF) affords a centralized RF receiving system for the range. Antennas associated with the sets are located on the roof of the building and on poles in an adjacent antenna farm. The 66 transmitters for the communications system are also located in a special building with an adjacent antenna farm. A Collins microwave, multiplex system with 240 channels, installed in the Pt. Mugu Range Operations Building, ties together facilities at Pt. Arguello, Santa Cruz Island, Pt. Mugu, and San Nicolas Island. Instrumentation data as well as voice and teletype data are handled by the system. A Lenkurt multichannel carrier system also provides additional communications means among Pt. Mugu, Santa Cruz Island, and San Nicolas Island [22].

With such a multiplicity of electronic equipment at the range, an extensive frequency monitoring and control system is imperative. Facilities available at PMR for this function include fixed sites at Pt. Mugu, Laguna Peak, and Pt. Arguello; 40-ft semimobile vans; 27-ft mobile vans; 18-ft mobile vans; WV-2 (Super Constellation) aircraft; smaller aircraft; and the USNS *Skidmore Victory* instrumentation ship.

a. *Radar Tracking and Telemetry*

A variety of pulsed and passive CW radars are used at PMR for tracking, surveillance, and range safety. Sites are located on the mainland, offshore islands, tracking ships, downrange islands, and in aircraft. The various sites, types of radars, and uses are given in Table IV.

The principal CW tracking systems presently used at PMR are the COTAR target acquisition system at Pt. Mugu and the GERTS receiving system for precision tracking.

Fig. 31. AN/FPS-16 radars on San Nicolas Island, PMR. Courtesy North American Aviation, Inc.

The COTAR system, consisting of a 450-ft diameter antenna field and a 40-ft equipment van, is employed to position radar antenna by furnishing target azimuth and elevation angle data. No special equipment is required by COTAR since it uses signals from a standard telemetry transmitter. It can acquire targets through 360 deg in azimuth to a range of 2 to 500 nautical mi without provision for parallax cor-

TABLE IV. RADAR TRACKING FACILITIES AT PMR [14, 15, 21][a]

Type	No.	Location	Use
AN/FPS-16	4	Pt. Mugu	Precision tracking and range safety
	4	San Nicolas Island	Precision tracking and range safety
	2	Pt. Arguello	Precision tracking (Fig. 31)
	1	Kokee (Hawaii)	Precision tracking and range safety
AN/MPS-19	1	Pillar Point	
	2	San Nicolas Island	Air surveillance
AN/FPS-6A	1	Pt. Arguello	Height finder
AN/FPS-41	1	Kwajalein	Storm detection
AN/FPN-33	2	Pt. Arguello	Range safety
AN/APS-20	1	Pt. Mugu	Surface surveillance
	1	Santa Cruz	Surface surveillance
	1	San Nicolas Island	Surface surveillance
	1	Pt. Arguello	Surface surveillance
AN/SPS-5B	2	Pt. Arguello	Area clearance
AN/SPS-6C	1	Pt. Mugu	Air surveillance
AN/SPS-8A	1	Pt. Mugu	Height finder
	1	San Nicolas Island	Air surveillance
AN/MPS-26	2	Pt. Mugu	Range safety
AN/MPS-19	2	Pt. Arguello	Range safety
AN/MPQ-38	2	Barking Sands (Hawaii)	Precision tracking
SG-1B	1	Pt. Mugu	Surface surveillance
SCR-584/615	3	San Nicolas Island	Precision tracking
M33	1	Pt. Arguello	Precision tracking
	1	Vandenberg AFB	Precision tracking
	1	Barking Sands	Precision tracking
	1	Kwajalein	Precision tracking
	1	Eniwetok	Precision tracking
Cotar	1	Pt. Mugu	Precision tracking
	2	Vandenberg	Range safety
Verlort	1	Kokee (Hawaii)	Range safety
	1	Pt. Arguello	Range safety

[a] Other radars on Kwajalein include tactical radars for Nike-Zeus system supplied by the U.S. Army.

rection, and it can track 10 to 60 deg in elevation with a tracking accuracy of 3 mil in elevation. The maximum tracking rate is 18 deg-sec.

GERTS is a combination pulse/CW system, requiring two transponders in the target. An extremely accurate X-band pulse radar measures distance from and direction to the target. Range rates are obtained by means of X-band, CW Doppler measurements from receivers located at both ends of a 2000-ft baseline. The system has its own digital computer.

Telemetry data from missiles or Earth satellites can be received at PMR by a variety of sensors. There are permanent sites at Pt. Mugu,

Pt. Arguello, San Nicolas Island, and Hawaii; two specially outfitted ships; mobile vans; and aircraft. A central telemetry data processing facility is located at Pt. Mugu and operates from magnetic tape inputs, employing both digital and analog techniques.

A summary of telemetry stations at the range is given in Table V.

TABLE V. PMR TELEMETRY RECEPTION STATIONS

Location	Quantity	Type
Pt. Mugu	10	FM/FM & PAM station
	1	FM/FM & PAM/FM van
	2	PDM/FM station
	1	FM/PAM & PDM mobile van
Pt. Arguello*	3	FM/FM, PAM/FM, PDM/FM station
San Nicolas Island*	1	FM/FM receive/record station
	1	PAM/PDM/FM receive/record station
Kokee Park, Kauai	4	FM/FM, PAM/FM, PDM/FM station
Canton Island	2	FM, PAM, PDM/FM record station
	1	FM, PAM, PDM/FM display station
Kwajalein	1	FM/FM, PAM/FM, PDM/FM system
USNS Range Tracker	12	FM/FM receivers
Pt. Pillar	1	PCM/FM
USNS Longview	8	FM/FM receiver
USNS Richfield	4	FM receiver
	5	AM/FM receiver
USNS Sunnyvale	8	FM/FM receiver
USNS Watertown	6	FM receiver
USNS Huntsville	6	FM receiver

* PCM also available.

While not listed in the table, there are several instrumented aircraft that carry a number of FM/FM, PDM/FM, and PAM/FM receivers in addition to recording equipment and spectrum displays.

b. Optical Tracking

Optical tracking equipment at PMR includes fixed cinetheodolites, transportable cinetheodolites, mobile optical tracking units (MOTU), tracking telescopes and various movie and still cameras. All cinetheodolites are KTH-53 trackers, which can simultaneously photograph a missile in flight and the azimuth and elevation angular scales of the theodolite at a maximum rate of four times per second. The MOTU is easily moved from place to place. Each unit is presently equipped with one 35-mm and one 70-mm camera, with 10-ft and 3-ft focal length lenses, respectively. The tracking telescope such as LA-24 and ME-16 provide time-correlated black and white or color pictures taken at high frame rates. Ribbon-frame acceleration cameras include the RC-2 and CZR, on mobile, three-axis mounts. Other cameras, such as the Photosonic fixed camera mounted coaxially with the LA-24 tracking telescope, are used in conjunction with other optical equipment and for the correlation of data from other means of tracking [20]. The BC-4 is also used.

Latest addition to the optical tracking means available at PMR is the MOPTS or mobile photographic tracking station. It is a high-speed, multiple-camera, tracking mount for 16-mm, 35-mm, and 70-mm cameras; but it can accommodate larger ones. The MOPTS is manually moved through azimuth and elevation by a "stiff stick" control that permits the operator to maintain 8 sec of orthogonality while tracking. The mount can also be remotely slaved to tracking radars for control, and it can be adapted for use as a cinetheodolite. MOPTS is installed on a flat-bed trailer equipped with screw jacks for leveling [23].

Optical tracking stations at PMR are summarized in Table VI.

TABLE VI. PMR OPTICAL TRACKING LOCATIONS

Location	Quantity	Type
Pt. Mugu	7	Cinetheodolite
	9	Acceleration camera
	1	MOTU (Mobile Optical Tracking Unit)
	1	MOPTS (Mobile Photographic Tracking Station)
Pt. Arguello	3	Cinetheodolite
	1	LA-24 telescope
	2	Acceleration camera
	2	MOTU
Vandenberg AFB	4	Cinetheodolite
	2	MOTU
	3	Acceleration camera
San Nicolas Island	6	Cinetheodolite
	2	MOTU
Kwajalein	2	LA-24 telescope
	3	Cinetheodolite
	3	Tracking camera
	18	Photosonic fixed camera
Eniwetok	18	Eyemos camera
	10	CA3-2B aerial camera

4. SUPPORT ACTIVITIES

Support activities for the PMR are similar to those at AMR, except in the areas where topographical, climatic, or operational differences dictate a divergent or unique approach.

Technical support activities include miss-distance indication, impact prediction, vectoring and drone control, data processing and reduction, range timing, and meteorology, and range safety.

Indications of missed distances are provided by an optical system that uses eight-camera, missed-distance, indication pods mounted on the wing tip of drone aircraft. PMR timing signals are recorded on 16-mm film exposed for about 10 sec. Two electronic systems are being evaluated: the AN/USQ-6(XN-2) scalar and the AN/USQ7(XN-2) vector system.

The IRBM/ICBM Missile Location System (MILS) uses SOFAR and a splash detection system. The two SOFAR are the Broad Ocean Area (BOA) SOFAR and the Miniature SOFAR. The BOA is used to determine IRBM impact positions; it employs widely-separated, shore-based listening stations that time and record arriving signals. Each station is connected to underwater hydrophones, which can receive reliable signals from over 2000 nautical mi from a signal bomb with 4 lb of explosives. The Miniature SOFAR system uses four pairs of hydrophones, arranged with baseline dimensions from 50 to 100 nautical mi. They are connected by cable to a single shore station that receives, records, times, and correlates the signals to determine impact locations.

The splash detection system permits a high degree of accuracy in locating impact pointers. No bomb is required and the area covered is relatively small. A group of hydrophones is arranged on the perimeter of a given target area, and a single hydrophone is located at its center.

Vectoring and drone control at PMR is performed both by a voice control system and an automatic vectoring system. Beacons, either S-band or C-band (depending on the type of radar used) UHF receivers, decoder systems, and flight control stations for out-of-sight control are also employed.

Computers include: IBM 7090 and 7094 for general purpose data processing; IBM 1401; Bendix G15; Remington Rand 1206. In addition the AN/USQ-20 and AN/UYK-1 computers are also used.

Range timing is provided by timing centers located at Pt. Arguello, Pt. Mugu, and on San Nicolas Island, the latter center synchronized to the Pt. Mugu center by a radio link. The Pt. Mugu signal generator is synchronized to a cesium atomic frequency standard. Signals originate at the Pt. Arguello center, which provides 14-digit, binary-coded range timing; theodolite time-control signals; radar plotting-board timing; a 100-cps sine wave signal; and a 100-kc, secondary-frequency, sine-wave signal. An Instrumentation Data Transmission System (IDTS) synchronizes the signal generator at the three range timing centers. The timing signals, which can be correlated with WWV transmissions, are available to about 75 mi beyond San Nicolas Island over the extended Sea Test Range. A converted seaplane tender used for guided missile launching and tracking can be integrated into the timing system; and a AN/FRW-2 transmitter at Pt. Mugu can be used to send simultaneous command and timing signals to aircraft.

Meteorological facilities at the Pacific Missile Range include three fixed-location stations at Pt. Mugu, Pt. Arguello, and San Nicolas Island, plus two seagoing stations. These stations are designed to accurately forecast weather conditions that can affect scheduled test operations and to record weather data during test operations for later use in test evaluation. Meteorological equipment used at PMR includes the AN/GMQ-14A semiautomatic weather station, AN/UMQ-5 wind measuring set, AN/UMQ-3 handheld wind vane for remote sites, AN/GMQ-2 and AN/GMQ-13 cloud ceiling measuring sets, AN/GMQ-10 transmissionmeter for visibility measurements, AN/GMD-2 and AN/GMD-2 rawinsonde sets, AN/FMQ-2A and AN/FMQ-1 radiosonde sets, AN/AMQ-8 airborne aerograph set, and AN/ASH-14 airborne microwave refractometer set. In addition, synoptic weather data for altitudes up to 100,000 ft are obtained by meteorological sounding rockets such as Loki and Arcas.

The range safety system at PMR provides the missile safety officer with real-time missile trajectory and impact prediction information from a number of sensors, computers, and displays. Missile trajectory data are supplied mainly by AN/FPS-16, AN/MPS-19, and COTAR radars. During missile lift-off, position and attitude are monitored by means of AN/FPN-33 radars, Mk 51 optical sky screens, and vertical-wire sky screens located near the individual launching pads.

Data gathered by the various centers are transmitted by land line to the Range Operations Building at Pt. Arguello for computation and display in the Missile Flight Safety Center. The major displays in this center are 30-in. × 30-in. and 150-in. plotting boards that show missile present position and predicted impact location; strip chart recorders angle-vs-time information; and an elapsed-time and countdown indicator.

Protection for mainland and offshore island installations is afforded by AN/FRW-2 command destruct transmitters with 10-kw amplifiers.

Nontechnical support activities throughout the range are performed by Naval personnel, including members of the Marine Corps, contractor personnel, and, in some instances, by native personnel at such down-range stations as Kwajalein and Eniwetok atolls. These activities and corresponding facilities are under the jurisdiction of the Commander, Pacific Missile Range, and include air-sea rescue units, transportation of supplies and personnel, cafeterias and messhalls, fire protection, security and community (where applicable) police, facilities maintenance, and organized recreational activities.

Additional support facilities available at Vandenberg AFB include a liquid oxygen plant, missile assembly buildings, a water treatment plant, technical support buildings, and a photographic processing laboratory.

C. White Sands Missile Range

White Sands Missile Range (WSMR), New Mexico (Fig. 32), is the largest inland missile and rocket test center in the United States and the only overland test center of the three national ranges. Originally established by the U.S. Army Ordnance Corps as the White Sands Proving Ground, on 9 July 1945, this Army-operated facility became a National Range in 1952 and received its present designation in 1958. On 26 September 1945, the first firing from the range occurred with the launch of a modified Tiny Tim rocket adapted to simulate the Wac Corporal, a larger, more powerful rocket configuration then being developed. Later the same day two more missiles were fired. (And seven days following these flights the first atomic explosion took place at Trinity Site, now in the heart of the upland missile impact area.) Since its establishment, the operational capability of White Sands Missile Range has grown to more than 2000 missile firings a year.

On 16 April 1946, the first captured German V-2 was flight tested, and 24 February 1949, the range achieved a milestone when a V-2 first stage with a Wac Corporal second stage reached an altitude of nearly 250 miles during a Project Bumper firing.

WSMR is operated for the military services and other government agencies conducting missile tests and is headed by an Army officer with Navy and Air Force deputies. The varied missions typify the range's role as a national missile test center operated by the Department of Defense.

The Army program tests and evaluates various missile systems, including the Nike series of surface-to-air weapons and the Sergeant and Pershing surface-to-surface missiles. The Navy came to WSMR in 1945 as a co-worker with the Army on V-2 experiments; later, it initiated its own program with the Viking and Aerobee upper-atmosphere research rockets. Present Naval activity concerns the Talos and Typhon surface-to-air missile systems. The Navy also fires Aerobee missiles for Air Force and NASA projects.

The U.S. Air Force Development Center, located at Holloman Air Force Base halfway along the eastern boundary of WSMR and inte-

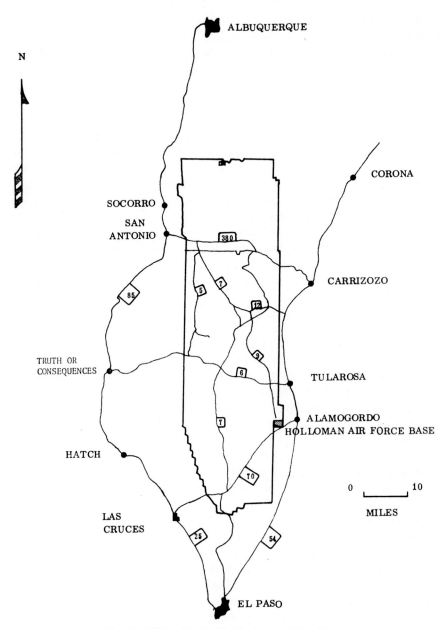

Fig. 32. White Sands Missile Range, New Mex.

grated with it for range purposes, is a user of the range and is charged with the testing program for many Air Force missile and research projects, including air-to-air and air-to-surface missile systems. The Marine Corps has been represented at the Range since 1945; and its mission is evaluation and testing, with the Army, of missiles used by the Corps, including Honest John, Little John, Hawk, and Lacrosse.

Test activities of NASA at White Sands are centered on development of the Apollo manned spacecraft. Initial tests used mockups of the spacecraft, which approximated it in weight and size, using Little Joe 2 research test vehicles. The NASA Propulsion System Development Facility also includes static test stands for the Lunar Excursion Module of the Apollo lunar spaceship.

1. Geographical Factors

White Sands Missile Range, 32° 23′ N 106° 20′ W, elevation 4,200 ft, is located in southern New Mexico. It is 100 mi long and 40 mi wide. The nearest towns are Alamogordo, to the east of the Headquarters area, and Las Cruces, west of the Range's southern tip. El Paso, Texas, and Albuquerque, New Mexico, are also nearby. The 4,000 mi² of range is in a nearly flat desert located in the Tularosa Basin. The terrain is rugged, with elevations over 9,000 ft in the St. Andres Mountains, which are within the western border of the range. The Sacramento Mountains form a natural boundary to the east. At the northern end of the range are the Oscura Mountains and Los Pinos Mountains. Contained within the range is the White Sands National Monument, just west of Holloman Air Force Base and off U.S. Highway 70, which traverses the southern portion of the range.

Like the missile ranges at Colomb Béchar, Algeria, and Woomera, Australia; WSMR is in an area where weather and geography permit the maximum use of optical and radar instrumentation.

The climate varies from very hot summers to mild winters; however, the temperatures in the summer are offset by low relative humidity. The mean temperature in January is 45°F, with minima of 4°F and maxima of 73°F. The average summer temperature is 85°F, with highs of 107°F in July and lows of 58°F in the same month. The mean rainfall is 8 in., and the annual snowfall averages 3.5 in. Thunderstorms occur most frequently over the range during July and August. Extremely low temperatures during the winter are rare.

The winds are primarily from the west, northwest, and southeast; but it is calm 31.9 per cent of the time. Wind velocities, regardless of direction, seldom exceed 17 mph, which is the mean for winds from the west. Unlimited ceilings or ceilings above 10,000 ft are available 90 per cent of the time, and visibility is 10 mi or better at least 96 per cent of the time.

2. LAUNCHING FACILITIES

A variety of launching facilities for liquid and solid propellant missiles is located at White Sands, tailored to the requirements and exigencies of a range that conducts its experiments entirely over land.

Flight test activities at WSMR are concentrated in six primary launching areas. With the exception of the Small Missile Range, these are located along a hard-surfaced road that forms the southern border of the range, which is contiguous with Ft. Bliss, Texas, and connects the main post area with Highway 54 to the west. These areas are:

(1) *Army Launch Area 1.* A complex of 13 launching sites, two of which are equipped with a special gantry service tower. One is a special, underground launcher. A central fire control building serves the area. ALA 1 is used primarily for prototype or research and development missiles, sounding rockets, and research test vehicles.

(2) *Army Launch Area 2.* A complex of four concrete launching pads with a centrally located blockhouse, assembly building, and power and communications building.

(3) *Army Launch Area 3.* A complex of nine special launching areas used primarily for engineer-user flights of missiles in early production. This area also has the NASA Little Joe 2 facility described below.

(4) *Army Launch Area 5.* A complex designed solely for launching the Nike Zeus antimissile missile.

(5) *Small Missile Range.* Located 6.5 mi to the northwest of the Army blockhouse at ALA 1 and near Highway 70, this area features three concrete pads, flight control building, and various instrumentation sites. It is used for the flight testing of surface-launched or air-launched missiles for which detailed data on launching or separation are desired. Sounding rockets such as Loki Dart are also launched from this area for gathering meteorological data.

(6) *Navy Launch Area.* A complex of seven launching sites and two fire control buildings (Navy blockhouse and LLS-1 *Desert Ship*). It is used for testing Navy missiles and has an Aerobee launching tower as well.

Target drones for surface-to-air missiles can be launched from any of these complexes. The Pogo and Pogo-Hi are standard range target missiles, but numerous others are available and have been tested. Sounding rockets of various types are also regularly launched for obtaining meteorological data and in support of scientific programs.

Impact zones for these launching complexes lie in four large areas downrange. They are the 30-mi, 50-mi, 70-mi, and 90-mi areas. In addition, a 40-mi extension on the northern boundary accommodates special firings. Within each are located precisely surveyed targets, survey control points, instrumentation sites, and airstrips. Hard-surfaced and unimproved roads throughout the impact areas permit access to range instrumentation sites and targets. These impact areas are also used for missiles and research test vehicles fired from the Hueco Range at Ft. Bliss, to the south of WSMR, and at the Green River, Utah, range of the U.S. Air Force, 470 mi to the northwest.

The Army blockhouse (Fig. 33), from which most tactical missile

Fig. 33. Army blockhouse at Army Launch Area 1, WSMR. U.S. Army photograph.

and rocket firings on the range are conducted, is an above ground concrete shelter, built to withstand a direct missile hit. It is located at Army Launch Area 1. A personnel ladder from a flat and low exterior level leads to the top of the pyramid-shaped high level and permits access to meteorological and prelaunch warning apparatus. One side of the lower level contains a view port with sliding metal doors. Naval launches are conducted from a separate blockhouse.

Most of the launchers at White Sands are self-propelled or transportable, being tactical devices developed for the armed services or simple launchers for prototype rockets and missiles. Typical are a 30-ft, truck-mounted launcher for the Honest John rocket; a truck-mounted

FIG. 34. Radio frequency hazard test on the Hawk missle prior to launching at WSMR. Operator and monitoring equipment on vehicle circle the trailer-mounted launcher to detect stray rf energy that might prematurely detonate the squibs in the missile motor. U.S. Army photograph.

launcher for the Lacrosse missile; a wheeled erector-launcher for the Sergeant missile; a self-propelled, full-tracked, vehicle for the Mauler missile; a trailer-mounted Hawk launcher, seen in Fig. 34; and a tube mount on a tank for the Shillelagh missile. These launchers are usually emplaced on concrete launch pads. An exception is the Army's most transportable missile launcher, a Bazooka-like tube that utilizes the human shoulder as a launch pad and fires the 22-lb, infrared-seeking, Redeye missile [24].

Most elaborate of the missile launching facilities at White Sands is located at Army Launch Area 5, where the Nike-Zeus surface-to-air antimissile missile is fired. While an early prototype launcher (illustrated in Fig. 35) was first used, a tactical launcher is now available. This vertical, straight-rail launcher with missile *in situ* is stored in a steel and concrete cell 16 ft × 20 ft and 60 ft deep. Launchings are conducted from an underground launch control building.

A unique launch facility is the U.S. Navy LLS-1 *Desert Ship* (Land Locked Ship). The interior of the LLS-1 has a shipboard appearance, and special missile launchers controlled from the LLS-1 simulate the movements of the ocean. It is equipped to launch the Talos guided missile (Fig. 36), which is carried by the Navy's guided missile cruisers. The Naval complex also contains a standard Aerobee launching tower [24].

The Air Force Missile Development Center, at Holloman Air Force Base on the east side of the range, also has an Aerobee launching tower and launch and test facilities for such air-to-air and air-to-surface systems as Sidewinder, Falcon, Genie, Mace, and Matador. Launchers for meteorological sounding rockets are also available.

Other launch facilities at Holloman include balloon launchers and a sled track nearly 7 mi long. The former have launched special research balloons to conduct upper atmosphere and cosmic ray studies, and balloons with manned gondolas that have reached altitudes of 20 mi. Balloons carrying 500-lb charges were used in Project Banshee, a series of high-atmosphere blast behavior tests conducted by the Department of Defense. The rocket sled track is used to investigate the hazards of space exploration, particularly the effects of acceleration and deceleration on both animals and missile components, and has been used to propel man up to 632 mph. Unlike other rocket launches, a test is successful without the vehicle, or sled, leaving the ground [25–27].

NASA's launching facility at WSMR is located at Army Launch Area 3, which used to be the pad from which Redstone missiles were fired. In 1962, NASA, after a survey of other national ranges, decided that the launching and tracking facilities at WSMR were optimum for its Little

Fig. 35. Nike-Zeus antimissile missile launcher at Army Launch Area 5, WSMR. U.S. Army photograph.

Fig. 36. Talos missile on launching pad at Navy launching facility, WSMR. Shown to rear is the concrete blockhouse *LLS-1* Desert Ship. U.S. Navy photograph.

FIG. 37. Little Joe 2 and launcher at Army Launch Area 3, WSMR. The converted Redstone missile service gantry is behind the Little Joe 2. Sandbag revetments protect cameras and other instrumentation. Courtesy NASA.

Joe 2 research test vehicle launchings in support of the Apollo lunar spaceship program.

The former Redstone launching pad is too small to accommodate both the Little Joe 2 launcher and a special launcher from which Apollo command module abort tests are made. And, since the Little Joe 2 vehicle represents more of a potential hazard than did Redstone, a new pad was constructed 1100 ft from the blockhouse—almost twice as far as the Redstone. Other changes in the area included extending the Redstone service gantry tracks out to the new pad and the provision of a utilities tunnel to it [28].

The Little Joe 2 launcher, shown in Fig. 37, is a steel structure weighing 100,000 lb and trainable through 140 deg of azimuth and from 75 deg to 90 deg in elevation. It is basically a pivot frame mounted on double-flange, crane-type trucks. A support platform provides fittings for securing the vehicle to the launcher and screwjacks for tilting it to the required elevation angle. Also attached to the support platform is a cable mast that helps stabilize the upper portion of the vehicle with a support arm for the payload umbilical cable. The launcher is remotely adjusted in azimuth and elevation [28, 29].

The existing Redstone service gantry was modified to meet the Little Joe's needs by adding 33 ft to its height and extending the working platforms 10 ft out from the structure (Fig. 37). A white room was also incorporated into the gantry. Other features of the launching complex include a vertical assembly and checkout building near the pad and a quickly assembled and disassembled, "Teki" hut used on the pad as a checkout building for the Apollo command module. This hut is removed prior to launching. Interior arrangements in the existing blockhouse were altered to conform to Little Joe requirements.

By 1963, the concern for extremely strict range safety regulations so far as missile overflights of populated areas had abated. Coupled with the fact that missile reliability had increased to a very high degree from the early days of WSMR when a German V-2 impacted in a cemetery in Juarez, Mexico, across the Rio Grande River from El Paso, Texas, launchings of fairly long-range missiles were begun from areas outside WSMR into its impact area. Thus, research test vehicles for re-entry nosecone studies, such as the Athena rocket, can be launched from the range at Green River, Utah, for re-entry and impact into WSMR, making use of its extensive instrumentation.[5]

3. COMMUNICATIONS AND TRACKING [30]

The communications system at WSMR reflects its unique position as the only national range entirely located inland. It is not required, like

[5] Long-range tactical missiles such as the Pershing are also fired from outside WSMR into its impact area. A firing site on Black Mesa, near Blanding, Utah, is used as well as a launching area at Ft. Wingate, New Mexico. The Pershing has also been fired from Ft. Bliss, Texas, into the impact area. The Sergeant missile has been fired from Datil, New Mexico, into the same area. Athena re-entry vehicles launched from Green River, Utah, also impact at WSMR.

other ranges, to maintain communications links over extensive areas of open water. Thus, the geography of WSMR eliminates the need for submarine cables and troposcatter systems and reinforces the reliance placed on microwave, UHF and VHF radio, telephone, and teletype systems. This system includes over 35,000 mi of wire and cable, 50 microwave channels, and 200 rf channels.

Two microwave systems are presently in use at WSMR: one, a special purpose, frequency-division, multiplex system, connects telemetry ground stations and relays FM/FM telemetry signals received from missiles by uprange ground stations to control and data centers for recording and real-time display; the other, a general-purpose, frequency-division, multiplex system, connects various range centers with facilities in remote areas such as Stallion Site and North Oscura Peak. The microwave relay station at Alamo Lookout is a center for cross connecting individual channels between the terminal stations.

Multichannel, UHF transmissions from six base locations throughout the range permit ground-to-air communications. Both UHF and VHF facilities are provided for drone control and to support surveillance of missile flights. Mobile radio is also used at WSMR. Military police and security vehicles, maintenance trucks, fire fighting units, ambulances, and other vehicles are radio-equipped.

The range telephone system has dial control offices at the Lower Range Instrumentation Center, King 1, Stallion Site, and Rhodes Canyon, with a total of 1600 lines. A universal numbering system is used and interexchange routing is automatic. This automatic system is augmented by manual exchanges at North Oscura Peak and Oscura Range Camp, and PBX facilities are provided at some of the launching and instrumentation areas.

Teletype stations at WSMR are provided for operational support and are arranged into networks for range scheduling, air weather dissemination, data transfer, and support operations.

WSMR is one of the world's most extensively instrumented ranges, having over 650 sensors, optical, radar, and telemetry, within its boundaries.

a. Radar and Telemetry

The chain radar system employed at WSMR utilizes AN/MPQ-12, AN/MPQ-18, and AN/FPS-16 radars to provide in-flight missile performance data and position data for target acquisition, missile flight safety, and post-flight analysis. Data from an AN/FPS-16 can be used to position the MPQ radars, and MPQ radar data can be used to place an AN/FPS-16 on target. The AN/FPS-16 radars in the chain system function with an accuracy of 0.1 mil in both azimuth and elevation, and 5 yd in range on all targets having an S/N ratio of 20 db, in either the skin- or beacon-track mode. A mobile version of the AN/FPS-16, designated the AN/MPS-25, is also available at WSMR; and it is used for special projects in locations not having fixed installations. It can also be used to replace fixed units being serviced or modified.

Aerial surveillance of the range is maintained by an AN/FPS-8 surveillance radar and an AN/TPS-10D height finder radar at Elephant Mountain and an AN/FPS-33 surveillance radar and an AN/FPS-6 height finder radar at the north end of the range. This system provides complete coverage of the range and also permits surveillance of adjacent areas. Data from the surveillance radar can be used by associated, 12-channel, track-while-scan equipment at Elephant Mountain, which can simultaneously track 12 targets; these data can also be used by the chain radar system and can be plotted in real time or used to position individual tracking radars in the chain on targets being tracked by the surveillance radars.

Radar tracking sites are shown in Fig. 38.

The nerve center for radar tracking, as well as other types of tracking, communications, and telemetry, is the range control station, more familiarly known as C-station at WSMR. This building viewed in Fig. 39, is located about 4 mi south of the Army blockhouse at ALA-1.

Position, velocity, and time data on missiles in flight are gathered from an active Dovap system and a Miran system. This later electronic trajectory system is a pulse technique, while Dovap relies on the Doppler shift principle. Velocimeters in the passive mode are also used to obtain radial velocities during the boost phase of small missiles. It operates on the principle of comparing the reflected Doppler shift of a CW signal. The resulting Doppler beat is directly proportional to the radial velocity of the missile.

Also a part of the range's tracking system is the 60-ft diameter Amrad radar with a peak power output of 10 Mw. A part of Project Defender, this powerful tracking instrument is used to obtain precision radar cross sections of objects and to make Doppler shift measurements of Athena re-entry bodies launched from Green River, Utah.

Two principal telemetry systems are used at WSMR. These are AM/FM, using 1 to 18 channels, with up to 18 subcarriers on one channel, and the PDM/FM system with 30 channels sampled 30 times a second or 45 channels sampled 20 times a second. Standard FM/FM systems are also available. All telemetry systems conform to IRIG standards.

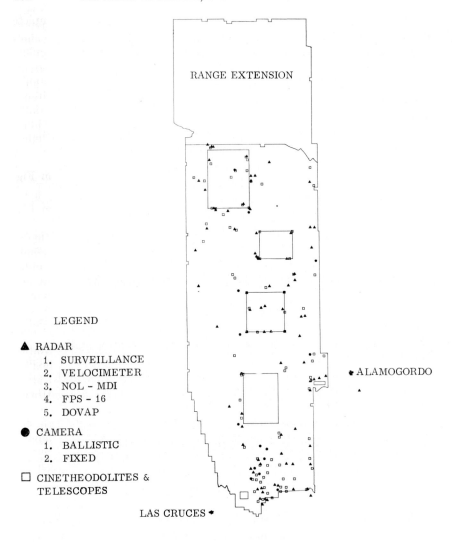

RANGE EXTENSION

LEGEND

▲ RADAR
    1. SURVEILLANCE
    2. VELOCIMETER
    3. NOL - MDI
    4. FPS - 16
    5. DOVAP

● CAMERA
    1. BALLISTIC
    2. FIXED

□ CINETHEODOLITES &
   TELESCOPES

ALAMOGORDO

LAS CRUCES ✦

EL PASO

FIG. 38. Radar and optical tracking sites at WSMR.

b. Optical Tracking [31]

WSMR is one of the best equipped ranges in the world for optical instrumentation. Its sites are precisely located and contain a variety of cameras. In addition, the prevailing favorable weather conditions at the range also make optimum use of optical tracking possible. A unique

FIG. 39. Range Control Station, C-station, a5 WSMR. U.S. Army photograph.

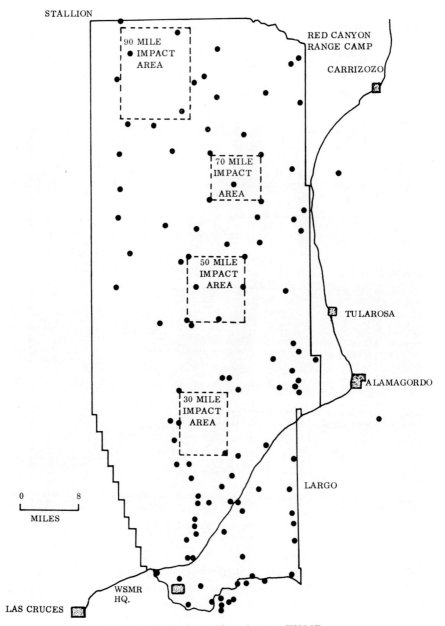

Fig. 40. Optical tracking sites at WSMR.

feature of the instrumentation system is that certain sites are remotely controlled because their locations are considered hazardous. The remote control is accomplished by television cameras, monitored at a safe distance by the site operators. Optical sites throughout the range are located in Fig. 40.

Cinetheodolites at the range consist of both the Askania and the

Fig. 41. Cinetheodolite tracking site G-80 at WSMR showing Contraves instrument in astrodome structure. U.S. Army photograph.

Contraves. They are housed in three types of structures to protect them from the weather when not in use: astrodome shelters, cinderblock buildings, and concrete silos. The astrodomes are mounted on concrete bases or elevated, environment-controlled pedestals (Fig. 41) and are slaved to the cinetheodolites. In use, a section of the astrodome rolls back to expose the instrument; it is similar in appearance and function to the familiar dome covering astronomical telescopes at observatories.

Within the cinderblock buildings, instruments are mounted on hydraulic lifts, which rise through openings in the roof to position the cinetheodolites for use. The concrete silos are air conditioned and are specially designed to minimize optical ground clutter. The instrument is mounted on the top floor and covered by a roof in four sections that slide back on rails.

Mobile cinetheodolite units are also available for extending and assuring coverage. These are installed on concrete pads throughout the range area as required.

A variety of tracking telescopes is also available at WSMR. The

Fig. 42. Small Missile Tracking Telescope at Largo site, WSMR. U.S. Army photograph.

range's first tracking telescope was a primitive instrument consisting of a pair of 20-× Japanese binoculars; two Eyemo, 35-mm movie cameras; and a pair of refracting telescopes mounted on a modified M-45 antiaircraft machine gun mount. Even this instrument proved itself when it revealed an unsuspected tumbling of the V-2 missile at peak altitudes during a firing on 5 December, 1946.

The tracking telescope network is designed around a few, permanently installed instruments supplemented by mobile models. Standard tracking devices include the Igor and Roti used at most large missile ranges, installed in astrodome structures of the type used to protect the cinetheodolites. One of the permanently installed instruments is the Telescope 4, a Newtonian optical system with a 16-in. primary mirror and variable focal length of 125 in. to 400 in. It is mounted on a 90-mm gun mount and uses a Mitchell 35-mm camera for photographic recording.

Other instruments include the specially designed Small Missile Tele-

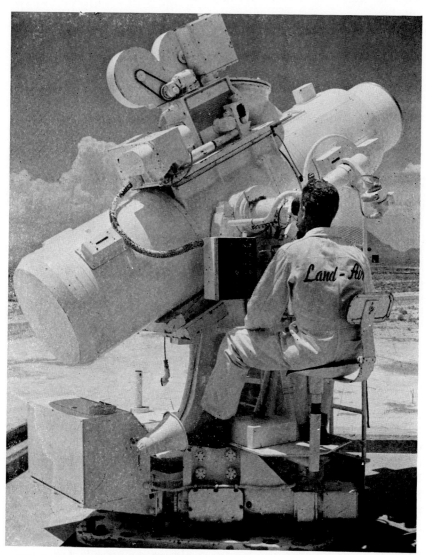

Fig. 43. Terminal Trajectory Telescope (Tetra) used at WSMR. Courtesy Land-Air, Inc.

scope (Fig. 42). It has a corrected Cassegrainian optical system with a focal length of 100 in. and a 30-in. aperture. A 70-mm camera records the image, while a 35-mm camera records the azimuth and elevation dials to give an accuracy equivalent to that of a cinetheodolite. The Tetra (Terminal Tracking Telescope) illustrated in Fig. 43, is made of Igor components but uses an 18-in. primary mirror and a 70-mm image recording camera. The Medium Focal Length Tracking Telescope has a modified Baker reflector corrector and focal lengths of 40 in., 60 in., and

FIG. 44. Wild BC-4 ballistic camera at WSMR tracking site Nan. U.S. Army photograph.

80 in. Images are recorded by a 70-mm camera, and the high-speed tracking mount is installed on a trailer that permits it to be transported quickly to previously prepared sites. The Gorid (Ground Optical Recorder for Intercept Determination) and Midor (Miss Distance Optical Recorder) are also used at WSMR. Gorid is a modified version of the Small Missile Telescope, with improved aided-tracking mechanism. Midor has a 5-in. by 5-in. format rather than the usual 70-mm. Thus, it can record greater miss distances.

A number of smaller, mobile tracking telescopes are also available. These instruments use a variety of lens, camera components, and mounts;

generally, they are used only when specific requirements call for additional coverage beyond the primary capability.

Ballistic cameras or phototheodolites at WSMR are highly accurate instruments located with great precision. Practically all stations are now furnished with astrodomes to protect instruments when they are not in use. The primary camera used is the Wild BC-4 (Fig. 44), but the Askania PI, Princeton ballistic camera, and a special astroballistic camera developed by technicians at WSMR are also used. These cameras furnish position versus time data but are occasionally used to record intercept events and acceleration data.

Fixed cameras at the range consist of ribbon-frame, high-speed, and very high-speed motion picture instruments.

The ribbon-frame cameras are mounted on three-axis mounts and are not able to track missiles. They are synchronized by a coded signal from the central timing generator. Ribbon-frame cameras at WSMR include: Bowen-Knapp RC-1, RC-4, RC-5, and CZR-1; Clark, and 5M. These instruments are used extensively on either side of the line of fire at the Small Missile Range and at the rocket sled track at Holloman AFB.

High-speed cameras include the Mitchell 35 mm and 70 mm; Bell and Howell 16 mm; Photosonic 4B, 5B, 10A. The very high-speed cameras are the Photosonic 1A, 1B; Fastax WF-3, WF-5, WF-8; Fastair 16 mm; Fairchild HS-101, HS-70A; and Warrick G. Cameras of these types are also mounted parallel to the center of the beam on some tracking radars.

4. Support Activities

Support activities at WSMR are not unlike those of AMR and PMR, except for activities related to their geographical differences. White Sands is not required to operate and maintain a fleet of supply or tracking ships, nor must it be concerned with the political aspects its activities may have on foreign governments (other than nearby Mexico, in which the rogue V-2 impacted). Conversely, it must consider factors peculiar to an inland range; for example, the precautionary measures taken for the safety of mainland populations, particularly those in the extended impact area at the extreme northern end of the range and those in relatively nearby Las Cruces.

The range user has scientific, technical, logistics, and services backup in a variety of categories. Scientific and technical support activities include flight simulation, environmental and climatic testing, computer services, frequency coordination, static propulsion testing, meteorological

FIG. 45. Guided missile flights are simulated on analog computers at WSMR. U.S. Army photograph.

support, drone control, timing generation and distribution, television coverage, and pictorial services.

Typical of the technical support available to range users is a modern flight simulation laboratory, viewed in Fig. 45, with its digital and analog computers and associated equipment. This facility offers the means for simulating a wide variety of missile flights. Problems in trajectories, for example, are handled by the IBM 704 digital computer and the CRG 103 digital differential analyzer, while problems in flight dynamics and guidance are handled by the Electronics Associates 16-231-R, 16-131-R, and 16-24-D analog consoles. The laboratory also has a random noise generator that utilizes a radioactive source to produce true white noise signals, which, in turn, are used to simulate the random errors of radar tracking [32].

Environmental and climatic tests in the field and laboratory are conducted and provide information on a missile's ability to withstand heat, cold, sun, rain, dust, sea air, fungus, and altitude pressure changes. Shock and vibration facilities at WSMR duplicate the effects of air turbulence, overland transportation, and rough handling; in essence, the effects of all hazards can be produced that may be experienced by a missile or allied components in transit, in storage, or in flight. Environmental and weather factors, such as salt spray, dust, or high humidity can be simulated in special indoor chambers.

Computer services at White Sands provide mathematical analysis and technical consultation on matters relating to computer applications, and techniques for transmitting, handling, and processing data. Present computers include a Philco 2000, Univac 1130A, Burroughs E101, LPQ-20, IBM 704, IBM 610, IBM 650, and a Remington Solid State 80. Other equipment includes tape-to-tape format converters; spectrum analyzers; film readers for Dovap; other readers and telereaders for films, dials, and paper oscillograph records; and a Zeiss stereocomparator. A special analog telemetry data reduction system with readout on x-y plotters, pen recorders, or oscillographs. A similar system is available for digital telemetry [33].

Frequency coordination utilizes fixed, semifixed, mobile, and airborne monitoring stations. The fixed stations have high-gain, directional antennas for overall surveillance signal detection and are located in areas where a concentration of electromagnetic radiation is indicated. The semifixed stations have similar capability and are located in semitrailers. The mobile stations have less equipment but are more versatile; they make field intensity measurements to locate interference sources and can act as fixed stations when more extensive equipment is not required. The

Area Frequency Coordinator is responsible for coordinating military use of the frequency spectrum in an area with a radius of 150 mi.

Static firing facilities exist for both liquid and solid propellant motors. Specifically these include a 500,000-lb thrust stand for liquid propellant engines up to 8 ft in diameter. It is instrumented to provide up to 250 channels of data. Tests are controlled from an integral blockhouse. Large storage tanks for various types of propellants are also integral to the stand. However, the stand can also accept engines with vertical propellant tanks in place.

A 100,000-lb thrust stand for liquid propellant engines up to 8 ft in diameter is also available. Engines may be mounted in either the vertical or horizontal position.

The solid propellant test facility consists of five concrete bays. Motors with thrusts of 5000 lb, 250,000 lb, and 300,000 lb can be fired. The motor mount in the 250,000-lb bay can measure thrust components in six different directions. Firings are controlled from a blockhouse sited 150 ft from the bays. The means for approximately 75 channels of information are available, with provisions for adding supplementary ones.

NASA also has an extensive testing area at WSMR. In an 87-mi² area some 20 mi to the west of the Little Joe 2 pad at ALA-3, on the western slopes of the Organ Mountains, is the Propulsion System Development Facility (PSDF), shown in Fig. 46. The area is designed for static testing the propulsion systems for the service module and the lunar excursion module of the Apollo manned spacecraft [28].

As shown in Fig. 46, PSDF consists of three major areas: an administrative area, a service module test area, and a lunar excursion module test area. The service module area has two stands. Test Stand 1 will accept only the service module while Test Stand 2 is designed to accept the service module with the command module atop it. Tests are conducted from a control center 1000 ft from the two stands. Each stand has a reinforced concrete building adjacent to it for housing required ground support equipment and propellant transfer equipment. Instrumentation for the stands is contained in an underground bunker near Test Stand 2. It is connected to the control center by a concrete tunnel.

Other facilities in the area include a preparation building; propellant storage areas; explosive storage areas; static water supply, treatment pond and trenches; and a fire station.

The lunar excursion module area has three test stands, two of which have an altitude simulation capability. Also located in this area is a storage area for hypergolic propellants that are used in this test area as well as the other.

The administrative area, some 4 mi to the west, has an administration building, storage facilities, emergency medical center, garage and main-

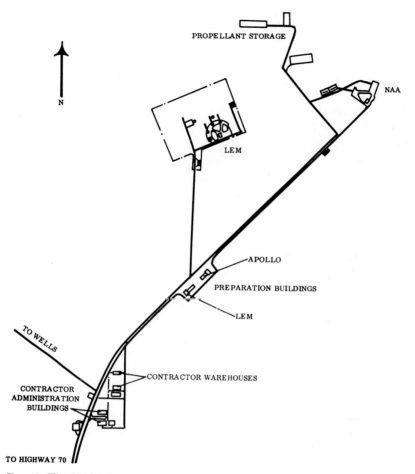

FIG. 46. The NASA Propulsion Systems Development Facility at WSMR, showing LEM (Lunar Excursion Module) test site and NAA (North American Aviation) facility.

tenance shops, an electrical power substation, static water supply, and a helicopter pad.

Meteorological facilities at WSMR are used to collect data on atmospheric parameters before, during, and after a missile launch. Included in these facilities are mobile aerovane anemometers, which can be quickly set up, with masts extending up to 50 ft above the surface; and a 500-ft tower with 10 levels where temperature, relative humidity, and wind speed and direction are measured. The U.S. Air Force furnishes meteorological service for WSMR.

Other facilities and techniques are employed to gather wind measurements above the levels of surface instrumentation, including the double theodolite technique, which utilizes a balloon, a recorder, and two theodolites at the ends of a surveyed baseline. A double-theodolite computer reads and computes elevations and azimuth information from the theodolites and produces a graph profile of the wind observations. This technique permits measurement of data from the surface to 40,000 ft. Measurements to 90,000 ft are provided by Rawin Sets AN/GMD-1 and 2 and by radiosonde transmitters attached to balloons. A wiresonde unit, with helium-filled balloon and ground recorder, is used at the point and time of interest to provide accurate temperature and humidity data up to 1500 ft above the surface. Sounding rockets, such as the Loki, Arcas, and Nike Cajun, are used to collect data at altitudes beyond the reach of balloons.

Range recovery services at WSMR are similar to those employed at Woomera, Australia; and Colomb-Béchar, Algeria. Indeed, the methods at these three ranges are dictated by the terrain, which is amazingly similar. Aircraft, helicopters, and trucks are used to transport personnel and material into and out of the impact area. However, since WSMR is so extensively instrumented and since targets are precisely located, it is unnecessary to employ homing beacons in the impacted vehicle. Recovery is facilitated by the Sotim (Sonic Observation of the Trajectory and Impact of the Missiles) system and the Arip 2 (Automatic Rocket Impact Predictor 2) [34]. The Sotim is, as the name implies, a sonic measuring system; while the Arip 2 is a system used primarily to plot the impact of high-altitude, unguided, multiple-stage rockets. Basically it is a system that takes wind data from different sensors, including the 500-ft tower at WSMR, and continuously computes and predicts the point of impact.

Timing signals are generated and distributed by a synchronized time signal generator (Fig. 47) using the WWR primary frequency standard and provides simultaneous signals to WSMR, Holloman, and the Small Missile Range. Both WSMR and Holloman use microwave links to relay timing signals to range instruments, while the Small Missile Range signals are directly tied into a mobile radar tracking installation.

Air support at WSMR is provided jointly by the Army and the Air Force. The range has its own airfield, Condron Field, located 7 mi to the southeast of the headquarters area. This field is staffed by personnel from the Air Force Missile Test Center at Holloman AFB. Auxiliary dirt airstrips are also scattered throughout the impact area. The Air Force provides logistic maintenance and other support for the various Army and Navy aircraft stationed at Holloman and used for range purposes. In addition, the Air Force also provides high-performance aircraft

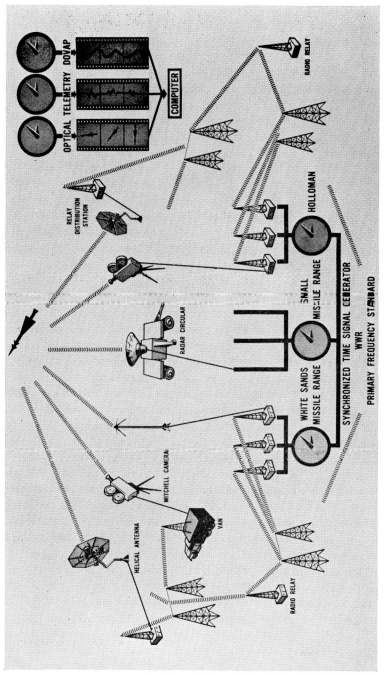

FIG. 47. Range timing system at WSMR. Courtesy U.S. Army.

for WSMR for "tracking," reconnaissance, and photographic missions as requested by the range. Contractor-operated private aircraft are permitted to land at both Condron Field and Holloman AFB, provided prior arrangements are made.

Nontechnical support services include Military Police and civilian security guards, who maintain order and provide the security of classified defense activities; logistic services such as supply, transportation, storage, and maintenance of range equipment; and service facilities which cater to the comfort, convenience, and well-being of personnel.

D. Wallops Station

Wallops Station is the only launch facility entirely owned and operated by the National Aeronautics and Space Administration. Established in 1945 as the Pilotless Aircraft Research Station of the National Advisory Committee for Aeronautics to gather aerodynamic data at transonic and low supersonic speeds, it is one of the oldest rocket launching facilities in the United States [35].

Usually thought of as a center for sounding rocket activity, Wallops Station has other missions as well. These include aeronautical research, primarily in the areas of flight characteristics of vehicles and spacecraft and aerodynamic data related to aerospace flight; components and systems development; small scientific satellite launchings; and assistance to other U.S. government agencies, as well as cooperation and training of scientists and aerospace technicians from other countries. In addition, Wallops Station also engages in tracking and data acquisition from scientific satellites launched elsewhere, e.g., Tiros and Echo.

Typical of these programs are the high-speed re-entry studies using the seven-stage Trailblazer test vehicle; sub-orbital flights of manned space capsules; Nike Cajun sounding rockets with scientific payloads; and the launching of Explorer 9 satellite by a Scout space carrier vehicle.

Wallops Station has a versatile and flexible capability, which can be tailored to meet the needs of individual tests. Using the Atlantic Ocean as the major part of its range, rockets launched from Wallops Island have reached apogees of 3500 mi and velocities faster than 20,000 mph. More than 5000 research vehicles had been launched from the range by 1964, with approximately 300 being launched each year, including experiments for subsequent launches from Cape Kennedy.

1. GEOGRAPHICAL FACTORS

Wallops Station (37° 50′ N 75° 29′ W) is located approximately 40 mi southeast of Salisbury, Maryland, on the Delmarva peninsula of the

Virginia coast. The entire installation is actually composed of three areas: (1) Wallops Island proper (5 mi long and $\frac{1}{2}$ mi wide), which is located south of Chincoteague, Virginia; (2) Wallops mainland, on the Virginia mainland, two miles west of Wallops Island; and (3) Main Base, which is located west of Chincoteague at the former Chincoteague Naval Air Station (Fig. 48a). Despite its proximity to such population centers as Washington, Norfolk, and Richmond, it is relatively isolated by the waters of the Chesapeake Bay, Delaware Bay, and the Atlantic Ocean [35].

The climate of Wallops Station is continental for approximately half the year and oceanic for the other half. From October through March the area is cooled by continental air from the west; from April through September the climate is more balmy, influenced as it is by the Bermuda High with air from the east. The average annual precipitation is 37.03 in., with a maximum monthly precipitation of 9.84, recorded in August 1958, and a maximal daily precipitation of 2.5 in., which occurred on 25 August 1958. The average mean temperature is 58°F, with a mean maximum of 63°F and a mean minimum of 53°F. The extreme maximum temperature recorded since 1946 was 99°F and the minimum was −6°F. The average wind velocity for Wallops Station is 12 mph, with the highest wind speed recorded at 110 mph during Hurricane Donna in 1960. Prevailing winds are southerly in the summer and northwesterly in the winter. Sea breezes occur most frequently during late spring and early summer.

2. LAUNCHING FACILITIES [36]

Wallops Island proper has five launching areas with various equipment available for different rocket configurations. Each area is tied into the Range Control Center by an integrated instrumentation system shown in Fig. 48b.

Launch Area No. 1 contains a blockhouse and a straight-rail launcher and a gimbal-mounted tower for the Aerobee 150A, 300A, and 350 sounding rockets, Fig. 49A.

Launch Area No. 2 is a large, concrete pad with several launch positions. Pad 2A has a tubular-boom launcher and a straight-rail launcher, while pad 2B has several small launchers. Pad 2C is equipped with an I-beam launcher. The tubular launcher is a 28-ft mast, supporting a boom with variable length up to 34 ft. The cable-supported boom is hinged to the mast, which is anchored to the pad and supported by three steel-pipe legs. Trainable in azimuth, it can be used to launch a variety of rocket configurations, including multistage vehicles weighing up to 14,000 lb (Fig. 49b). The I-beam launcher is similar in appearance

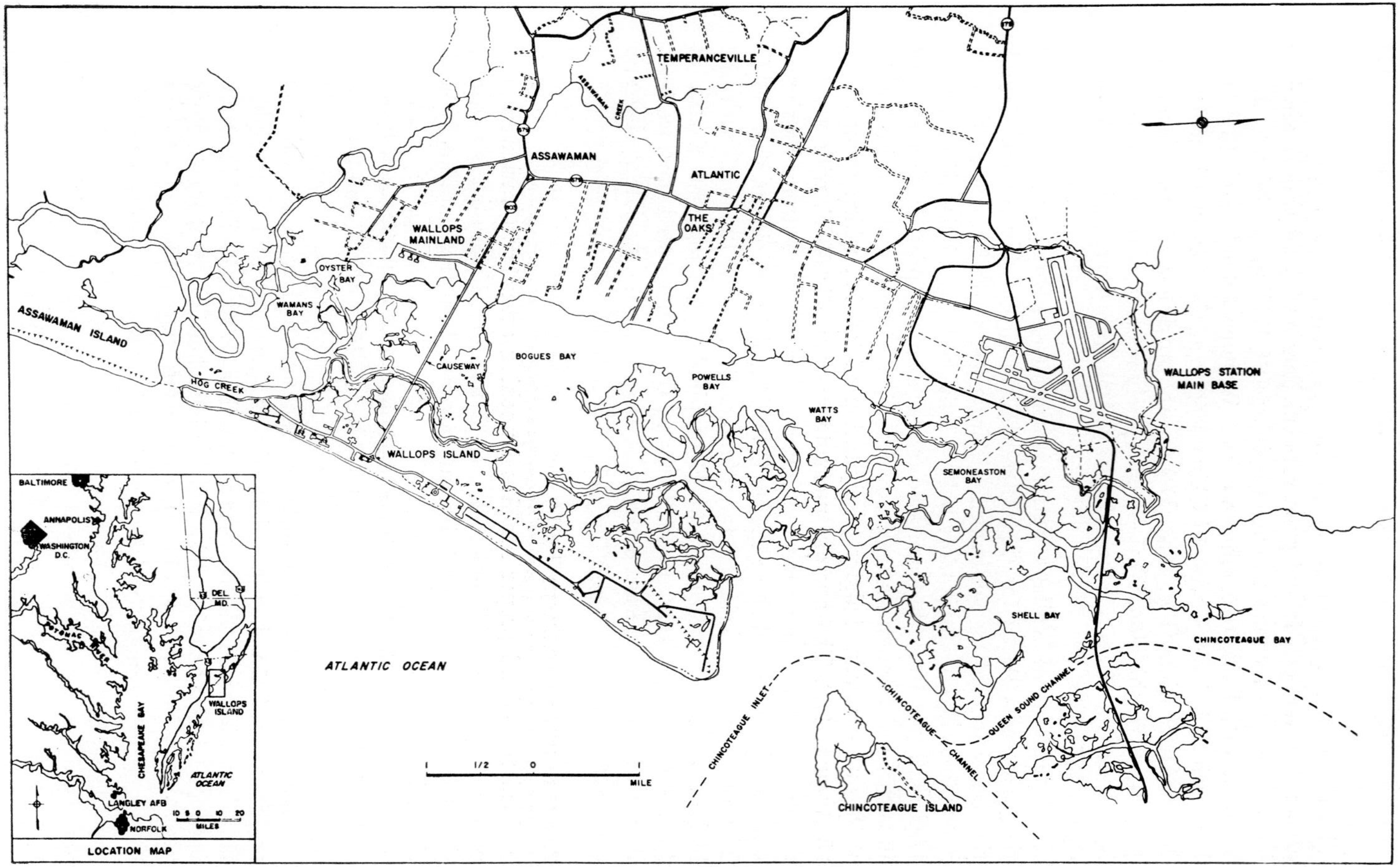

FIG. 48a. Wallops Station, Virginia. Courtesy NASA.

FIG. 48b. Wallops Island launch facilities looking north from Aerobee tower in the foreground. Courtesy NASA.

to the tubular launcher, except that the boom is a modified I-beam and that the mast is supported by three steel cables instead of pipes. The launcher also has a versatile capability and can be used to launch multistage rockets weighing as much as 7300 lb, including those boosted by the Honest John motor.

Portable launchers in the area include a 5-in. naval gun, seen to right in Fig. 49b, and launchers for the Arcas, Loki-Dart, and Viper-

Fig. 49a. Aerobee launch tower at Pad No. 1 on Wallops Island. Courtesy NASA.

Falcon rockets. The gun is used to launch the Hasp sounding rocket; the Arcas launcher, shown being loaded in Fig. 50, is a tube mounted on a free-volume cylinder and movable in both azimuth and elevation. It is designed specifically for the Arcas sounding rocket. An 11-ft long, rifled tube is used to launch the Loki-Dart and Judi-Dart rockets; and a 20-ft long, rail-type launcher is used with the Asp and Viper-Falcon rockets.

Launch Area No. 3 has a concrete pad, blockhouse, terminal building,

FIG. 49b. Adjustable, tubular boom launcher at Wallops Island Launch Area No. 2. A Nike Apache sounding rocket is shown on the launcher. To the right rear can be seen a modified U.S. Navy 5-in. gun used as a launcher for the Hasp sounding rocket. Courtesy NASA.

equipment carrier, and a wheeled Scout launcher service structure which moves on curved rails. The Scout launcher proper is a 40-ft, welded, steel beam pivoted to the structure base (Fig. 51). The 110-ft tall, open-framework, service structure has three winches, a 1000-lb capacity elevator, and other launching equipment.

FIG. 50. Arcas sounding rocket launcher at Wallops Island Launch Area No. 2. Courtesy NASA.

Also located in the same area is a newer type of Scout launcher. It is hinged at the base and is 85 ft tall in the vertical position. Associated with it on the pad are a 70-ft long rocket transporter and a 120-ft long, mobile shelter that can be heated or air conditioned. The upper structure of the launcher is positioned by two screwjacks powered by a 20-hp motor. The base of the launcher can be turned through 140 deg to adjust azimuth launching angles. Using this system, the preparation time is cut approximately in half, thus speeding up the rate at which Scouts can be launched (Fig. 52).

Launch Area No. 4 contains a concrete pad, tubular-boom launcher, Little Joe 1 launcher, straight-rail launcher, and a pad terminal building. The Little Joe 1 launcher was used to boost the Mercury capsule through ballistic trajectories during the development phase of Project Mercury. The Little Joe 2 launcher, for the Apollo capsule, is located

FIG. 51. Scout launcher and service structure at Wallops Island Launch Area No. 3. Only the first and second stages of the rocket are shown on the launcher. Courtesy NASA.

at White Sands Missile Range; see Section II, C. The tubular launcher is similar to the one in Launch Area No. 2.

Launch Area No. 5 also has a concrete pad, tubular launcher, and a pad terminal building. The launcher is similar to the one on the pad at Launch Area No. 2.

The immediate launch areas at Wallops Island, in addition to the blockhouses, terminal buildings, and launchers, also have fixed camera stations, meteorological towers, and television camera stations.

FIG. 52. New Scout launcher at Wallops Island Launch Area No. 2. This hinged structure permits horizontal checkout of the rocket and subsequent erection in much shorter periods of time. Courtesy Ling-Temco-Vought, Inc.

3. TRACKING AND COMMUNICATIONS [36, 37]

a. Communications

Wallops Station does not employ an extensive communications network, such as those used at AMR and PMR, to maintain contact with range activities several thousands of miles apart. Most downrange Wallops communications are handled through tie lines to a switchboard at Goddard Space Flight Center and through dial access to other telephone networks. Teletype and datafax circuits connect Wallops with the Goddard Center and other stations in the NASA administrative teletype network. The principal local facilities include a dial PBX at Wallops Station; an interconnected satellite PBX on Wallops Island; a 15-channel intercommunications system; a paging and countdown system; and two-way HF, VHF, and UHF radio. The radio communications system permits radio hookup to any telephone location through a central control operator.

b. Radar Tracking and Telemetry

Several radar systems are employed at Wallops Station for data acquisition. They include the Mod II; AN/FPS-16 (Fig. 53); Spandar, long-range S-band; AN/FPQ-6; and MSQ-1A. The Spandar is located on Wallops mainland and has a 60-ft parabolic reflector mounted on a 95-ft tall tower (Fig. 54). Maximum range is 5000 miles; with skin tracking on a 10 ft^2 target, the range is approximately 600 mi. The mobile Doppler radar, a Resdel 10A velocimeter, uses an audio beat frequency to determine target velocity. The average range is 40,000 ft to 50,000 ft. Two experimental 60-ft diameter dish radar sets are also available for tracking.

Wallops Station utilizes PDM/FM, PAM/FM/FM, FM/FM, and FM/AM telemetry systems, using AM, FM, and combination AM/FM receivers with high-gain, directional antennas. The main tracking antenna (Fig. 55) is a high-gain array with a gain equal to that of a 60-ft parabolic reflector. It has a phase-monopulse circular array, operating in the 225–260 Mc/sec range, with 33 elements and a diameter of 40 ft. Several mobile stations are available as well as a 136-Mc/sec system for satellite tracking. Ancillary telemetry equipment includes preamplifiers, tape recorders, and quick-look, real-time display at the Telemetry Receiving Building on the Main Base (Fig. 56).

A special, 60-ft parabolic reflector, covering the frequency band 220 Mc/sec to 10,000 Mc/sec, is scheduled for operation in 1965.

Continuous wave acquistion is provided at Wallops Station by a 73-Mc/sec Doppler velocity and position system (Dovap). Two stations are located on Wallops Island, including a single-station Dovap located at the south end of the island. Its use requires a target-borne transponder and, since the system is ambiguous, a precisely known point from which to integrate the Doppler cycles. The system is used for special projects only rather than regular operations. Also using the Doppler technique is the Single Station Doppler (SSD), an economical and portable means of supplying trajectory information. Designed for use in remote areas where range facilities are limited or nonexistent, the SSD is housed in one trailer. The system uses a Doppler effect to obtain radial velocity of the rocket with respect to the ground station and measures the radius

Fig. 53. Radar tracking and telemetry tracking antennas at Wallops Station. To the left is a quad-helix telemetry antenna, while an AN/FPS-16 tracking radar is seen on the center building. A Mod II and SCR 584 radars are shown on the building to the right. Courtesy NASA.

Fig. 54. A 60-ft diameter receiving antenna atop a 95-ft pedestal at Wallops Station. Courtesy NASA.

vector cosine with a two-axis radio interferometer. Integration of the radial velocity gives the radial distance. With the data acquired, it is possible to compute the rocket position with respect to the ground station [38].

Fig. 55. Thirty-three element, automatic tracking, telemetry receiving antenna at Wallops Station. Courtesy NASA.

c. *Optical Tracking*

Photographic and optical systems in use at Wallops Station employ a wide range of equipment, the principal components including an Igor tracking telescope, T-5 tracking telescope, RT-2 telespectrograph, Mitch-

FIG. 56. Aerial view of the Main Base at Wallops Station. Facilities include: 1, main gate; 2, communications and receiving building; 3, high-gain telemetry building; 4, special projects office; 5, procurement and supply; 6, fiscal office; 7, range center; 8, engineering building; 9, photographic laboratory; 10, airfield; 11, central heating plant; 12, control tower; 13, damage control station; 14, hangar; 15, dormitory; 16, Wallops Station headquarters; 17, technical services area; 18, explosives storage area C; 19, sewage disposal plant; 20, explosives storage area B; 21, explosives storage area A; 22, shipping and receiving buildings. Courtesy NASA.

ell high-speed motion picture cameras, Fastax cameras, 70-mm cameras, and Milliken cameras. The mobile, long-range T-5 with an Iflot camera provides sequential data on distant targets. The Igor, stationed 13 miles downrange, is equipped with 35-mm or 70-mm cameras and provides long-range performance and sequential data of vehicles in flight. The RT-2, equipped with a mosaic objective grating, obtains spectrographic data during re-entry body studies. Data are recorded on an 11-in. format. The Mitchell camera is used for medium- and high-speed photography; and the Fastax to record vehicle performance at lift-off. The programmable, 70-mm cameras, which permit larger image sizes and greater detail than the 16-mm and 35-mm cameras, are employed in the launching area to collect sequential data. Other equipment includes a variety of hand cameras and a small selection of aerial cameras. Camera stations are located on Wallops Island and Wallops mainland.

d. Range Timing

The range timing and programming system provides the following:

(1) Correlation of data from various fixed and mobile instrumentation sites by supplying precise time-of-day information, which is recorded along with the collected data on oscillographs, tape recorders, cameras, digital recorders, etc.

(2) Automatic control functions associated with the launching and in-flight monitoring of test vehicles.

The timing system is synchronized with Greenwich Mean Time to an accuracy of 1 msec. Two, precision, time-of-day codes, a fast code (time word every second) and a slow code (time word every minute) are generated. Maximum resolution is 1 msec. Program time, synchronized with time of day, is provided from 100 min before launch to 100 min after launch. Events can be programmed with a resolution of 0.1 sec at any point in the program count.

4. Support Activities [36]

Technical support activities at Wallops Station include range surveillance, meteorological support, range recovery, range safety, data reduction, and support service shops. These are located primarily at the Main Base, as shown in Fig. 56.

Airspace and range area limits are determined in advance from flight path and flight determination computations. Range surveillance is accomplished by aircraft and ground-based surveillance radar.

Meteorological support at Wallops Station is provided jointly by the Aeronomy and Application Office and contracted U.S. Weather Bureau personnel, who provide operational meteorological services, including atmospheric measurements, basic data reduction for ballistic applications and climatic analysis, weather forecasting, and briefing.

Measurements of wind, cloud-base heights, temperature, pressure, associated hydrometers, and density include the use of ground-based and upper-air flight instrumentation.

Surface data are acquired by utilizing standard indicating or recording instruments such as anemometers, ceilometers and ceiling lights, rain gages, barometers, thermometers, and psychrometers at fixed and mobile sites. Special, instrumented, 250-ft and 300-ft meteorological towers, primarily used for ballistic applications, are located at the north and south launching areas respectively. Marine instrumentation is housed aboard a downrange ship for similar atmospheric measurement needs.

Upper atmospheric data is gathered by using various techniques and AN/GMD-1, AN/GMD-2, optical theodolites, tracking radars and associated balloon- or rocket-borne flight equipment. The use of small meteorological rockets and radiosondes aboard ship provide similar data downrange.

Range safety at Wallops Station is provided with numerous data systems for evaluation and control of errant vehicles.

During the early trajectory skyscreens assure that the vehicle remains within predescribed limits. Two types of skyscreens are available: vertical-wire and closed circuit television each positioned so that they define azimuth and vertical limits when aligned with the launch site. The vertical wire skyscreens are portable and can be located so that they provide the proper limits for any planned launch site. The closed circuit television system uses fixed camera sites and tripod-mounted cameras to provide prelaunch surveillance and display vehicle pitch attitude during the initial launch phase. Video pictures from each camera are transmitted by a microwave link to the Range Control Center at the Main Base and are displayed on monitors.

Soon after launch radar track is acquired and data are displayed at the Range Control Center on 30-in. × 30-in. x-y plotters. These data usually are present position of the vehicle in real time; however, velocity, flight-path angle, and heading derived by an analog computer are available if desired. Also displayed at the Main Base are real-time telemetry data, which are utilized by range safety in the evaluation of vehicular deviations.

A command destruct system, AN/FRW-2, operating in the 406 Mc/sec to 549 Mc/sec band, is used for vehicle command and destruct purposes.

Data reduction at Wallops Station is performed by an IBM 650 computer, a digital plotter, a Doppler chronograph, film reader, and strip-chart reader. The IBM 650 will be replaced by a real-time data computer system in the near future. This system will be composed of two major components: (1) a real-time data collection and display system, and (2) a central, general purpose computer capable of real-time input-output and general data processing. It will provide vehicle impact prediction and expand the data reduction capabilities of the range.

Wallops Station also has its own downrange tracking ship, the 940-ton *Range Recoverer,* acquired from PMR. It is used for the downrange reception of telemetry data, surface surveillance, and recovery operations.

Logistics support at Wallops Station includes liquid propellant storage areas, magazines for storage of rockets and other explosives, compressed air facility, water treatment plant, heating plant, vehicle assembly shops, and heavy equipment storage areas.

Nontechnical support services include transportation, security, transient housing, cafeterias, medical facilities, and post office.

E. Eglin Gulf Test Range

The Eglin Gulf Test Range (EGTR) is operated by the Air Proving Ground Center (APGC) of the U.S. Air Force's Air Force Systems Command at Eglin Air Force Base, Florida. While this base has been an active Air Force installation since 1941, its use as a missile and rocket launching center dates from 1958. Prior to this time, it was used mainly as a bombing and gunnery range and for the static testing of aircraft engines.[6] Today, however, its activities include the testing of a wide variety of guided missiles and aircraft, the training of pilots in bombing and gunnery, the training of air commandos, and the study of upper atmospheric physical processes by rocket probes.[7]

EGTR's range was designed with an integrated capability of supporting the launch of three guided missiles against three target drones. The various tracking and command instrumentation is sited with this requirement in mind; thus, the range has a greater operational flexibility than other ranges that expanded by evolution and the needs of the moment [39].

[6] During the crucial days of early 1944, before the invasion of Europe, Eglin Field, as it was then called, built its first missile launching sites for the German V-1 "buzz bomb." Under utmost secrecy, V-1 launching ramps were constructed using plans made from aerial photographs of sites in France. Various bombing techniques were tried upon them, until low-level attacks with heavy bombs were found to be most destructive. This solution was then used in Project Crossbow, the systematic reduction of V-1 sites before D-Day.

[7] The range is also a part of the U.S. Air Force meteorological rocket network that includes stations at Thule, Greenland; Churchill Research Range, Canada; Wallops Station, Virginia; Antigua, British West Indies; and Ascension Island.

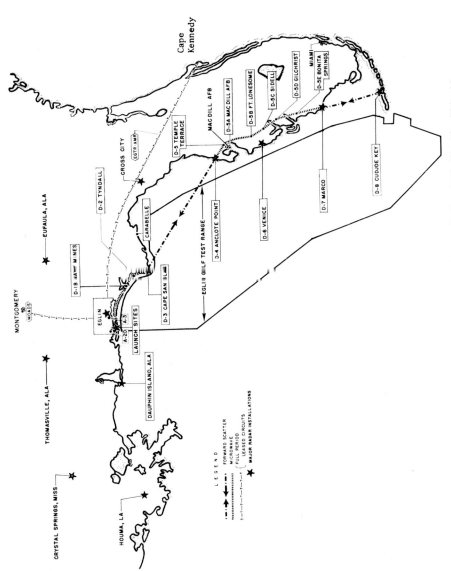

Fig. 57. Air Proving Ground and Gulf Test Range, Eglin AFB, Florida. U.S. Air Force photograph.

The major portion of Eglin Main, that area of the base located on the mainland, consists of a series of 10 auxiliary landing strips and 10 well-instrumented bombing and gunnery ranges. Sited outside the boundary of Eglin AFB proper is a chain of long-range tracking radars and communications systems in an arc reaching from Houma, Louisiana, through Thomasville, Alabama and Tampa, Florida to Key West, Florida. These sites and other major installations at EGTR are shown in Fig. 57.

While a variety of air-to-surface and air-to-air missiles as well as surface-to-air missiles such as Bomarc has been fired from the range, the number and variety of sounding rockets regularly launched for meteorological and scientific purposes are greater. Typical rockets include the Astrobee 200; Aerobees 150, 200, 300, 500; Exos; Arcas-Robin; Loki; Nike-Apache; Nike-Cajun; Nike-Javelin; and various combinations of the Nike M-5E1 booster such as Honest John-Nike; Nike-Nike-Nike (Cree); and Honest John-Nike-Nike.

1. Geographical Factors [40]

Eglin Air Force Base (Fig. 57) is situated on the southern coast of the panhandle area of Florida about halfway between Pensacola and Panama City. Including the impact area in the Gulf of Mexico, the range has some 44,000 mi². The sounding rocket launching facilities (Fig. 58), are located on Santa Rosa Island (30° 23′ N 86° 42′ W, elevation 14 ft), which is separated from the mainland by Santa Rosa Sound. This launch facility is some 14 mi from Eglin Main. Santa Rosa Island is a low, flat, sandy reef, which varies in elevation from 0 ft to 14 ft. Some 40 mi long it is little over a mile wide at its western end but is narrower toward the east. Vegetation is sparse though some areas of the island are marshy or thinly covered by scrub pine trees.

Temperatures in January have a mean daily maximum of 62°F and a mean daily minimum of 43°F. However, extremes of 10°F have occurred. The mean daily maximum in July is 90°F and the mean daily minimum for the same month is 73°F. Extremes during the summer have reached well above 100°F between May and August. The weather during the summer is sultry despite the land and sea breezes because the relative humidity ranges between 70 per cent and 80 per cent. There are also frequent thunderstorms during the summer. The prevailing winds during the summer are from the north, but the wooded terrain of the mainland to the north acts as a windbreak and morning fogs are not uncommon in the launching areas of Santa Rosa Island.

Fig. 58. Aerial view of sounding rocket launching sites on Santa Rosa Island, Eglin AFB. Aerobee launching tower is seen in foreground, while payload checkout building and assembly shops are seen in the background with smaller launch pads. U.S. Air Force photograph.

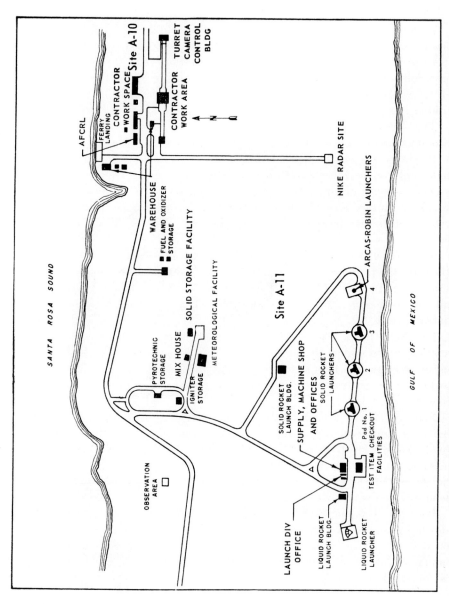

FIG. 59. Sounding rocket launching sites and facilities on Santa Rosa Island, EGTR. U.S. Air Force photograph.

2. Launching Facilities [40–44]

With the exception of the launching site for the Bomarc surface-to-air missile, which is some 6 mi to the west, the locations of rocket launching sites and facilities on Santa Rosa Island are shown in Fig. 59.[8] In general, these consist of five launching pads, along the shore about 100 yd from the Gulf; solid propellant storage areas; machine shops and assembly buildings; liquid propellant storage areas; and administrative areas.

The Aerobee launcher is a 162-ft, open-truss, steel tower that can accommodate rockets up to 24 in. in diameter with fin spans up to 120 in. It can be turned through 90 deg of traverse for adjusting the firing angle and can be elevated 10 deg to adjust for range. Other facilities associated with this launcher include a launch control building, liquid propellant storage and handling equipment, and a helium pressurization system (5000 psi) located 1000 ft to the east. This area is shown in Fig. 58.

Solid propellant rocket pads Nos. 1, 2, and 3 are separated from each other by 250-ft intervals, and pad No. 1 is 1300 ft from the Aerobee launch pad. Pad No. 4 is 500 ft east of pad No. 3. Launchers on these pads are remotely aimed in azimuth and elevation from the launching control station.

Pad No. 1 has a boom-type launcher that can accommodate vehicles weighing up to 40,000 lb. This launcher can be positioned through 100 deg of azimuth and elevated to 89 deg by means of a winch and cable. Rockets launched from this unit include the Nike-Nike-Nike (Cree) and Honest John-Nike.

Pads No. 2 and 3 have modified Jason launchers, shown in Fig. 60. They are similar in appearance and operation to the launcher at pad No. 1, being a 28-ft, straight-rail, boom type. They can handle rockets weighing up to 10,000 lb and are trainable through 100 deg of azimuth and 90 deg of elevation. Typical rockets launched from these positions include Nike-Cajun, Nike-Apache, Astrobee-200, and Nike-Javelin.

Pad No. 4 is the Arcas pad. The standard, 7-ft long, Arcas gun launcher is shown in Fig. 61. It is mounted on a turntable that permits 360 deg of traverse and 90 deg of elevation, both adjustments being manual. Two of these units are available at the pad. Tube-type launchers for the Loki and 2.75-in. meteorological rockets are also available at the pad.

8 Many air-to-surface and air-to-air missiles are tested and launched at EGTR, but these are fired for the most part at specified areas over the Gulf of Mexico or at several gunnery and bombing ranges on the mainland.

Fig. 60. Launcher at pad No. 2, Santa Rosa Island, EGTR. Rocket is an Astrobee 200. U.S. Air Force photograph.

FIG. 61. Arcas launcher at pad No. 4 on Santa Rosa Island, EGTR. U.S. Air Force photograph.

The launching control station for the four solid propellant rocket pads is located 800 ft to the rear of pad No. 3.

3. TRACKING AND COMMUNICATIONS

Downrange tracking sites on the Florida west coast and at Cudjoe Key, shown in Fig. 57, are linked to Eglin AFB by forward scatter and microwave radio systems, while full-period telephone circuits link the base to Cape Kennedy, to Goddard Space Flight Center, and to the Montgomery Air Defense Section (MOADS), in Alabama [39, 41, 43].

The microwave system from Eglin site A-20, Eglin Range Control Center and Project Mercury tracking station no. 17, to site D-3 at Cape San Blas, then across the Gulf of Mexico to site D-4, at Anclote Point on the west coast of the Florida peninsula, is a dual system with two frequency diversity microwave chains in parallel which provide two, 528-kc, baseband-frequency spectra with a total capacity equal to more than 240 voice channels. The system from site D-4 to site D-7 has a capacity of one, 528-kc, baseband-frequency spectrum. The forward scatter equipment provides tropospheric communications in the 2000-Mc/sec frequency band for the 180-nautical mi distance between sites D-3 and D-4.

Information transmitted over this communications system includes drone and missile radar and telemetry data; signal and control channels; countdown and frequency monitoring channels; and air-to-ground, intercommunications, and public address channels. All physical data are transmitted in digital form. In addition, timing and synchronizing signals on a double-sideband, FM carrier channel are also provided.

There are also seven fixed networks, three for the missile sites, three for the drone sites, and one for the frequency monitoring and interference control service. These are essentially party lines that service the entire range.

Operations at the Eglin Main area are served by several communications means: submarine cables linking it to Santa Rosa Island; voice communications from several sites to the Base Communication Central provided by a local 24-channel microwave system; and HF radio to control marine support operations and for long-range air-to-ground communications.

a. Radar and Telemetry

The EGTR radar system for test data acquisition, with seven major instrumentation sites, can track and beacon-control airborne objects throughout the range. Types of radars at each site are given in Table VII.

TABLE VII. MAJOR EGTR RADAR SUBSYSTEM ACQUISITION RADAR SITES

Site	Quantity	Type	Range (nautical miles)
Eglin (A-20)	2	AN/FPS-16	500
	2	AN/FPS-16	200
	1	AN/MPQ-31	2,500
Eglin (A-3)	2	AN/MPS-19	200
Cape San Blas (D-3)	3	AN/FPS-16	200
	1	AN/MPS-19	200
Anclote Point (D-4)	1	AN/FPS-16	500
	1	AN/FPS-16	200
	1	AN/MPS-19	200
Marco (D-7)	2	AN/FPS-16	500
	2	AN/MPS-19	200

Surveillance radar coverage of the Eglin test area is a cooperative effort of the various Air Force commands in or adjacent to the area and the Federal Aviation Agency. They are given in Table VIII.

TABLE VIII. EGLIN TEST AREA SURVEILLANCE RADAR SITES

Site	Quantity	Type	Range (nautical miles)
Dauphin Island, Alabama	1	FPS-7	200
Eglin (A-19)	1	FPS-20	200
	1	FPS-6	200
Cape San Blas (D-3)	1	FPS-20	200
	1	FPS-6B	200
Cross City, Florida	1	FPS-20	200
	2	FPS-6A	200
MacDill AFB, Florida	1	FPS-20	200
	1	FPS-6	200
Miami, Florida	1	ASR-1A	200

Site A-20, on the mainland, is the center of tracking activity at the Eglin range and is also the range control. It has range control consoles as well as facilities for range safety, optical and radar instrumentation, and other support systems. In addition to the four AN/FPS-16 and AN/MPQ-31 radars located on the roof of the control building, there are an Agave telemetry installation and several optical tracking instruments.

Another important instrumentation site is A-3, which is the drone control station for the range. It has control units for three simultaneous drone flights and contains tracking radars, several types of optical instruments, and three AN/FRQ-2, UHF, command transmitter systems. A special feature of this station is a chain telemetering control system that automatically selects the telemetering station receiving the best data and displays them for the drone controller on his console. Drones are also controlled from site D-4 at Anclote Point. This station has special microwave transmitters for command guidance of drone aircraft used at the range.

Site A-6 is the major telemetry station for the range. It has four missile-data receiving sets and four drone-data receiving sets. The range's telemetry system provides for the simultaneous reception of data from six airborne objects at three sites and from three objects at two sites. Most sites have redundant equipment installations to insure the receipt of data, and they are located so that an airborne object is always within sight of at least two of them. Separate telemetry systems are provided for missiles and drones, though a patching facility permits a highly flexible configuration of equipment.

A typical telemetry station has several helical antennas, of which some are single-helix while others are trihelix. These are remotely positioned from data supplied by tracking radars. The four telemetry modes used at the range are FM/FM, PDM/FM, PAM/FM/FM, and PDM/

FM/FM. Discrimination is provided for 54 channels of FM sub-carrier frequencies for any standard IRIG frequency and 36 channels of PAM or PDM decommunication.

In addition to the range stations, additional equipment is used in the local Eglin area, including a 60-ft diameter, parabolic antenna located at site B-4, auxiliary airstrip No. 5, with a high-sensitivity, narrow-beam, telemetry tracking capability, a gain of 28 db, and an automatic tracking capability.

Mobile vans and telemetry aircraft are also used at Eglin. The vans have capabilities similar to those of the fixed range stations except that fewer channels are available. Telemetry aircraft equipped for airborne telemetry reception can be controlled from individual range site control rooms. Two rf channels of telemetry data can be handled by the airborne stations through magnetic tape recorders.

b. Optical Tracking [42, 45]

Rocket launchings from Santa Rosa Island are covered primarily by Contraves and Askania cinetheodolites located on the island and at two sites on the mainland. Other optical tracking stations, with the same types of cinetheodolites, are located near the various auxiliary landing fields and bombing and gunnery ranges on the mainland. Also there are three sites provided with skyscreens to provide visual trajectory safety limits for missiles during launching. These are located on Santa Rosa Island and the mainland.

Ballistic cameras or phototheodolites used at the range include the Wild BC-4 and several modified aerial cameras such as the K-17, K-24, and K-47. The BC-4 are mobile units housed in vans.

The standard acceleration cameras used at Eglin are Bowen CZR-1's. They are arranged at 450-ft intervals and a distance of 1000-ft from the flight line of the missile trajectory. Other stations are located at greater distances and greater intervals for backup purposes. In addition, high-speed Mitchell cameras are used to obtain acceleration and velocity data. They are deployed in conjunction with the CZR-1 cameras. Fastax cameras, both 16 mm and 35 mm, are also used.

The TPR mobile tracking telescope, with 100-, 200-, and 300-in. focal length optics and a 24-in. aperture is also used for tracking missile flights to provide qualitative data such as attitude, roll rate, fin opening, etc. Ultrahigh speed Beckman-Whitley 189's and 192's are also used for documentary purposes.

4. Support Activities [43]

Technical support activities include data handling and processing; range timing; frequency monitoring and interference control; technical laboratories; marine facilities; and machine and fabrication shops.

The data handling and processing at EGTR utilize three primary data chains, which can be used interchangeably with S- or C-band radar. Primary data are obtained from tracking radar in digital form and then are fed to a data converter that changes them into the proper computer format, after which they are ultimately made compatible for real-time use of the IBM 7094 computer. Other computers include the LPG 30 and IBM 650.

The range timing system at EGTR consists of a master control station, at A-20, and slave stations at downrange sites. The master station has a timing generator with a 1-Mc/sec frequency standard as do the slave generators at the downrange sites. The stability of this standard is one part in 10^9 per day. The system also has provisions for calibration with timing signals from WWV. The signals produced by the master station and the slave stations are highly accurate and coordinated. This accuracy is produced by a synchronizing signal generated at A-20 and transmitted by various communications links to the slave generators. The synchronization is within 50 μsec. Even if this synchronizing signal is lost, the system maintains an accuracy of at least 0.25 msec in 8 hr.

Two other timing subsystems are also used at the range. These include a Vitro M2 and M3 system and a ballistic camera system. The M2 code provides pulses at a moderate rate for display and other purposes, while the M3 code is a lower rate and is used for camera pulses and time identification. The Vitro system can be slaved to the EGTR timing system with little bias since both are synchronized with WWV.

The ballistic camera timing system has its own time signal generator that produces a 32-bit, straight-binary code with 0.1 msec resolution. This generator is also synchronized with WWV or with VLF stations. It provides shutter-open, shutter-close, and strobe-flash pulses, which are recorded digitally at each station on magnetic tape with precise time information.

The frequency monitoring and interference control (FMIC) subsystem monitors and records rf signals for post-test analysis and detects, records, and locates signals that could interfere with a mission. Equipment for the subsystem includes various broadband receivers to monitor the rf spectrum, recorders, and displays. This equipment is grouped into five types of installations: fixed telemetry, FMIC buildings, semi-mobile vans, chase vehicles, and aircraft. Downrange sites and aircraft communicate with master site A-6 by the FMIC, fixed-net, air-to-ground segment of the EGTR communications system.

In addition to its extensive means for direct or simultaneous support of test operations, EGTR also offers the capabilities of several scientific and technical laboratories: mathematical services, metrology, precision measurement equipment, infrared (ir), photo-physics, instrumentation, photographic, and climatic. The mathematical services laboratory provides mathematical and data reduction services and furnishes computation support for APGC units. The metrology laboratory makes a wide range of measurements in electricity, light, radio frequency, microwaves, and sound. The precision measurement equipment laboratory repairs, calibrates, and certifies precise measurement equipment. The ir laboratory develops and evaluates ir instrumentation and equipment and performs radiation measurements in the ir region of the electromagnetic spectrum.

The photo-physics laboratory provides facilities for photo-optical instrument design, prototype manufacturing, optical calibration and test, and maintenance and modification of nonstandard photographic and optical instrumentation equipment. The instrumentation laboratory is responsible for the design, and installation, and coordination of use of equipment for the overall airborne instrumentation system. The photographic laboratory provides motion picture and still photography support for APGC test missions and meets the photographic requirements of Eglin AFB and other range users. As the name implies, the climatic laboratory simulates global environments that range from hot desert to polar regions; and it can produce a variety of climatic conditions, including sand storms, rain, salt spray, and high humidity.

Marine facilities are used to support missile testing over the Gulf of Mexico in recovery operations, transportation of diving personnel, range clearing, salvage operations, and air rescue training programs. Secondary missions are support operations in local emergencies and search and rescue work. Vessels used in this work include a 110-ft ferry, yard salvage derrick, range clearance and missile retriever, intercoastal range clearance boat, and two MR-41A's. The ferry is used to transport vehicles and materiel to sites and areas not readily reached by overland transportation. The derrick, with an 18-ton capacity crane, is a control vessel during over water tests and is also used to tow targets. The 63-ft, modified crash rescue boat is used primarily as a clearance and retriever vessel. The intercoastal range clearance boat, as its name implies, is used to control range clearance on intercoastal waters. The 170-ft MR-41A has modern communications and navigation equipment; a 360 deg,

12-ton capacity stern crane, and a 23-ft, diesel-powered, auxiliary boat. It is utilized for a variety of purposes.

The machine and fabrication shops include a machine shop, sheet metal shop, tubing and cable shop, wood shop, paint shop, welding shop, plating and heat-treating shop, and a fabric shop. They produce aircraft and missile components, ground support equipment, and components and end items for instrumentation and experimental equipment.

Housing and other amenities for military personnel as well as nontechnical support and administrative facilities are located at Eglin Main.

F. United States Naval Ordnance Test Station

The U.S. Naval Ordance Test Station (NOTS) began in 1943 as a firing range for the California Institute of Technology, which was a contractor for the Navy. As this field of development grew, the need for greater facilities increased in proportion. The range and its associated facilities were relocated in the Indian Wells Valley near Inyokern. In April 1945, the U.S. Navy Bureau of Ordnance assumed responsibility for the installation. Three years later NOTS incorporated the former Foothill Boulevard Ordnance Plant, at Pasadena, into its complex as the Underwater Ordnance Department. Today the Pasadena Annex, as it is called, includes the Foothill Boulevard Plant, the Morris Dam Test Range, the Long Beach Test Range, and the San Clemente Island Test Range [46].

NOTS, with its different ranges, well-equipped laboratories, and excellent technical support facilities, is capable of performing a wide variety of operations in the design, development, testing, and engineering evaluation of guided missiles, rockets, and spacecraft components. The principal areas of research and activity at NOTS are in the fields of underwater missiles, surface-to-air missiles, air-to-air missiles and rockets, air-to-surface missiles, aircraft fire control systems, deep-sea studies, and space research [47]. Typical of the missile systems with which NOTS has been concerned are Sidewinder, Shrike, Walleye, Polaris, Subroc, Tartar, Terrier, Redeye, Rat, Asroc, Weapon A, and the Mk 46 torpedo. In addition, NOTS scientists have also conducted cetacean research with dolphins.

1. GEOGRAPHICAL FACTORS

With headquarters at China Lake, California, NOTS is located (Fig. 62) on the northern edge of the Mojave Desert ($35° 41'$ N $117° 41'$ W, elevation 2215 ft). It occupies some 2000 mi^2 of the Indian Wells Valley, a broad, sandy desert and dry lake bed extending 30 mi north and south and 15 mi east and west. This valley is completely surrounded by mountains, with the Sierra Nevadas to the west, the Coso Mountains to the north, the Argus Mountains to the east, and the El Paso Mountains to the south. It has a typical desert climate with low humidity and tem-

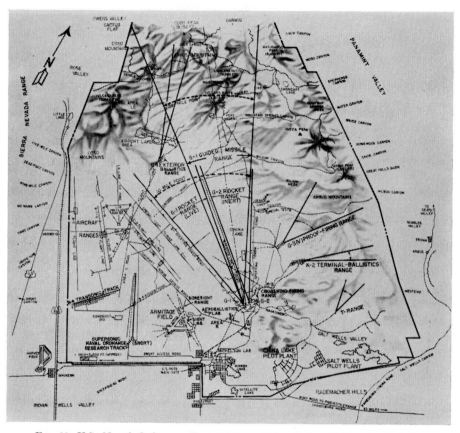

FIG. 62. U.S. Naval Ordnance Test Station, China Lake, California. U.S. Navy photograph.

peratures ranging between 20°F and 80°F in winter and 60°F to 115°F in summer.

The Randsburg Wash Activities of NOTS are situated in another valley 23 mi to the southeast. This level desert valley is ringed by mountains with heights up to 3500 ft on three sides but open to the east. The area encompasses 350 mi² and has firing ranges over 15 mi long. Because of its isolation, it is particularly adaptable for tests involving classified weapons.

San Clemente Island, 60 mi south of Long Beach, California, is 21 mi long and 5 mi wide. The island lies generally northwest to southeast, with an underwater demolition team training area and shore bombardment area in the south and an airfield and missile launching facility in the north. The island has elevations of 1500 ft and steep cliffs bordering the ocean.

2. Ground Ranges [47–49]

There are four permanent ranges and one temporary range maintained by NOTS. These include ranges at China Lake, Randsburg Wash, San Clemente Island, and a temporary range at Walker Lake, Hawthorne, Nevada. Ranges within the China Lake area are shown in Fig. 63.

The two primary ranges at China Lake, G-1 and G-2, are located 4 mi northeast of the headquarters area (Fig. 64). The two ranges overlap downrange and share tracking instrumentation in this area.

G-1 has a length of 37 mi, the first 10 mi of which is flat with the remaining 27 mi being hilly or mountainous. The nominal line of fire for guided missiles is on an azimuth of 356 deg, with firing limits of 640 mils. For rockets with high-explosive warheads the firing azimuth is 333.5 deg, with limits of 400 mils. Launching equipment at G-1 consists of an X-7 launcher and an X-8 launcher for the Terrier surface-to-air missile (Fig. 65) and a concrete pad on which other launchers can be emplaced. The area also has an earth-barricaded, concrete pad for use with the marine Mk 3 Mod 0 launcher and fire control system. All firings are controlled from a fire control building located to the rear of the X-8 pad. Other facilities at the range include a three-bay missile assembly building, two checkout buildings, range timing station, radar and optical tracking instrumentation, telemetry receiving stations, missile repair shops, and three storage buildings for inert materiel.

G-2 offers 10 mi of flat desert range and is used for flight-testing inert-head rockets and for determining exterior ballistics of rockets during launch, midflight, and impact. It also is available for developmental work on rocket launchers. The central feature of G-2 is a large concrete pad with an Mk 108 universal launcher. The range centerline is on an azimuth of 356 deg with firing limits of 400 mils. Other launchers can be mounted on the pad as required. Sited near the pad is a 6 deg ramp launcher, 524 ft long, consisting of two standard-gage, 75-lb railroad rails. It is used for rockets needing preaccelerated firings. Depending

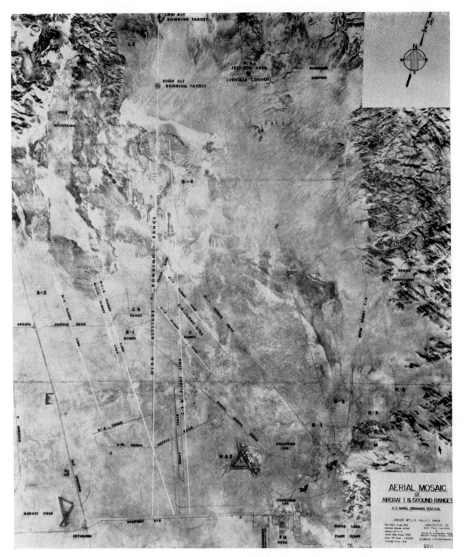

FIG. 63. Range areas at NOTS. U.S. Navy photograph.

upon the propulsion units used with the sled on the track, velocities are possible from 500 ft/sec to 1300 ft/sec with resulting accelerations of $17g$ to $72g$. Other facilities at G-2 include a magazine area for explosive components (primarily for use of G-1 range), fire control center, inert storage facilities, a temperature cycling building for rockets, and an optical and radar tracking system. This range has a limitation in that no rockets with explosive heads can be fired on it.

K-2, 8 mi northeast of the headquarters area, is a special range used primarily for the terminal ballistic testing of rocket warheads and fuzes. It has a 1500-ft long, two-rail track launcher, which fires into a 2-mi

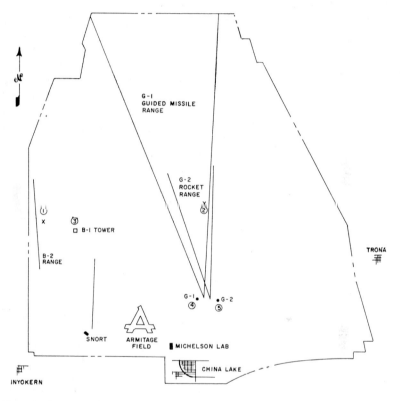

FIG. 64. G-1 and G-2 ground ranges at NOTS. Installations include 1, Midas West tracking station; 2, Midas East tracking station; 3, B-1 range control; 4, G-1 range control; 5, Doppler velocimeters at G-2 range. U.S. Navy photograph.

wide and 5-mi long range. These rails are set 2 in. apart and alignment is held within 0.021 in. over the length of the track. The azimuth of fire is 83.5 deg. Slipper-mounted sleds with test rockets *in situ* can be accelerated to velocities of 3000 ft/sec and 125*g* using solid propellant rocket motors for propulsion units. If a braking device is needed, retro-rockets are used. A magnetic pickup system on the rail permits velocity data to be recorded. A smaller, 200 ft launcher is also located on the left side of

Fig. 65. The X-8 launcher for Terrier missiles at G-1 range, NOTS. U.S. Navy photograph.

the muzzle end of the 1500-ft launcher. Other facilities available at K-2 include a fire control building, a standard 4-channel PCM timing station, a temperature cycling building for rockets, inert and explosive storage, and a machine shop. A 100-ft spotting tower is also located behind the launcher as well as at five downrange spotting posts. Barricaded camera stations are sited at distances out to 800 ft, and there is a large barricaded camera and instrumentation building located 500 ft to the side of the launcher at its muzzle end. Tracking instrumentation available is primarily photographic.

One of the most versatile and interesting launching facilities at NOTS is at the K-3 range. Located some 5 mi northeast of the headquarters area, this range is comprised of 2.84 mi of precisely laid railroad track of a standard gage. This range simulates the cross-wind firing of missiles from moving ships. Launchers for such weapons as Rat, Asroc, and Weapon A are mounted on a special, 4000-lb flatcar that is towed by a prime mover equipped with a 155-hp engine. This combination can produce velocities up to 70 knots; and, because of the configuration of the track pattern, permits launchings in cross-wind, dead-ahead, and astern directions, to ranges of 13,500 yd, impacting into the downrange G range areas. Third-rail circuits at preselected launch points actuate firing devices.

Other features of the range include the spiraled and super-elevated curves for the track, portable blast aprons, and portable firing barricades and camera stations. There is no permanent tracking instrumentation at the range, but it is provided as required.

The G-3 (Victor) range is located 9 mi northeast of the headquarters area and is used for firing rockets with high-explosive heads from portable launchers, for studying terminal ballistics, and the production lot testing of rockets. The range is only 2 mi long but has a mountain backstop. The firing azimuth is 56 deg with lateral limits of 400 mils. While a 350-ft monorail, I-beam launcher base is in place at the range, the rail has been removed; and the only launcher used is a 3-in., smooth-bore gun launcher. Rocket heads and fuzes are launched from it using sabots. There are only three small utility buildings available, and no permanent instrumentation, power, or water supply.

The Randsburg Wash Activities of NOTS provide several specialized ground ranges. They are, for the most part, used for rocket and missile fuze evaluation tests.

This rocket range has two 300-ft towers (Fig. 66), 700 ft apart, from which a number of full-scale or small-scale models can be suspended at heights up to 150 ft above the ground. Since the towers are wood and the suspension cables are hemp ropes, there is very little radar return from them or reflections from the ground. The launcher is an open-truss, steel structure 150 ft tall mounted on railway trucks so that it can be moved to and from the target (Fig. 67). This tower is topped by a 70-ft, variable-angle launcher. At the 50-ft level there is an armor-plated control room from which the launcher can be fired. Junctions boxes at 700-ft intervals

FIG. 66. Wooden towers for target support at rocket range, Randsburg Wash Activity, NOTS. U.S. Navy photograph.

along the launcher tower track provide power outlets, timing signals, and camera control signals.

Phototheodolites are positioned throughout the area to provide camera coverage. A special, mobile instrumentation van is available for radio telemetry.

Also located in the Randsburg Wash area is the Gun Target Range. The gun-line facility consists of a concrete pad and firing control build-

Fig. 67. Steel, 150-ft, mobile rocket launcher at rocket range, Randsburg Wash Activity, NOTS. U.S. Navy photograph.

ing with some 24 different Army and Navy guns (Fig. 68). The control building, a three-story, reinforced concrete structure with bulletproof windows houses the test conductors and associated control equipment. On the roof of this building is an infrared, burst-time indicator that is accurate to 1 msec. Cameras and weather instrumentation can also be emplaced on the roof. Targets for this range are suspended from two

360-ft high wooden towers located 640 ft apart and 3200 yd downrange. Targets as large as a stripped-down B-29 bomber can be suspended between them at heights up to 250 ft above the ground. Other facilities

FIG. 68. Gun-line area, Randsburg Wash Activity, NOTS. U.S. Navy photograph.

include storage structures; a 36-ft, mobile, steel spotting tower; and a temperature-controlled building in which rockets can be cycled from −65°F to 165°F.

Similar ranges are in the area for howitzers and for the vertical firing of projectiles with base-impact fuzes.

A variety of radar and camera instrumentation coverage is provided in the Randsburg Wash area, and there is an administrative area consisting of a headquarters and laboratory building, carpentry shop for target mockups, machine shop for minor weapons repair and range maintenance, instrumentation shop for the modification and maintenance of electronic and photographic instruments, and a large warehouse for the storage of inert components. A special magazine area provides storage for explosive components and rocket motors.

3. Special Ranges [47–49]

Among the various activities of NOTS are several types of ranges that find use in the missile and space programs of the United States. These are at China Lake as well as other locations.

a. CT-4

A special range to study the effects of atomic and high-explosive items to thermal response is located some 15 mi northeast of the headquarters area. CT-4 range is also used to study weapon fragment size, velocity, and distribution; nuclear contamination particle size, concentration, and distribution; blast pressure, magnitude, and distribution; and the effects of various firefighting techniques. It is especially useful in determining the effects of the cook-off of simulated atomic weapons and other high explosive items in gasoline fires.

The range is a circular area 12,000 ft in diameter, the center of which is the test area or ground zero. In concentric rings around the ground zero point are three zones: at 50 ft is the firefighting area; at 150 ft the fragmentation area; and at 12,000 ft the contamination-array zone. The entire area is instrumented with radiological contamination collector stations.

Instrumentation at the range includes special devices, such as Celotex panels to catch metal fragments, copper-indenter pressure gages, thermocouples, and air samplers, as well as high-speed cameras and closed-circuit TV. Firings and tests are conducted remotely and controlled from a building 1 mi from the range. There is a radiological safety control center located near the range control building. Shop buildings for the range are sited some nine miles from it.

b. Air Ranges

There are three air ranges at NOTS: the high-altitude range, B-1; the special test range, C; and the boresighting range. B-1 consists of several flight lines, with the high-altitude range being the major one. It is 25 mi long and is shown in Fig. 63. These flight lines are used for several purposes: (1) air-to-surface missile, rocket, and bomb evaluation tests; (2) angle-of-attack aircraft calibration tests; (3) dispersion patterns of ammunition; (4) airborne radar calibration; (5) aircraft bomb-director and fire-control system evaluation.

The high-altitude flight line is instrumented along its entire 25 mi with special camera stations, flight analyzers, and cinetheodolites. Other flight lines of the range are instrumented as needed for specific trials. Radar coverage is also provided.

C range is an air-to-surface missile range with three flight lines. The major one is designated C-3-W, is more than 20 mi long, and has its target located at the midpoint so that firings can be made from either half of the flight line. The target is especially designed for visual and radar detection, and the range is used primarily for training aircraft pilots in the various special weapons delivery techniques, mine drops, night flare tests, and guided missile homing runs. Since the range is primarily a training area, instrumentation provides real-time data rather than precision position data. Cameras are available for obtaining documentary flight coverage. Manned spotting towers located 1 mi from the target provide accurate impact data to the range control station where it is placed immediately on a plotting board.

As the name implies, the boresighting range is used for the alignment of cameras, gyroscopes, radars, guns, etc., on aircraft. The range is some 6000 ft long and has a concrete pad on which aircraft are parked. A special screw lift provides a means of leveling the aircraft by raising and lowering its tail. The device can also swing the craft through an arc at the rate of 1 ft/min. Boresight targets are positioned on the perimeter of a circle 1000 yd from the pad. The centerline marker is at 2000 yd. The arrangement permits alignment accuracies to 2 mrad. Camera coverage is provided as needed.

c. Indoor Ranges

The controlled-atmosphere range is a heavy-wall, steel tube 34 ft long and 3 ft in diameter. It can be evacuated to a pressure of 1 mm or filled with various mixtures of gases. It is used to study the flight characteristics of hypervelocity fragments. Windows in the tube permit visual and photographic observation.

The aeroballistic laboratory (Fig. 69) is a fully instrumented and enclosed range used for making experimental determinations of the aerodynamic and ballistic characteristics of rockets and missile models. Guns with 40-mm, 3-in., 5-in., and 8-in. tubes are used to launch inert models in free flight through the 500-ft long building. Photographic coverage is provided by 22 pairs of ballistic cameras precisely located at 4-ft intervals. The range also has the means for obtaining spark shadowgraphs. In addition, there is a 400-ft pressurized tube 4 ft in diameter and used for studying the flight characteristics of models at hypersonic velocities. The pressure in the tube can be varied between 1 mm and 10 atm.

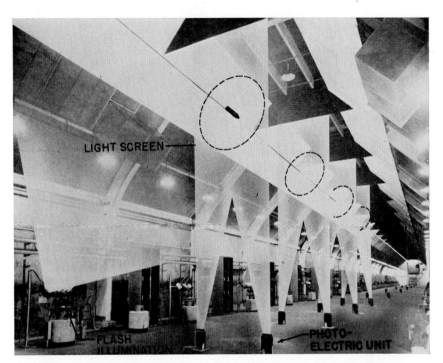

FIG. 69. Aeroballistic Laboratory range at NOTS. U.S. Navy photograph.

d. Static Test Facilities

The most elaborate of the static firing stands for rocket motors at NOTS are those designed for the Polaris; they are located 8 mi northeast of the administrative area. These are collectively called the Skytop facility (Fig. 70), and they could be adapted for other motors than the Polaris [50]. While there are three of these stands, a description of Skytop 2 is typical of the capabilities offered by the facility. It has both a horizontal and vertical test stand. The former is capable of withstanding a steady load of 1,000,000 lb or a momentary impulse of 10,000,000 lb. The latter will sustain a force of 200,000 lb in either an upward or a downward direction.

The horizontal pad is 30 ft × 75 ft and 6 ft thick. An asphalted work area about the pad increases the overall area to 120 ft × 180 ft. The

Fig. 70. Skytop static test facility, NOTS. U.S. Navy photograph.

pad is designed to accept different thrust buttresses to permit greater flexibility in testing. Some 50 ft behind this pad is an underground equipment and cable termination room designed to withstand impulsive overpressures of 30 psi.

The vertical stand is of conventional design and has a water-cooled flame deflector. Initially built to test fire the Polaris second stage motor and thrust vectoring equipment, the stand contains an altitude simulation device of the secondary-throat type of exhaust diffuser. This chamber, which permits simulated altitudes to 100,000 ft, is attached to the test motor by a flexible mounting.

Also available at the Skytop 2 facility is a second horizontal test bed with an altitude simulation feature. However, the motor chamber of this stand is enclosed in a steel chamber that can be evacuated by a mechanical pump; it, too, has an exhaust diffuser.

A smaller static firing stand is used for complete missiles such as the Sidewinder. The missile is restrained by a nylon harness. Likewise, small liquid propellant engines with thrusts from 50 lb to 50,000 lb can be tested in a small but completely instrumented static firing facility 3 mi northeast of the administrative area at NOTS (Fig. 71).

e. San Clemente Island

A number of rocket and missile testing facilities are located on San Clemente Island, which is the center for underwater missile research and testing. However, there is considerable activity in the testing of air-to-water, underwater-to-air, and surface-to-surface missiles. Entire weapons systems can be tested from aboard ship, aircraft, submarine, or from launchers on shore. Firings from the shore take place in easterly and southeasterly directions from firing positions located on the northeastern shore of the island.

One of the most interesting features of this facility is the Variable Depth Launch Facility from which Polaris missiles are fired. It consists of three underwater launchers of the type shown in Fig. 72. These launchers are completely instrumented, with cables being supported by floats. Towers nearby are equipped with underwater cameras, including closed-circuit television.

Missiles fired from these launchers can be captured in a net suspended over the launching point.

Fire control equipment and telemetry stations for the launchers are located on shore as are additional camera stations and the range timing central.

Other shore facilities include two launch pads with gun-mount launches for the Asroc and Zuni guided missiles, a bombing range for aircraft; a

FIG. 71. Small liquid propellant engine static firing area, NOTS. U.S. Navy photograph.

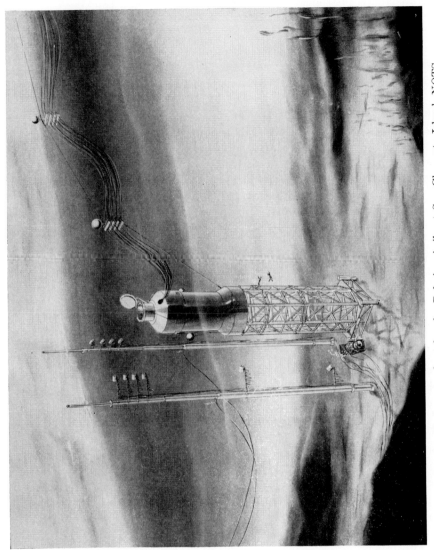

FIG. 72. Underwater launcher for Polaris missile at San Clemente Island, NOTS. U.S. Navy photograph.

captive cable range for war-shot testing of torpedoes; storage and as-sembly buildings; and messing and housing for range personnel. The entire island is instrumented for both radar and camera tracking.

f. Long Beach Test Range

This area covers some 476 mi^2 off the coast of Long Beach, Cali-fornia. It is used for underwater missile tests made under operational conditions. Missiles can be fired from submarines or other tactical craft; or they can be fired from a special Deep-Depth Launching and Test Facility, called Trygon, which consists of a converted LCU equipped and instrumented for the underwater launching of naval ordnance and the study of ocean acoustics. The unit has a fully instrumented and stabilized platform that can be lowered to depths of 600 ft for the underwater launching of missiles or for studying sonar transducer characteristics.

Shore facilities at Long Beach include torpedo and missile assembly shops, a pier, and boat houses.

g. Morris Lake Test Range

Located at Morris Dam in the mountains above Azusa, California, is a NOTS facility for test launching full-scale underwater weapons. Since the lake is formed by a dam, this body of water provides a range 1000 yd long, 200 yd wide and 150 ft deep. The central feature of this range is a variable-angle launcher (Fig. 73) for injecting missiles into the water at various angles of entry. The shore end of the 300-ft long, 22-ft wide, 35-ft high structure pivots on a rail-mounted carriage, while the op-posite end rests on two steel barges. The bridge supports two launching tubes, fired by compressed air; and it can be adjusted in elevation to produce entry angles between 0 deg and 40 deg.

The range also has a floating, variable-angle launcher mounted on a pontoon barge. It is used for launching small missiles or scale models and has a tube that will accept models of 5-in. or 8-in. diameter. One end of the tube is hinged to the barge while the other can be adjusted to per-mit launching angles from 0 deg to 18.5 deg. Propulsive power for the launching tube is supplied by the detonation of a mixture of hydrogen and air. The tube also has a feature that permits prespinning models to angular velocities as high as 18,000 rpm.

Two other specialized launchers are available for studying the effect of various entry angles on underwater missiles. These are the rocket launching tower and the slingshot launcher. The former is a tower 166 ft high with platforms at the 100-ft level and 150-ft level. Models up to 5 in. in diameter can be launched from the tower, and vertical drops can be made as well. The latter facility provides steeper entry angles

Fig. 73. Variable-angle launcher for underwater missiles at Morris Lake Test Range, NOTS. U.S. Navy photograph.

than can be created by other launchers. Missiles are attached to a cable and hoisted to heights of 180 ft above the lake. Elastic cables anchored to barges provide the propulsive power for driving the missiles into the water. The launcher will take missiles as heavy as 350 lb and can launch them at angles between 65 deg and 90 deg at velocities up to 200 ft/sec.

A simulated submarine impact facility consisting of a rail launcher mounted on a barge is also located at the range. This launcher can be lowered to a depth of 9 ft between two pontoon barges. Torpedoes and depth charges are launched down the rail and exploded against a steel plate target, which can be varied in angle up to 90 deg.

An underwater cableway, 2500 ft long, and at a depth of 60 ft, provides a means of studying torpedo propulsion systems and hyrodynamics over a limited range.

h. Walker Lake Range

This range is used as needed for testing free-flight missiles and rockets with ranges up to 12,000 yd. It is located on the southern shore of Walker Lake, at Hawthorne, Nevada, 260 mi north of China Lake. There are two concrete pads from which missiles can be launched. One pad is oriented so that the line of fire is on an azimuth of 328 deg, while the other is on an azimuth of 22 deg. Range support includes a large missile assembly building, and both pads and the downrange areas are surveyed for cinetheodolites and ballistic cameras. Portable vans containing range timing equipment and firing controls are provided as needed.

4. SUPERSONIC ROCKET SLED TRACKS [48, 51]

NOTS provides three supersonic rocket sled tracks that are used extensively in the development and testing of guided missiles, high-performance aircraft, and spacecraft components.

The Snort (Supersonic Naval Ordnance Research Track) track is 5 mi west of NOTS headquarters and is used for a variety of purposes including captive tests of rockets, missiles, model or full-scale aircraft; aeroballistic tests of projectiles, missiles, and rockets fired under simulated launch conditions; standard and VT (variable time) fuze development tests; inertial guidance systems developmental and calibration tests; aircraft damage tests; and ejectable items trials from aircraft and missiles.

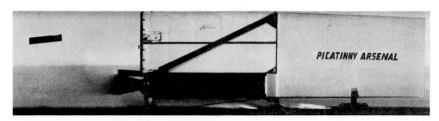

FIG. 74. 155-mm howitzer projectile being caught in flight by a sled on Snort track, NOTS. U.S. Navy photograph.

One of the more unusual uses of the track is shown in Fig. 74. Here a 155-mm howitzer projectile is being caught in flight by a rocket sled equipped with a container of cushioning material. Thus the shell is recovered intact after firing for analysis. The howitzer is mounted parallel to the track and fired electrically by the sled at a predetermined point.

Physical characteristics of the tracks are given in Table IX.

TABLE IX. SUPERSONIC SLED TRACKS AT NOTS

Item	Snort	B-4 Total	B-4 3rd Rail	G-4
Length (ft)	21,560	14,560	11,000	3000
Gage (in.)	56.6	56.6	13	33.87
Heading	N 3° 02′ 35″ W	N 72° 3′ 45″ W		N 9° 10′ 58″ W
Rail size (lb/yd)	171	75		171
Alignment (in.)	0.036 vertical 0.060 horizontal	0.062 vertical and horizontal		0.036 vertical 0.060 horizontal
Brake	water	sand		retrorocket (if required)
Artificial rain device	yes	no		no
Camera coverage	yes	yes		yes
Telemetry	yes	yes		yes

On all the tracks a variety of sleds are available, the particular sled used depending primarily on the nature of the experiment. Typical of the Snort sleds is the liquid propellant rocket engine shown in Fig. 75. Solid propellant sleds are simpler in design and use "off the shelf" motors for propulsion units. A typical sled of this type is shown in Fig. 76.

The Snort track is programmed for firing and control in much the same fashion as the launcher for a large space carrier vehicle. The programming system, Fig. 77, provides automatic countdown timing, control signals for starting and stopping relay-controlled apparatus, and safety interlocks. Electronic instrumentation associated with the track includes telemetry systems (both radio and landline as well as on board recording), timing station, and time-position system. The radio telemetry is primarily an FM/FM system although PDM/FM and PAM/FM are available. The system can receive seven carriers in the 200 Mc/sec band. The timing system at the track is an independent one deriving its 9-channel PCM/UHF signals from the range's central timing center.

The B-4 track, located 9 mi northwest of NOTS headquarters, is used basically for the same types of tests as the Snort track. Sometimes it is employed for dummy runs of tests to be made on Snort. The characteristics of the track are given in Table IX. Unlike the longer track,

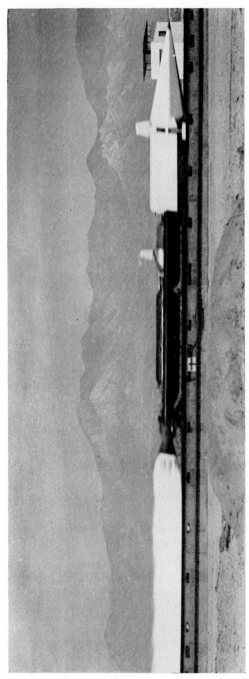

Fig. 75. Liquid propellant rocket sled on Snort track at NOTS. U.S. Navy photograph.

Fig. 76. Solid propellant rocket sled used to accelerate Mercury space capsule on Snort track, NOTS. U.S. Navy photograph.

B-4 has a third rail, which extends from a point 4000 ft from the breech end to a point 11,000 ft down-track. Instead of a water-trough brake, the B-4 facility uses loose sand between the rails. With light vehicles, accelerations as high as 125g are possible with velocities of 1980 ft/sec. Track programming and instrumentation is much the same as that of Snort, although less elaborate.

Fig. 77. Snort track programmer at NOTS. U.S. Navy photograph.

The G-4 track, 15 mi northwest of the NOTS headquarters area, differs from Snort and B-4 in that it is essentially a terminal ballistics track, overlooking an expanse of desert that permits free-flight of objects from the track. Since it is used for terminal ballistics tests, G-4 has no track braking device, but individual sleds can be equipped with retro-rockets as needed. Track programming and instrumentation is limited, but basically the same as that of B-4 and Snort, being housed in portable vans.

5. Tracking, Communications, and Telemetry [48, 52, 53]

a. Radar Tracking

Despite the decentralized range facilities at NOTS, it is one of the better instrumented of U.S. missile ranges, both electronically and optically. The Metric Electronics Branch of the range's Test Department operates the tracking radars, Doppler velocimeters, and Midas system, which are the primary tracking means.

Radars used at NOTS include the AN/MPS-26, C-band (Fig. 78); Mod 7298, S-band; M-33, X- and S-band; and SCR-584, S-band. While permanently installed radars are available at G-1 range, mobile sets are available as required at other sites. Other tracking instruments include the Resdel 10A Doppler velocimeter and the Midas (Missile Intercept Data Acquisition System).

The Midas system can track two airborne objects simultaneously and provide continuous space coordinates for aircraft control and range safety purposes. There are two Midas sites located 12 mi apart, one at B-2 range and one at G-2 range. Each site is an arrangement of antenna-pair interferometers, which measure the phase difference of signals from target telemetry transmitters or special beacons. The output of system is in digital form in a format acceptable by the IBM 7090 computer. The overall accuracy of the Midas is said to be 50 parts in 1,000,000 in direction cosines.

b. Communications

Both wire and radio communications tie the range operations together. Most communications are handled by two-way, FM radio with two independent channels. A repeater station located on a mountain within the area effectively increases the range of the network by using a transmitter of higher power than any of the stations within the network. During critical tests the repeater station is operated by landline from G-1 or G-2 range control center as a means of assuring positive control. The network has about 160 stations in fixed, portable, and mobile installations.

Ground to air communications are furnished by two UHF channels.

China Lake, San Clemente Island, and the Pasadena Annex (as well as their associated facilities) are linked by a single-channel FM network utilizing mountain-top repeater stations.

Two portable, closed-circuit television links are available for remote coverage of ordnance tests. There are also six microwave data links for

Fig. 78. Mobile AN/MPS-26 tracking radar used at NOTS. U.S. Navy photograph.

transmitting data from the two Midas sites to the G-1 fire control center and to the computer facility.

NOTS also employs a radio control and flight termination system. It is used both for the rf command of drones such as the QF-9F, KDA, KDB, and Q2C and for the destruct signals to missiles equipped with such devices. The ground station consists of two AN/FRW-2 systems that are interconnected in such a way that switching between transmitters occurs in only 100 msec in case of a malfunction in the transmitter on the air. A single AN/FRW-2 station installed in a trailer is available for use in remote locations.

c. Optical Tracking

A variety of optical instruments is used at NOTS facilities. Some are permanent emplacements, but there are hundreds of surveyed positions in which mobile instruments can be installed as required. For example, there are 132 sites in which cameras can be emplaced for coverage of firings at the G-2 range.

Primary optical tracking instrument at NOTS ranges is the cinetheodolite. There are three types used: the Askania Kth 41, Mk 2 Mod 0, and Mk 5 Mod 0. The Mk 2 Mod 0 is a Navy modification of the Askania Gth 40, while the Mk 5 Mod 0 is a Naval Gunfire Factory instrument. There are both permanent and temporary installations for these instruments. Approximately half of the cinetheodolites available at NOTS are housed in air-conditioned astrodomes similar to those shown in Fig. 41. Many of the stations are tied into a target acquisition system that permits remote slaving to other range instrumentation, particularly the Midas.

The other standard tracking instrument used at NOTS is the M-45 machine-gun mount fitted with a variety of camera and lens combinations. These mobile units (Fig. 79) are towed by trucks equipped with a power supply, timing system, and radio communications. The usual combination is a 96-in. lens on one mount and a 48-in. lens on the other. Cameras employed are the 35-mm Mitchell or Photo-Sonics, 70-mm Mitchell or Photo-Sonics, and the 16-mm Mitchell, Milliken, or Photo-Sonics. (These same cameras with lens up to 150-in. are also used as boresight cameras with tracking radars.)

The primary ballistic camera used at NOTS is the Bowen CZR-1. It is especially useful in taking qualitative photographs at the various sled tracks, as shown in Fig. 80. With a 10-in. lens as the usual lens, the camera is fixed in a precision, three-axis mount that is mobile. The mount also has a control cabinet and timing receiver. To be installed at a remote position it needs only a surveyed position and a 208-volt power supply, which is available from the prime mover. The Bowen RC-2, also on a three-axis, mobile mount, is utilized as well.

Other cameras used at NOTS include the familiar Fastax, Fairchild, and Eastman high-speed cameras; the K-24, modified aerial camera;

Fig. 79. M-45 camera tracking mount used at NOTS. U.S. Navy photograph.

Warrick FN; Fastair, TSR, and Test Vehicle Recorder 1A and 1B, ribbon-frame cameras.

d. Range Timing System

The primary timing system at NOTS is a 9-channel, PCM code supplemented by a simpler 4-channel system. The frame rate repeats at a 10 kc rate and is transmitted on a 500 Mc/sec carrier. This PCM system provides a highly accurate (2 μsec) time base for both optical and radar instrumentation. The range central is located in the instrumentation operations building at range G-1; and there are additional central stations located at G-2, K-2, the three sled tracks, Randsburg Wash ranges, and San Clemente Island.

FIG. 80. CZR-1 ribbon-frame camera photographs of seat ejection test on Snort track at NOTS. U.S. Navy photograph.

At the beginning of each day's activity at NOTS the secondary frequency standards at each of these stations is checked against the primary standard at the instrumentation operations building. The primary standard is calibrated as required with the singles from WWV or WWVH.

Typical timing codes include a 100 pps binary code for integrated time referenced to test zero (usually squib firing pulse) for ribbon-frame, Fastax, Mitchell and other cameras where motion is continuous at some point; 100 pps displayed on Askania film next to film frame code to assure positive identification of the presence or absence of a pulse in the code; 20 pps as a convenient rate for telemetry records; and 4 pps for actuating the shutters of Askania cinetheodolites.

Receiving and demodulating equipment for some 200 sites is available at NOTS.

Each instrumentation receiving site has a directional, high-gain antenna; UHF pulse receiver; PCM demodulator; and associated control boxes and timers.

A limited IRIG standard format timing is being generated and transmitted by a 141-Mc/sec link; as equipment becomes available to receive the IRIG format, it will be phased into the timing system and the PCM operation will be dropped.

e. Telemetry

Both radio and landline telemetry are utilized at NOTS. The radio telemetry receivers and associated equipment are housed in a central telemetry building of the headquarters area. Landline terminal equipment is located in vans and the radar building at G-1 range. Mobile landline equipment is also available.

NOTS employs three types of radio telemetry: FM/FM, PAM/FM, and PDM/FM. The telemetry building contains six 12-channel, FM/FM receivers (Fig. 81); and two time-division receiving stations (Fig. 82). Four of the six FM/FM stations are used for real-time recording and presentation of data. Each of these stations is comprised of two phase-locked, crystal-controlled receivers; a 7-track magnetic tape recorder; a 12-in. recording oscillograph; and a 12-sub-carrier discriminator bank. The two other stations are used for tape playback records. The antenna system consists of a quad helix with remote drive, two bifilar manual-track helices, and two 28-ft, parabolic antennas.

Each of the time-division stations includes a helical antenna, a pre-amplifier, multicoupler, receiver, 7-channel magnetic tape recorder, and decommutator. The decommutators can handle 45 channels at all standard IRIG rates.

FIG. 81. FM/FM receiving stations at NOTS central telemetry building. U.S. Navy photograph.

In addition to these telemetry facilities, there is also a Microlock station for receiving satellite telemetry.

6. SUPPORT ACTIVITIES

a. Technical and Scientific Research Laboratories

Together with its variety of testing facilities and ranges, NOTS has a diversified capability in both fundamental and applied research in propellant chemistry and in aeronautical and astronautical engineering. For the most part these activities are located in the Michelson Laboratory (Fig. 83) at China Lake. Within this huge building (10 acres of floor space), named for Albert A. Michelson (a former Navy scientist and Nobel winner) are physics laboratories, a standards laboratory, a fuze design and evaluation laboratory, and an electronics laboratory. Other technical support activities at NOTS include a materials engineering laboratory, an engineering evaluation laboratory, a hyperballistics field laboratory, and a warhead-test field laboratory.

Perhaps the most complete research and development facility available at NOTS is in the field of propellants and explosives. Here there is an integrated complex consisting of basic research laboratories and

Fig. 82. Time-division telemetry receiving equipment at NOTS. U.S. Navy photograph.

several applied research laboratories as well as prototype manufacturing lines and extensive storage facilities.

b. Data Processing and Reduction [54, 55]

Data processing for all NOTS activities is provided by a NOTS designed redactor called Nodac (Naval Ordnance Data Automation Center). It is a general-purpose, data-processing system comprised of three major components: (1) analog handling equipment, (2) digital handling equipment, and (3) conversion equipment for changing continuous data to dicrete form and vice versa. Outputs from Nodac are applied to an IBM 7090 for reduction. Nodac accepts several forms of electrical data, e.g., FM/FM, PDM, PDM/FM, PCM, and PAM/FM. Noncomputer format digital information can also be processed by the system.

c. Aviation Support

The Naval Aviation Facility at Armitage Field, near China Lake, provides complete air support for NOTS various requirements. Over 30

FIG. 83. Michelson Research Laboratory at NOTS. U.S. Navy photograph.

different types of aircraft are available, and the runways are long enough to accommodate all jet and cargo aircraft. In addition to purely aviation support, the Naval Air Facility also has a 50-ft tower with two fully instrumented bays for testing fire control systems and radar. Another service offered by this facility is a drone control and maintenance unit that supplies target drones for tests of guided missiles at NOTS various ranges.

d. Meteorological Support

The primary meteorological information collection system at NOTS consists of a double Rawin complex with data reduced by the IBM 7090 computer. Basically the system involves an AN/GMD-1B Rawin set sited at either end of a 31,781-ft baseline. One station, tower No. 8 on C-1 Tower Road, is designated North Station and is regularly used for single station readings. The South Station is located on the roof of the Instrument Laboratory in the G-1 Range area. Both stations have AN/GMD-1B antennas, frequency time recorders, and FM communications as well as wire. The North Station, in addition, has a baseline check set. The Balloon sonde is released from the North Station and telemetry is usually recorded there.

G. Fort Greely

A small sounding rocket range is located at Ft. Greely, Alaska, formerly Big Delta Air Base. Since 1955 it has been a part of the US Army, Alaska and is used by the U.S. Army Arctic Test Board, Chemical Test Group, and Cold Weather and Mountain School. It is one of the stations in the synoptic meteorological network of ranges that includes ranges at AMR; PMR; WSMR; Wallops Station, Virginia; Eglin AFB, Florida; Tonapah Test Range, Nevada; and Churchill Research Range, Canada [39].

1. Geographical Factors

The range is 4 mi southwest of Ft. Greely (64° 00′ N 145° 44′ W, elevation 1274 ft) and within a 25-mi × 30-mi danger area as shown in Fig. 84. Ft. Greely is in the Tanana Valley of Alaska at a point where the Delta River joins the valley and about halfway between the Canadian border and the confluence of the Tanana and Yukon Rivers. The area contains many small ponds, streams, and marshes. Some 40 mi to the south glaciated summits about Mt. Hayes rise to heights of 13,000 ft. The climate at Ft. Greely is similar to that of Ft. Churchill, Canada, but somewhat milder—even though it is considerably farther north. The mean temperature during January is −5°F, while the July mean is 59°F. However, the mean daily minimum temperature in January is −13°F, and the mean daily maximum temperature in July is 69°F. Snowfall in January averages 7 in. and there is an average annual precipitation of 11.5 in. Surface winds, especially in January, at mean speeds of 17 mph, present problems in launching the smaller rockets such as Arcas.

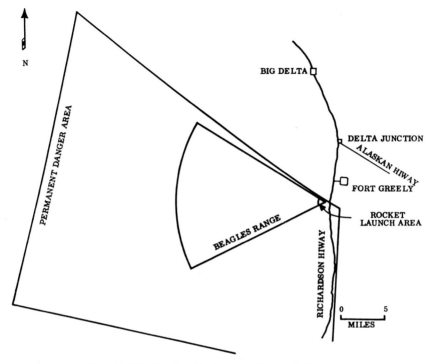

Fig. 84. Ft. Greely, Alaska, sounding rocket range.

2. Launching Facilities

The launching pad and its associated facilities are situated on a high bluff overlooking the Delta River. Unlike most pads, which are made of concrete, the one at Ft. Greely is made of wood and has a standard, gun-type launcher for the Arcas rocket and a launcher for the Loki rocket as well. Launch control center for the pad is in a trailer that is a component of the range's radar tracking system.

3. Communications and Tracking

Radar tracking at the range consists of a modified M-33D fire control system. As a reserve tracking system the range also utilizes two AN/GMD-1 radars, with one station in the launch area and one located at Ft. Greely, some 4 mi away. Even though this presents a rather short baseline, the system of double receivers permits an alternate means of tracking, and the antenna in the launch area is often used as an aid in radar acquisition. The AN/GMD-1 antennas are also used to receive telemetry data from the rockets.

4. Support Activities

A wooden structure is located at the launching site for the shelter of generators and convertors for the radar. A special metal structure is located some 350 ft from the launching pad for ready-storage of an operating quantity of various types of meteorological sounding rockets. A similar building located at the launch site contains a rocket assembly room and a firing room for Loki and Arcas launchers.

Technical support facilities and amenities for range operating personnel are in the cantonment area of nearby Ft. Greely.

H. Tonapah Test Range

The Sandia Corp. operates the Tonapah Test Range for the Atomic Energy Commission (AEC). It was opened in 1957, and by 1959 the Sandia Corp. had established a full-time staff at the range. While it is primarily used by the AEC for gathering ballistic information on bombs dropped from aircraft and data from sounding rockets and research test vehicles, it is sometimes made available to other government agencies and their contractors on a noninterference basis. The AEC tests only nonnuclear weapons components at the range. Typical of the rockets launched are the Deacon-Arrow, Arrow-Dart, and Arcas [40]. The range supports X-15 flights from Edwards AFB by supplying meteorological profiles to 200,000-ft altitudes.

1. Geographical Factors

The range (37° 45′ N 116° 45′ W, elevation 5330 ft) is located 32 mi southeast of Tonapah, Nevada, and is on the northern edge of the Las Vegas Bombing and Gunnery Range. However, it is not a part of the range nor of the Nevada Test Site, operated by the AEC. The range proper (637 mi²) is situated on Cactus Flat in a vast valley with many dry lake beds. Some 10 mi to the west lie the Cactus Mountains, while the Kawich Range is 20 mi to the east. The area is semidesert with sparse vegetation of sagebrush and mesquite.

Normal summer temperatures at the range are in the low 90's (°F), but minima of 70°F are not uncommon. Winter temperatures have reached an absolute low of −15°F, but the mean maximum is 50°F. In the late winter, spring, and early summer gusts of wind from 35 mph to 45 mph occur. Average annual precipitation is only 5 in., and the area has a low humidity all year.

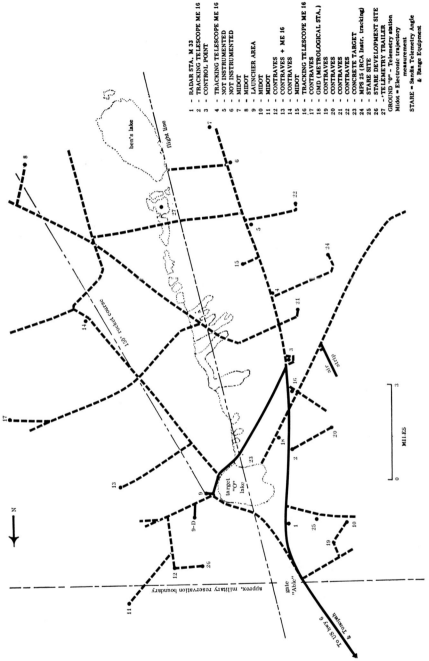

FIG. 85. Tonapah Test Range, Nevada. Courtesy Sandia Corp.

2. Launching Facilities

The control center for the range is located some 5 mi from the rocket launching pad, as shown in Fig. 85. It is only 400 ft² and accommodates a maximum of four test controllers at a time. The rocket launching complex consists of a concrete pad and a 400-ft² concrete blockhouse some 300 ft from the pad. The launch pad (Fig. 86) has two boom-type, straight-rail launchers; a high-altitude Sampler rocket launcher and two small, tube-type launchers. Direction of fire is almost due west. Research test vehicles with first stages consisting of the Honest John motor can be launched from the pad as well as a variety of smaller rockets, some designed by the Sandia Corp. Rockets with altitude ranges up to 300,000 ft can be fired from the range.

3. Tracking and Communications

Tonapah has 28 permanent instrumentation sites and four more are planned. Optical instrumentation includes eight Contraves cinetheodolites modified for one-man operation; five 16-in. Newtonian tracking telescopes, with 117-in. focal lengths, on tracking mounts (ME-16); LA-24 tracking telescope; and a variety of smaller cameras such as Photosonic, Warrick, CZR, Mitchell, and Fastax. Locations of these sites are shown in Fig. 85.

Principal radars used at the range are the AN/MPS-25, a mobile version of the AN/FPS-6 with a range of 500 mi; and an M-33, X-band instrument especially modified by Sandia with a range of 100,000 ft. Other equipment includes a Tacoda (Target Coordinate Data) instrumentation acquisition system; five FM/FM telemetry receiving stations; and five Midot (Multiple Interference Determination of Trajectory) interferometer angle measuring stations. A special interferometer, angle-measuring, phase-comparison, distance-measuring system (Stare) is also being developed for the range tracking system.

The range timing system, called Digitime, consists of a master station and 27 substations, with few exceptions all transistorized. Each substation is independent but is synchronized by a special 1000-cps signal transmitted from the master station over VHF radio, telephone, and wire circuits. This signal phase-locks a tuning fork oscillator, which has sufficient stability that loss of the signal will not cause a serious drift in local oscillator frequency [53].

In addition to the 1000-cps synchronizing signal, the master station also continuously transmits two reset signals to the substations. A 2000-cps signal reverses phase each second. The substations detect this phase reversal and apply it to the three highest speed decades in the counter,

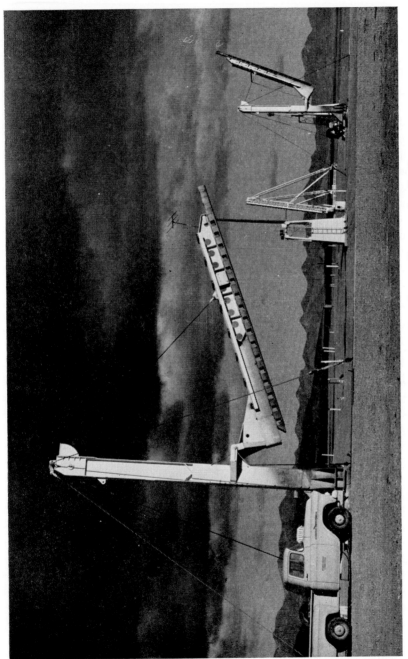

FIG. 86. Launching pad at Tonapah Test Range. Courtesy Sandia Corp.

resetting them to 0.000 sec every second on the second in case they have picked up or lost counts during the previous second. A 2500-cps signal reverses phase at test zero, resetting all counter decades on the range to zero.

Each station in the network generates six digits of time in the binary coded decimal format, thus the maximum time count is 999.999 sec and resolution is to the nearest millisecond.

These various tracking devices are arranged around what amounts to three separate ranges. The primary target area is a range instrumented with ballistic cameras that provide coverage for the release of test vehicles from aircraft. Behind this range area are the tracking cinetheodolites, tracking telescopes, radars, and rf tracking systems that provide ballistic coverage for drops from medium and high-altitude aircraft. This system of instrumentation also covers the rocket launching range, which is considered to be a third range.

All camera and telemetry stations are tied together by underground cables, and the range also has a commercial teletype station as well as HF, VHF, and UHF radio networks. Most vehicles operating throughout the range are equipped with two-way radios.

4. Support Activities

Adjacent to the launcher pad is a 3000-ft² rocket and payload assembly building.

Power for major activities on the range is supplied by diesel generators, while smaller generators are used for remote installations and sites.

An AN/GMD-1 meteorological station provides weather data for the range.

Other supporting facilities at the range include three specially instrumented artillery pieces that are used for acceleration and deceleration tests of nuclear weapons components. Cameras associated with these weapons permit high-speed photography at the muzzle, while other range instrumentation provides for complete trajectory coverage.

I. Edwards Air Force Base

Although Edwards AFB has no missile launching facilities, it is covered in this chapter because of its intimate association with other aerospace facilities and ranges and because it is the site from which rocket-powered airplanes are flown. Those activities at the base most immediately concerned with astronautics are described in this section; however, this does not do justice to all the technological facilities available, particularly those in the field of aeronautical engineering.

The base has seen a variety of military uses since its inception as a bomber and gunnery range in 1933. During World War II Edwards was used as a training base for bomber crews and was the site of the famous *Muroc-Maru*, an exact replica of a Japanese cruiser of the Mogambi class. This steel target was constructed in the dry lake bed—complete with earth waves rippling from the bow—and used as a bombing target. On 1 October, 1942 the base was the site of the first flight of the first U.S. jet aircraft, the XP-59A Bell [56]. In 1950 the name of the base was changed from Muroc AFB to Edwards AFB in honor of Captain Glen Edwards, who was killed flying an experimental aircraft at the base in 1948.[9]

Encompassing 301,209 acres of the Antelope Valley on the western edge of the Mojave Desert in California (Fig. 87), Edwards AFB (34° 56′ N 117° 53′ W, elevation 2300 ft) is the center of much research and development as well as testing of large rocket engines for space carrier vehicles. It is 100 mi north of Los Angeles and generally bordered by Highway 466 to the north, Highway 6 to the west, and Highway 395 to the east.

The base proper is situated on the western edge of Rogers Dry Lake, a geological phenomenon that is ideal for aircraft testing because of its surface. The lake bed is some 65 mi^2 and completely devoid of any vegetation. Its rock-like surface consists of a clay hardpan, which varies in thickness from 8 in. to 18 in. and weighs 111 lb/ft^3. For 10 months out of the year no rain falls on it. During the winter months what rain that does fall collects on the surface and is whipped back and forth by the wind. With the beginning of warmer weather, the surface water drains off leaving the surface as smooth as it has been for millennia.

The hard surface of the lake bed can support loadings of 250 psi.

The lake bed is used as a landing facility for various types of developmental aircraft and the X-15 rocket-powered aerospace research plane. There are eight runways crisscrossing the lake bed, six oriented generally east and west and two stretching north and south. Runway 35 north is 7.5 mi long. These runways are bordered with asphalt strips to make them visible from the air.

In the summer the daily temperature ranges from 62°F to 96°F, and in the winter it varies between 32°F and 59°F. Because of the large number of cloudless days, visibility is excellent over the range and there are on the average 350 flying days out of the year. The average annual rainfall is only 4 in., and the humidity is low all year.

[9] The name Muroc derives from Corum, spelled backwards, the name of early settlers of the area in 1910. When the post office objected to the name Corum for the desert community, one of the founders submitted Muroc.

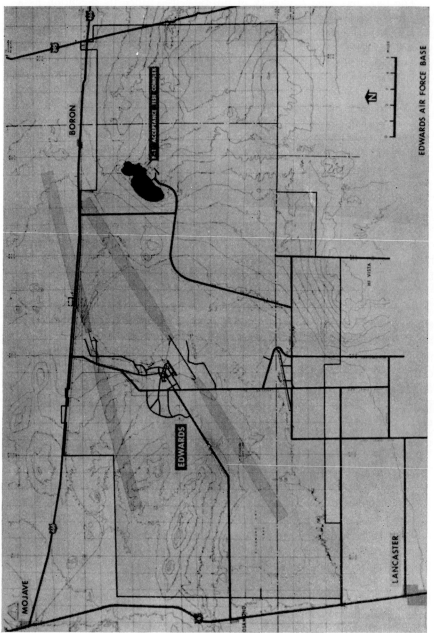

Fig. 87. Edwards AFB, California. Courtesy NASA.

The Air Force Flight Test Center (AFFTC), which conducts research and development and flight testing activities, is the host agency at Edwards AFB. Since 1946 the center has tested every U.S. Air Force aircraft except one, and AFFTC facilities are also available to aircraft manufacturers on a fee basis for the testing of commercial airplanes. The U.S. Army Aviation Test Activity and the NASA Flight Research Center are also at Edwards AFB.

1. Aerospace Facilities [57]

The complex at Edwards AFB consists of several major installations or operational areas. These include the Main Base, North Base, South Base, bombing range, high-speed track facility, the Baker-Nunn satellite camera tracking station, and the propulsion test facilities on Leuhman Ridge.

a. Main Base

The Main Base houses the administrative, logistic, and technical support facilities for Edwards AFB as well as the U.S. Air Force Experimental Flight Test Pilot School and NASA's Flight Research Center. The central feature of this area is the main airfield with its long row of hangars for the various contractors using facilities of the base. Also located on Main Base is the propulsion branch test facility, which has among other activities four enclosed gas turbine engine test cells for engines with thrusts up to 50,000 lb. One of the cells is designed for turboprop engines with power ratings up to 25,000 hp.

Also found on the Main Base are the housing and other amenities for range operating personnel.

b. Bombing Range

The bombing range is an area 17 mi long and 6 mi wide extending from west to east with its western boundary on the edge of Rogers Dry Lake some 7 mi to the southeast of Main Base. Within this range are several bombing targets that are accurately surveyed and surrounded by impact observation camera platforms. Also within it is the Haystack Butte hazardous propellants test site. Numerous cinetheodolite and ballistic camera stations are located both within the bombing range and without its perimeter [57].

c. High-speed Track Facility

The high-speed track facility is located at the south end of Rogers Dry Lake but not within the lake bed. It is used to test a variety of guided missile and aerospace components. Since the track facilities can produce accelerations and velocities in several regimes up to 100 g and Mach 4.0, the track is particularly useful in testing parachute recovery systems, seat ejection and canopy jettisoning devices, missile and aircraft radomes, as well as complete rockets and missiles or their guidance systems.

The 20,000-ft track consists of two precisely aligned, 171-lb/yd crane rails set at the standard railway gauge of 56.5 in. The rails are continuously welded and supported by sleepers at 3-ft intervals along the entire length of the track. The last 6000 ft of the track contains a concrete water trough, which is the braking system. A variety of solid propellant and liquid propellant sleds is available for use with the track. In general, the nature of the experiment or test dictates the type of rocket sled to be employed. In some tests as many as five stages of solid propellant rocket motors have been used and ejected after burnout. The track is instrumented for telemetry and optical coverage.

d. Nunn-Baker Camera Tracking Station

The Nunn-Baker camera site at Edwards AFB is one of a network of five Earth satellite tracking cameras controlled by the National Space Surveillance Control Center in Cambridge, Massachusetts. The central feature of the station is the camera itself (Fig. 88). It is the same Nunn-Baker camera used by the Smithsonian Institution's satellite tracking system. The camera is 10 ft tall, 6 ft wide, and weighs 6000 lb. Fixed in an equatorial mount, the camera can be pointed with extreme precision and can photograph satellites as faint as a star of the 17th magnitude.

The optical system of the camera consists of a 31-in. spherical primary element and a three-element corrector. The relative speed of the optical system is $f1$. Exposures are made on a 55-mm film, and the camera can traverse the horizon in 93 sec. The camera has a field of view of 5 deg by 30 deg and an accuracy of approximately 3 sec of arc. It has photographed the 6-in. diameter Vanguard 1 satellite at an altitude of 2400 mi [58].

e. Propulsion Systems Static Test Area

Located on Leuhman Ridge 10 mi east of Main Base is the U.S. Air Force's Rocket Propulsion Laboratory and NASA's High Thrust Test Area, seen in Fig. 89. Stationed around the perimeter of this 3200-ft ridge are ten major static test stands for large liquid propellant rocket engines linked to a central control center. There are also a number of horizontal and one major vertical stands for solid propellant motors and two Minuteman underground silos [59].

United States Air Force activities on Leuhman Ridge involve four

main operations: liquid propellant systems testing, solid propellant systems testing, propellants studies, and technical support. NASA's major activity is the F-1 liquid propellant engine acceptance test facility. Typical propulsion systems that have been tested at these facilities include those for the Thor, Atlas, Titan, Minuteman, Blue Scout, X-15, and Saturn 5.

Air Force test stands can accept complete missiles such as the Atlas and Thor as well as individual stages or single engines. One of the stands is especially adapted to provide various environmental conditions for the missile undergoing test. A covered structure over the stand permits the creation of a variety of temperature environments. The propulsion system can be fired at the induced temperature with the missile in the structure.

While most of the solid propellant test stands are simple horizontal fixtures, built into the hillside, Leuhman Ridge has one of the largest vertical solid propellant test complexes in the world. It is designed specifically to test the 120-in. diameter motors for the Titan 3 space carrier vehicle. In addition to the two test stands, the complex also has an instrumentation and control building, a storage area, and an administrative and assembly area. One of the huge test stands permits horizontal as well as vertical firings, while the other is restricted to vertical firings only. Since motors will be fired in a nozzle-up position, no elaborate, water-cooled flame deflectors are required. Each tower is 85 ft tall and 36 ft² at the base. The two stands are juxtaposed so that their walls are at 90 deg angles to each other. Located near the stands is a thrust vector control system fuel storage area, which is tied into the stands [60].

Also sited near these stands are two Minuteman underground launch silos, similar to those at Vandenberg AFB and other tactical Air Force sites. These units are used for test purposes only and are not operational, although they could be adapted to permit launchings if required. Tethered launchings with reduced propellant loadings have been made from the silos, but they are used primarily to conduct storage tests on the missile.

The Air Force also operates a special test area at Haystack Butte, within the bombing range, 7 mi to the southeast of Leuhman Ridge. Here tests and toxicity studies are made on highly poisonous chemicals that offer promise as liquid propellants for advanced missile and space carrier vehicle propulsion systems. Typical of the experiments are spill experiments to determine the toxic effects of such chemicals accidentally released upon the ground.

Figure 90 is a map of NASA's High Thrust Test Area. Located on Leuhman Ridge, it centers around six static test stands for the 1,500,000-lb thrust F-1 liquid propellant engine. Test stand 1A and 1B are used to static fire the complete engine, while test stand 2A is designed for test firing thrust chambers, propellant valves, and other components. These three stands are used for research and development model engine firings.

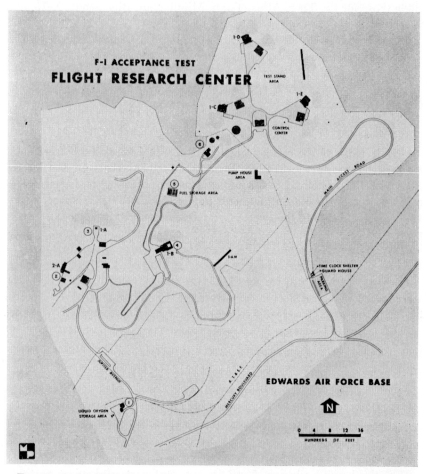

Fig. 90. NASA High Thrust Test Area on Leuhman Ridge at Edwards AFB. 1, liquid oxygen storage; 2, test stand 2-A; 3, test stand 1A; 4, test stand 1B; 5, fuel storage; and 6, water facilities. Courtesy NASA.

The 1A stand is a modified Atlas static test stand and accepts only one F-1 engine, operating at rated thrust. Stand 1B, shown in Fig. 91, is a dual-position stand and the largest of the three. This stand is 250 ft tall.

Each of the three other test stands, 1C, 1D, and 1E, located in an

FIG. 91. Static test stand 1B for the F-1 liquid propellant engine on Leuhman Ridge at Edwards AFB. Courtesy NASA.

FIG. 92. NASA High Thrust Test Area on Leuhman Ridge at Edwards AFB. Top, test stand 1E; lower right, test stand 1C; lower left, test stand 1D; control center is in the middle. Courtesy NASA.

area 1 mi to the northeast (see Fig. 92), is joined to a central control building by a concrete tunnel (for electrical conduits and other utilities). Each stand is also connected to a common water supply by a 66-in. diameter pipe. Water from all three stands is returned to a reservoir, which permits from 50 to 80 per cent of water used to be reclaimed, a very economical practice in the desert. The control center is a two-story, reinforced concrete structure. The first floor, which is underground, is a computer facility and equipment area; and the second story is the operations and control room. Each stand also has an associated reinforced concrete structure and high-bay work area. The reinforced-concrete structure is designed to withstand over pressures from explosions on adjacent stands but not on the associated stands. The test stands in this area are for the acceptance testing of engines.

Water for cooling the flame deflectors of the three stands is supplied by deep wells located some 10 mi from the test site. These wells supply water to a 3,000,000-gal tank through a 14-in. diameter pipeline. Pumps deliver the water to the individual stands at a rate of 150,000 gpm [61].

Also located in the High Thrust Test Area are eight observation bunkers, a liquid oxygen storage facility, a gaseous nitrogen storage area, and a fuel storage facility. Data from instrumentation on the test stands is transmitted by a microwave relay link to a computer facility at Canoga Park, California, 75 mi to the south, on a real-time basis. Within 1 hr after a test run, reduced data are back at the NASA Flight Research Center.

The Air Force also has a 160-ton per day air separation plant for liquid oxygen production located within the High Thrust Test Area south of stand 2A.

2. X-15 Aerospace Research Craft

The testing of high-performance aircraft at Edwards AFB dates back to 1946 when the X-1 rocket airplane was first tested. Experimentation with the X-series at the NASA Flight Research Center continues, with the X-15 being the latest. The concept for this unique aerospace test vehicle began in 1952 as a program of the National Advisory Committee on Aeronautics (the predecessor of the NASA).

The rocket-powered X-15 utilizes a flight corridor some 485 mi long by 50 mi wide, extending from Wendover AFB, Utah, to Edwards AFB. This corridor cuts southwest across the state of Nevada. The flight profile is shown in Fig. 93. At its maximum ordinate, the craft reaches an altitude of more than 250,000 ft and maintains proper attitude by reaction control jets. It is carried to an altitude of 40,000 ft by a B-52 mother

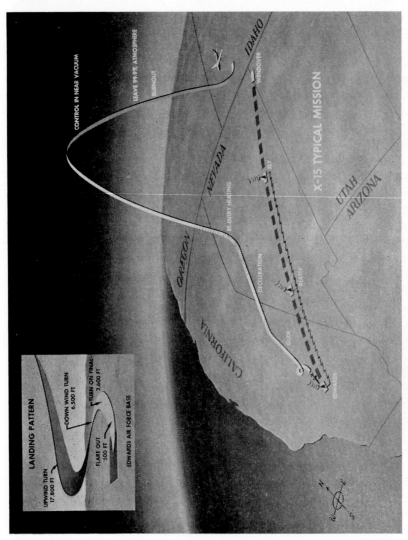

Fig. 93. X-15 flight profile. Courtesy North American Aviation, Inc.

ship before beginning its own powered flight with a 57,000-lb thrust, throttable liquid propellant engine.

A 500-mi tracking range supports the X-15 flight test program with ground tracking, data acquisition, and flight simulator facilities. Called the High Range, it provides direct support for X-15 flights. The terminal station is located at Edwards AFB and the two uprange stations are near the towns of Beatty and Ely, Nevada.

High Range ground stations support the X-15 flights in several ways by assisting in guiding and vectoring the B-52 launch plane to the required heading; monitoring the initial climb of the X-15; provding backup altitude and velocity information to X-15 pilots if on-board equipment does not function properly; monitoring flight path to provide homing and vectoring information if an emergency landing is required; providing escort rendezvous information; establishing long-range, UHF communications; providing real-time data; and furnishing space-trajectory data for aerospace research. To meet these requirements, each of the High Range sites has a Reeves Mod II instrumentation radar, a precision radar data recording system, telemetry receivers, and data monitoring and communications equipment.

Each of the S-band, 400-mi range radars is equipped with an 80-in. focal length boresight camera, which photographs the target and records tracking correction data. Other equipment used in conjunction with the radar includes analog-to-digital conversion equipment, a 14-track tape recorder, optical and brush recorders, and 35-mm cameras [37].

The timing system furnishes four timing signals to the High Range: a 1000-pps integrated time-of-day code; a 100-pps integrated time-of-day code; a 10-pps pulse rate; and a 1-pps time code. The signals are used by the precision data recovery system, telemetry recording system, plotting boards, velocity recorders, radar boresight cameras, and radar strip chart recorders. Two timing sources are available: the AAFTC precision timing system, which furnishes multiplexed signals received and demultiplexed at the Edwards range station, and a Model 6190 time-code generator located at the Edwards. Other timing equipment at the Edwards station includes a receiving station, time signal modulator, and a WWV standards frequency comparator. Time code converters are also located at the Beatty and Ely range stations.

High Range telemetry systems include an FM/FM, 15-channel, IRIG receiving station; a 90-channel, 225–260 Mc/sec PDM system; and a servo-driven helical antenna that can be operated in the automatic or manual mode.

The variety of critical test information provided by the telemetry system is monitored in real-time during test flights by ground observers to assist in mission performances. Real-time information on test parameters can be observed in bar-graph form on oscilloscopes in a data monitor console, permitting a quick assessment on several parameters; and, if a time-history presentation is needed, up to 12 channels can be plotted on a strip chart recorder.

Communications for the High Range is provided by an AN/GRC-27, UHF system, a data transmission system, leased telephone lines, and a microwave system. Ground-to-air communications are accomplished by UHF transmitting and receiving sets at each of the range stations. When transmitters are keyed at one station, the transmitters at the other station are keyed by a microwave link or telephone connection so that all stations simultaneously transmit the same information; and the UHF receivers at all three stations are also connected by microwave link or telephone line, thus permitting all stations to receive the transmission.

The data transmission system consists of a transmitter and a data receiver, the former being essentially an analog-to-digital converter that accepts data and converts them to binary-coded signals that are multiplexed for transmission over a single communications channel.

Leased telephone lines are used to connect the three range stations. Three circuits are used for data transmission and reception, one for station-to-station intercommunication, and one for range timing. The microwave system maintained by the Air Force supplies the Edwards station with telemetry signals from the Beatty and Ely stations. It also provides three additional information channels, and it can be used to back-up or augment the telephone circuits.

J. Mississippi Test Operations

A static test facility for future space carrier vehicle stages and engines is presently being constructed on 13,550 acres of land in southwest Mississippi. Easement rights to an additional 128,400 acres for a buffer zone, which extends into Louisiana, have also been purchased by the National Aeronautics and Space Administration. Although it is not a launching facility, it will embody many features found at a launch installation and will become an integral step in the development of large space carrier vehicles.

Located 10 mi south of Picayune (30° 25′ N 89° 37′ W) on the East Pearl River (Fig. 94) in a sparsely inhabited and wooded area with an elevation ranging from 4 to 30 ft above sea level, the site will have deepwater access from the Michoud space carrier vehicle fabrication plant at New Orleans, some 35 mi to the southwest. Designated the Mississippi Test Operations (MTO), the huge testing facility is an integral part of the George C. Marshall Space Flight Center at Huntsville, Alabama.

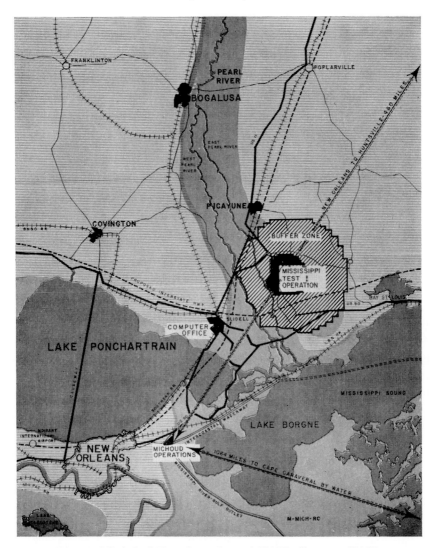

Fig. 94. Mississippi Test Operations of NASA. Courtesy NASA.

A vital link in support of the U.S. manned lunar space flight program, MTO will static test complete stages, components, and engines to be used in the Saturn 5 space carrier vehicle, which will launch the Apollo lunar spaceship.

The first phase of construction will include four test stands: two for the first stage of the Saturn 5 and two for the second stage. The first, or S-IC, stage will develop 7,500,000 lb of thrust, while the second, or S-II,

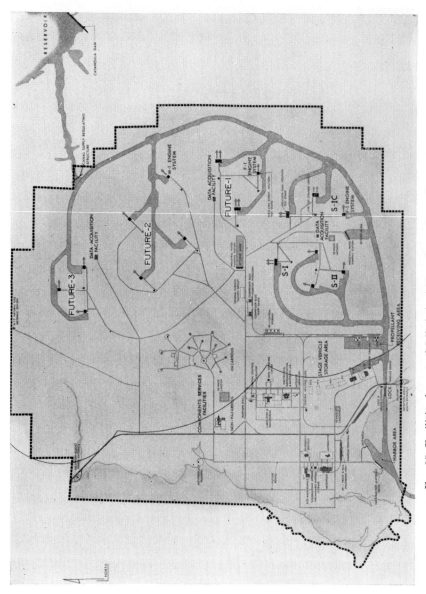

FIG. 95. Facilities layout of Mississippi Test Operations. Courtesy NASA.

Fig. 96. S 1-C static test stand at Mississippi Test Operations. Courtesy NASA.

stage will develop 1,000,000 lb. Later installations will include the construction of test stands for engines such as the F-1 and M-1 as well as the various stages for future space carrier vehicles of the post-Saturn type (Fig. 95).

The S-IC and S-II test complexes will consist of the test stands and various supporting installations. For example, the S-IC complex will

Fig. 97. S-II stage static test stand at Mississippi Test Operations. Courtesy NASA.

have, in addition to its two 290-ft tall test stands, a settling basin, two discharge basins and burn-off pits, fuel storage area, gas battery, liquid oxygen and barge docks, instrumentation towers, observation bunkers, and a test control center. Although each of the test stands is designed to accommodate two stages (Fig. 96), only one stage will be fired during each test.

The S-II static test stand, viewed in Fig. 97, will have facilities for

FIG. 98. Scientific and technical facilities at Mississippi Test Operations. Courtesy NASA.

FIG. 99. Loudspeaker with 5000 acoustical watts of power at Mississippi Test Operations. Courtesy NASA.

firing only one stage. For safety purposes this stand is located 6000 ft from the S-IC stand. The 185-ft tall structure will be topped by a stiff-leg crane for hoisting stages from barges and placing them in the stand.

In addition to the test complexes, MTO will also have scientific, technical, and service facilities such as an engineering laboratory, acoustical studies laboratory, electronics and instrumentation laboratory, site maintenance building, automotive repair building, warehouses and propellant storage structures (Fig. 98). Other buildings will house communications facilities and data handling installation. Also included among these will be special space carrier vehicle storge buildings. With 70-ft ceilings, these 200-ft tall structures will provide an area for minor repair and modification can be made to the various stages, and they will offer protection to such stages during hurricanes should they occur.

Perhaps the most unusual feature among the supporting activities at MTO is a large outdoor loudspeaker or exponential horn (Fig. 99), which is 30 ft long and 12 ft wide at the mouth. The horn is designed to determine the acoustical effects that various weather conditions will have on static firings and the degree to which such noise will disturb neighboring populations. Mounted on a turntable, it can be pointed in any direction. Operating through a frequency range of 30–400 cps, the horn produces 5000 acoustical watts of power and has a sound pressure level at the mouth of 150 db. It has been heard as far away as Bay St. Louis, which is 28 mi from MTO.

Heavy transportation requirements at MTO, such as those for the movement of space carrier vehicle stages, engines, and bulk propellants, will be met primarily by a waterway system consisting of 15 mi of improved river channel and an additional 15 mi of canals constructed within the area. The use of such a system results in a minimum need for road and railroad construction. However a 10.5-mi railway spur has been completed into the site.

All operations within MTO will be directed from a central control building, including the movements of cargo on the waterways system. The site is expected to be fully operational by 1966.

Note Added in Proof. In 1964 the Department of Defense designated the U.S. Air Force as cognizant command agency for both the Atlantic and Pacific Missile Ranges. The U.S. Air Force subsequently combined the facilities of Vandenberg AFB and Pt. Arguello, redesignating them the Western Test Range. AMR became the Eastern Test Range. Facilities at Kwajalein Island were turned over to the U.S. Army.

References

1. Anon. (1962). Cape Canaveral, Hope of the Free World. Report of the Committee on Science and Astronautics, U.S. House of Representatives, Eighty-Seventh Congress, Second Session, May 24, 1962 (Washington, D.C.).

2. Lange, O. H., and Stein, R. J. (1963). "Space Carrier Vehicles—Design, Development, and Testing of Launching Rockets." Academic Press, New York. See also Alexander G. (April, 1963). Cape facilities for Saturn 5/Apollo, Titan 3 keyed to delivery of flight ready boosters. *Aviation Week* **78**, 54–69.

3. Alexander, G. (May, 1963). First AMR tracking ship begins checkout. *Aviation Week* **78**, 28–29. See also Anon. (1963). Gen. H. H. Arnold, world's largest, most sophisticated missile tracking ship. *Marine Eng./Log* **68**, 42–43.

4. Scott, C. S. (June, 1962). Atlantic Missile Range instrumentation. *Missiles Space* **10**, 26.

5. Klass, P. J. (Aug., 1963). AMR cable enhances tracking reliability. *Aviation Week* **79**, 104–110.

6. Anon. (1963). "AMR Instrumentation Handbook. Vol. 1—Operational Systems." Tech. Doc. Rept. No. MTC-TDR-63-1, US Air Force Missile Test Center, Cape Kennedy, Florida.

7. Anon. (Fall 1962). Global range instrumentation. *Sperry Eng. Rev.* **15**, 2–10.

8. Mason, J. F. (Dec. 15, 1961). New pulse radars readied for space ranges. *Electronics* **34**, 26–28.

9. Anderson, K. R., and Crumley, T. D. (1962). MISTRAM—A "state of the art" tracking system. *Proc. IAS Natl. Symp. Tracking and Command of Aerospace Vehicles, San Francisco, February 1962.* Inst. Aerospace Sci., New York.

10. Glei, A. E. (1962). The design and operational philosophy of the ballistic camera system at the Atlantic Missile Range. *J. Soc. Motion Picture Television Engrs.* **71**, 823–827.

11. Economou, G. *et al.* (1958). Automatic exposure control for a high-resolution camera. *J. Soc. Motion Picture Television Engrs.* **67**, 249–251.

12. Schroeder, R. L., and Zvara, J. (Feb., 1963). NASA ground operational support system. *Missiles Space* **11**, 16–59. See also Anon. (Sept., 1963). USAF building expanded range center. *Aviation Week* **79**, 85–87.

13. Monroe, J. P. (Nov.-Dec., 1961). The Pacific Missile Range. *Ordnance* **46**, 361–364.

14. Anon. (1961). Finger-tip facts about PMR. Headquarters, Pacific Missile Range (Pt. Mugu, California).

15. Anon. (1963). Pacific Missile Range. "Range Manual. Vol. 3, Range Facilities and Instrumentation." Rept. No. PMR-MP 63-7, Headquarters, Pacific Missile Range, Pt. Mugu, California.

16. Bustamente, A. C., and Walker, W. E. (1961). Design of the Sandia High-Altitude Sampler Rocket System (HAS). Rept. SC-4589(RR). Sandia Corp. (Albuquerque, New Mexico).

17. Anon. (Jan., 1963). Preparing to launch at Pt. Mugu. *Naval Res. Rev.* **16**, 11.

18. Herron, E. A. (1962). Impact! *Skyline* **20**(2), 31–41.

19. Anon. (1962). Vandenberg, Sand-swept home of the giant missiles. *Skyline* **20**(2), 7–14.

20. Berger, C., and Howard, W. S. (1962). History of the 1st Strategic Aerospace Division and Vandenberg Air Force Base, 1957–1961. Strategic Air Command, US Air Force.

21. Anon. (1958). Range instrumentation of the Pacific Missile Range, MP-3-58. Headquarters, Pacific Missile Range Pt. Mugu, California.

22. Anon. (Summer, 1963). Collins at the Pacific Missile. *Signal* **11**, 1–8.

23. Beller, W. (Sept. 1963). AMF tracking mount on job at PMR. *Missiles Rockets* **13**, 26.

24. Anon. (Undated). A technical catalog of resources and capabilities. US Army Signal Missile Support Agency. (White Sands Missile Range, New Mexico).

25. Anon. (1960). Holloman track capabilities. Tech. Rept. AFMDC-TR-60-24. Air Force Missile Development Center (Holloman AFB, New Mexico).

26. Anon. (1961). AFMDC test track instrumentation and instrumentation support facilities. Air Force Missile Development Center (Holloman AFB, New Mexico).

27. Bushnell, D. (1958). AFMDC history of research in space biology and biodynamics at the Air Force Missile Development Center, 1946–1958. Air Force Missile Development Center (Holloman AFB, New Mexico).

28. Anon. (July 22, 1963). Apollo testing to begin at White Sands. *Aviation Week* 79, 281–295.

29. Anon. (Sept. 1963). Little Joe success will trigger Apollo test program soon. *Missiles Rockets* 13, 15.

30. Anon. (Undated). "White Sands Missile Range Technical Catalog," Vol. I. US Army Ordnance Corps, White Sands Missile Range, New Mexico. See also Anon. (1962). Instrumentation radar AN/FPS-16. US Army Signal Missile Support Agency (White Sands Missile Range, New Mexico).

31. Anon. (Undated). "White Sands Missile Range, Technical Catalog, Optical Flight Instrumentation," Vol. 2, Sec. 1. US Army Ordnance Corps, White Sands Missile Range, New Mexico. See also Durrenberger, T. E. (1960). Final report, development of Astro Makr II ballistic camera system. Integrated Range Mission (White Sands Missile Range, New Mexico). Also see Revyl, D. and Carrion, W. (July, 1962). Optical tracking methods and instrumentation: Research and Development at BRL. *J. Soc. Motion Picture Television Engrs.* 71, 505–508.

32. Anon. (1960). "White Sands Missile Range Technical Catalog, Army Mission Laboratories, Flight Simulation," Vol. 2, Sec. 7, Part 1. US Army Ordnance Corps, White Sands Missile Range, New Mexico.

33. Anon. (1960). "White Sands Missile Range Technical Catalog, Data Reduction," Vol. 2, Sec. 3. US Army Ordnance Corps, White Sands Missile Range, New Mexico.

34. Anon. (1962). Automatic Missile Impact Predictor 2. Prepared by Missile Meteorology Division, Tech. Rept. SELWS-M-14. US Army Electronic Research and Development Activity (White Sands Missile Range, New Mexico).

35. Anon. (Undated). "Wallops Station Handbook," Vol. 1, General Information. Wallops Station, Virginia.

36. Anon. (Undated). "Wallops Station Handbook," Vol. 2, Flight Test and Support Services Facilities. Wallops Station, Virginia.

37. Briskman, R. D. (1962). NASA ground support instrumentation. *Proc. IAS Natl. Symp. Tracking and Command of Aerospace Vehicles, San Francisco, February, 1962.* Inst. Aerospace Sci., New York.

38. Sheldon, J. C. (1962). Preliminary report on the single station Doppler-interferometer rocket tracking technique. NASA Tech. Note TN D-1344. Goddard Space Flight Center (Greenbelt, Maryland).

39. Altman, H. (March, 1962). Eglin Gulf test range. *Elec. Commun.* 37, 181–195.

40. Anon. (1961). Initiation of the meteorological rocket network, revised August 1961. IRIG Doc. 105–60, Secretariat Inter-range Instrumentation Group (White Sands Missile Range, New Mexico).

41. Brown, M. P. (July, 1961). Aerospace probing over the Gulf. *Astronautics* 6, 35–67.

42. Vickery, W. K. (1963). Firefly III, Sounding Rocket Launching Report, Launch

Facility, Vehicles, and Data Reduction, APGC-TDR-63-19, Vol. 1. Air Proving Ground Center, (Eglin AFB, Florida).

44. Brown, M. P. (Sept.-Oct., 1962). Testing our air weapons. *Armament Technol.* **3**, 3-5.

45. Schepler, H. C. (April, 1958). Photographic instrumentation at the Air Proving Ground Center. *J. Soc. Motion Picture Television Engrs.* **67**, 246-248.

46. Blenman, C. (Aug., 1963). US Naval Ordnance Test Station. *Underwater Technol.* **4**, 19-20.

47. The Inter-laboratory Committee on Facilities (1959). Facilities Review, West Coast Naval Laboratories, TS 59-200. (China Lake, California).

48. Anon. (1959). "Test Department Metric Photographic Instrumentation Handbook." US Naval Ordnance Test Station, China Lake, California.

49. Anon. (1956). Randsburg wash test activities, a facility for Ordnance testing under simulated tactical conditions, NOTS 1433. US Naval Ordnance Test Station (China Lake, California).

50. Thorn, W. F. (1963). Skytop 2 vertical static firing and altitude simulation facility. NAVWEPS 8068 (NOTS 3074), US Naval Ordnance Test Station (China Lake, California).

51. Anon. (1962). NOTS supersonic tracks. NOTS 1938 US Naval Ordnance Test Station (China Lake, California).

52. Anon. (1961). Instrumentation timing systems brochure. IRIG Doc. No. 103-59, Inter-range Instrumentation Group of the Range Commanders' Conference (White Sands Missile Range, New Mexico).

53. Anon. (1963). Instrument operations on test department ranges. NOTS TP 2692, US Naval Ordnance Test Station (China Lake, California).

54. Herman, R. W. *et al.* (1960). NODAC: the Naval Ordnance Data Automation Center. NOTS TP 2394 (NAVORD Rept. 7027), US Naval Ordnance Test Station (China Lake, California).

55. Hirschy, H. J. (1962). The NODAC Redactor: A general system description. NOTS TP 2995, US Naval Ordnance Test Station (China Lake, California).

56. Ball, J. (1962). "Edwards: Flight Test Center of the U.S.A.F.," pp. 15-16. Duell, Sloan, and Pearce, New York.

57. Anon. (1961). Technical support facilities. Directorate of Flight Test, Air Force Flight Test Center (Edwards AFB, California).

58. Hayes, E. N. (1962). The Smithsonian's satellite tracking program: Its history and organization. *Smithsonian Inst. Pub. 4482.*

59. Anon. (1963). Action on Leuhman Ridge. *Skyline* **21**(2), 22-229.

60. Anon. (Dec., 1962). AF begins building big Edwards solid rocket test facility. *Data Capsule* **3**, 1-4.

61. Momson, R. (Aug., 1963). Rocket test stand job demands CPM scheduling. *Contractors Engrs. Magazine* **60**, 12-19.

Note Added in Proof. An excellent pictorial history of the World War II guided missile development center at Peenemünde can be found in Ernst Klee and Otto Merk (1963), "Damals in Peenemünde." Gerhard Stalling Verlag, Oldenburg und Hamburg.

Author Index

Numbers in parentheses are reference numbers and indicate that an author's work is referred to although his name is not cited in the text. Numbers in italic show the page on which the complete reference is listed.

Subject Index